Modeling and Optimization in Science and Technologies

Volume 12

The book series *Modeling and Optimization in Science and Technologies (MOST)* publishes basic principles as well as novel theories and methods in the fast-evolving field of modeling and optimization. Topics of interest include, but are not limited to: methods for analysis, design and control of complex systems, networks and machines; methods for analysis, visualization and management of large data sets; use of supercomputers for modeling complex systems; digital signal processing; molecular modeling; and tools and software solutions for different scientific and technological purposes. Special emphasis is given to publications discussing novel theories and practical solutions that, by overcoming the limitations of traditional methods, may successfully address modern scientific challenges, thus promoting scientific and technological progress. The series publishes monographs, contributed volumes and conference proceedings, as well as advanced textbooks. The main targets of the series are graduate students, researchers and professionals working at the forefront of their fields.

More information about this series at http://www.springer.com/series/10577

Shengrong Gong · Chunping Liu
Yi Ji · Baojiang Zhong · Yonggang Li
Husheng Dong

Advanced Image and Video Processing Using MATLAB

Springer

Shengrong Gong
School of Computer Science
    and Engineering
Changshu Institute of Technology
Changshu, Jiangsu, China

Chunping Liu
School of Computer Science
    and Technology
Soochow University
Suzhou, Jiangsu, China

Yi Ji
School of Computer Science
    and Technology
Soochow University
Suzhou, Jiangsu, China

Baojiang Zhong
School of Computer Science
    and Technology
Soochow University
Suzhou, Jiangsu, China

Yonggang Li
College of Mathematics Physics
    and Information Engineering
Jiaxing University
Jiaxing, Zhejiang, China

Husheng Dong
School of Computer Science
    and Technology
Soochow University
Suzhou, Jiangsu, China

ISSN 2196-7326   ISSN 2196-7334   (electronic)
Modeling and Optimization in Science and Technologies
ISBN 978-3-030-08402-8   ISBN 978-3-319-77223-3   (eBook)
https://doi.org/10.1007/978-3-319-77223-3

Preface

Digital image processing mainly focuses on the research of signal processing, such as image contrast adjustment, image coding, image denoising and filtering. It is different that Image analysis emphasizes describing images with symbolic representations, analysis, interpretation, and recognition. Along with the boom in artificial intelligent and deep learning, digital image processing is going deeper and more advanced. People start the researches in simulating the human vision to see, to understand, and even to explain the real world using three techniques: image segmentation, image analysis, and image understanding. The Image segmentation is to extract the features such as the edges and regions for image analyzing, recognition, and understanding. Image analysis is to extract intelligent information from underlying features and their relationship using mathematical models and Image processing techniques. Image analysis and image processing are closely related. Although there may be a certain degree of overlapping, they are different in essence. Therefore, image analysis is more related to pattern recognition and computer vision. It is generally used to analyze the underlying features and superstructures by some mathematical models. The researches of image analysis are mainly focused on content-based image retrieval, face recognition, emotion recognition, optical character recognition, handwriting recognition, biomedical image analysis, video object extraction. The Image understanding is to further understand the meanings and scenario explanations by researching the properties and relations of the features and objects. The objects for image understanding are symbols from description; the process is similar to human brain.

Corresponding to image analysis, video analysis is to analyze the video frames of surveillance camera using computer vision techniques. It is also able to filter the background such as wind, rain, snow, fallen leaves, birds, and floating flags. It is so called object tracking in complex background. Due to the variant illusion, motion, occlusion, color, and complex background, the difficulty of object detection and tracking algorithm design is increased.

The steps in image and video analysis mainly include segmentation, classification, and explanation. The classification process normally extracts the features by SIFT and LBP. With the use of deep learning techniques, people start using deep feature by extracting automatically for image classification, scenario classification, and behaviors analysis.

Our purpose in writing this book is to present advanced applications in image and video processing. We believed that this book is distinguished from other MATLAB-based fundamental textbooks which only introduces the basic functions such as the transform, enhancement, restoration, coding, and resizing of image. Our book emphasized the advanced applications such as image dehazing correction, image deraining correction, image stitching, image watermarking, visual object recognition, moving object tracking, dynamic scene classification, pedestrian re-identification, behavior analysis with deep learning, and so on.

The book is divided into three parts:

Part I: The Basic Concepts

Chapter 1 briefly introduces the fundamental principles including the analysis techniques: scene segmentation, feature description, and object recognition. There are also some summaries about examples of advanced applications, such as image fusion, image inpainting, image stitching, image watermarking, object tracking, and pedestrian re-identification.

Chapter 2 introduces the functions of MATLAB toolboxes for image and video processing.

Chapter 3 presents the image and video segmentation methods of threshold, region-based, partial differential equation, clustering, graph theory, and cumulative difference-based motion region extraction.

Chapter 4 presents the feature extraction and representations, which includes Harris corner detection, SUSAN edge detection, the point feature detection algorithm SIFT and SURF.

Part II: Advances in Image Processing

This part includes the image processing techniques such as image correction, image inpainting, image fusions, image stitching, image watermarking.

Chapter 5 firstly introduces three filters for image denoise and blurred functions. Then, it mainly introduces the correction techniques of image dehazing, image deraining, and text image feature correction.

Chapter 6 presents the image inpainting techniques including the principle, structure, algorithm, and some example codes.

Chapter 7 firstly introduces the fusions types and their schemes and then mentioned a very important method: wavelet transform for image fusion. Finally, it discusses the evaluation of image fusion objectively and subjectively.

Chapter 8 introduces the image stitching techniques such as region-based, feature-based, and feature point method. The SIFT and Harris corner detection algorithms are also introduced in this chapter.

Chapter 9 briefly introduces the image watermarking in three different transforms which are spatial-domain-based, DCT-based and DWT-based watermarking techniques.

Chapter 10 introduces the object recognition techniques including face recognition, facial expression, and image-to-character extraction and recognition.

Part III: Advances in Video Processing and then Associated Chapters

Chapters 11–14 mainly introduce the video processing techniques of moving object tracking, dynamic scene classification based on TMBP, behavior recognition based on LDA topic model, person re-identification based on metric learning, lip recognition instance based on deep learning model, and deep CNN architecture for event recognition.

Chapter 11 introduces the object tracking techniques using Gaussian mixture model for background detection, and the RANSAC for feature points tracking. Further extend the mean-shift object tracking algorithm.

Chapter 12 introduces the dynamic scene classification and discusses the TMBP and LDA models for the classification.

Chapter 13 presents a person re-identification method by using the image understanding technique.

Chapter 14 presents the deep learning in image and video understanding.

For the convenience of the readers to evaluate the performance of the algorithms, we also give the common evaluation criteria in the appendix.

This book is written by Shengrong Gong, Chunping Liu, Yi Ji, Baojiang Zhong, Yonggang Li, Husheng Dong, Conghua Xie, Wei Pan, Yu xia, and Zhaohui Wang. Our M.Sc. researchers take participate in debugging most of the programs. They are Xinhua Dai, Ran Yan, Zongming Bao, and Pengcheng Zhou.

We gratefully acknowledge the professional suggestions of the reviewers and editors of Springer Press. We also thank to the support of Changshu Institute of Technology and Soochow University. We appreciate the support of National Natural Science Foundation of China (NSFC Grant No. 61170124, 61272258,

61301299), Integration of Cloud Computing and Big Data, Innovation of Science and Education (Grant No. 2017B03112), Provincial Natural Science Foundation of Jiangsu, China (Grant No. BK20151260, BK20151254), Six talent peaks project in Jiangsu Province, China (Grant No. DZXX-027).

Changshu, China   Shengrong Gong
Suzhou, China   Chunping Liu
Suzhou, China   Yi Ji
Suzhou, China   Baojiang Zhong
Jiaxing, China   Yonggang Li
Suzhou, China   Husheng Dong

Contents

Part I   The Basic Concepts

1   Introduction   3
   1.1   Basic Concepts and Terminology   3
       1.1.1   Digital Image and Digital Video   3
       1.1.2   Image Processing   6
       1.1.3   Image Analysis   6
       1.1.4   Video Analysis   8
   1.2   Image and Video Analysis   9
       1.2.1   Image and Video Scene Segmentation   9
       1.2.2   Image and Video Feature Description   10
       1.2.3   Object Recognition in Images/Videos   12
       1.2.4   Scene Description and Understanding   13
   1.3   Examples of Advanced Applications   14
       1.3.1   Image Correction   14
       1.3.2   Image Fusion   15
       1.3.3   Digital Image Inpainting   15
       1.3.4   Image Stitching   16
       1.3.5   Digital Watermarking   17
       1.3.6   Visual Object Recognition   18
       1.3.7   Object Tracking   20
       1.3.8   Dynamic Scene Classification   21
       1.3.9   Pedestrian Re-identification   22
       1.3.10 Lip Recognition in Video   22
   References   23

2   Matlab Functions of Image and Video   27
   2.1   Introduction to MATLAB for Image and Video   27
   2.2   Basic Elements of MATLAB   28

	2.2.1	Working Environment	28
	2.2.2	Data Types	29
	2.2.3	Array and Matrix Indexing in MATLAB	32
	2.2.4	Standard Arrays	34
	2.2.5	Command-Line Operations	34
2.3	Programming Tools: Scripts and Functions		35
	2.3.1	M-Files	35
	2.3.2	Operators	36
	2.3.3	Important Variables and Constants	38
	2.3.4	Number Representation	38
	2.3.5	Flow Control	39
	2.3.6	Input and Output	41
2.4	Graphics and Visualization		41
2.5	The Image Processing Toolbox		46
	2.5.1	The Image Processing Toolbox: An Overview	46
	2.5.2	Essential Functions and Features	47
	2.5.3	Displaying Information About an Image File	52
	2.5.4	Reading an Image File	52
	2.5.5	Data Classes and Data Conversions	53
	2.5.6	Displaying the Contents of an Image	55
	2.5.7	Exploring the Contents of an Image	57
	2.5.8	Writing the Resulting Image onto a File	58
2.6	Video Processing in MATLAB		58
	2.6.1	Reading Video Files	59
	2.6.2	Processing Video Files	59
	2.6.3	Playing Video Files	60
	2.6.4	Writing Video Files	61
	2.6.5	Basic Digital Video Manipulation in MATLAB	62
References			63

3 Image and Video Segmentation 65
3.1	Introduction		65
3.2	Threshold Segmentation		66
	3.2.1	Global Threshold Image Segmentation	68
	3.2.2	Local Dynamic Threshold Segmentation	69
3.3	Region-Based Segmentation		74
	3.3.1	Region Growing	74
	3.3.2	Region Splitting and Merging	78
3.4	Segmentation Based on Partial Differential Equation		88
3.5	Image Segmentation Based on Clustering		94
3.6	Image Segmentation Method Based on Graph Theory		97
	3.6.1	Introduction	97
	3.6.2	GraphCut and Improved Image Segmentation Method	99

3.7   Video Motion Region Extraction Method Based
        on Cumulative Difference   107
References   111

4   Feature Extraction and Representation   113
4.1   Introduction   113
4.2   Histogram-Based Features   115
        4.2.1   Grayscale Histogram   115
        4.2.2   Histograms of Oriented Gradients   117
4.3   Texture Features   121
        4.3.1   Haralick Texture Descriptors   122
        4.3.2   Wavelet Texture Descriptors   126
        4.3.3   LBP Texture Descriptors   131
4.4   Corner Feature Extraction   135
        4.4.1   Moravec Algorithm   135
        4.4.2   Harris Corner Detection Operator   137
        4.4.3   SUSAN Corner Detection Algorithm   141
4.5   Local Invariant Feature Point Extraction   144
        4.5.1   Local Invariant Point Feature of SURF   145
        4.5.2   SIFT Scale-Invariant Feature Algorithm   149
References   158

Part II   Advances in Image Processing

5   Image Correction   161
5.1   Introduction   161
5.2   Noise Reduction Using Spatial-Domain Techniques   161
        5.2.1   Selected Noise Probability Density Functions   162
        5.2.2   Filtering   168
5.3   Image Deblurring   173
        5.3.1   The Restoration of Defocus Blurred Image   174
        5.3.2   Restoration of Motion Blurred Image   176
5.4   Fisheye Distortion Correction Using Spherical Coordinates
        Model   180
5.5   Skew Correction of Text Images   186
        5.5.1   Feature Analysis of Text Images   187
        5.5.2   The Basic Idea of Hough Transform   187
        5.5.3   The Implementation Steps of Text Images Skew
                  Correction   188
5.6   Image Dehazing Correction   191
        5.6.1   Single Image Dehazing   191
        5.6.2   Dark Channel Prior   192
        5.6.3   Implementation Steps of DCP   194
        5.6.4   Refine Transmission Map Using Soft Matting   195

5.7   Image Deraining Correction...................... 200
       5.7.1   Related Work........................... 200
       5.7.2   Single Image De-rain with Deep Detail Network 200
       5.7.3   Implementation of Image Deraining with Deep
                Network 203
References 206

6   Image Inpainting.. 209
    6.1   Introduction 209
           6.1.1   Structure Oriented Image Inpainting Technology 210
           6.1.2   Texture-Based Image Inpainting Technology 211
    6.2   The Principle of Image Inpainting 211
    6.3   Variational PDE-Based Image Inpainting 213
           6.3.1   Image Inpainting Algorithm Based on Total
                    Variational Model 214
           6.3.2   Image Inpainting Based on CDD Model 219
    6.4   Exemplar-Based Image Inpainting Algorithm 222
    References 230

7   Image Fusion 233
    7.1   Introduction 233
    7.2   Fusion Categories.................................. 234
           7.2.1   Multi-view Fusion 234
           7.2.2   Multimodal Fusion 236
           7.2.3   Multi-temporal Fusion 240
           7.2.4   Multi-focus Fusion 242
    7.3   Image Fusion Schemes 243
    7.4   Image Fusion Using Wavelet Transform.................. 248
           7.4.1   Basis of Wavelet Transform 248
           7.4.2   Discrete Dyadic Wavelet Transform of Image
                    and Its Mallat Algorithm 249
           7.4.3   Steps of Implementation 250
    7.5   Region-Based Image Fusion 253
           7.5.1   Basic Framework of Regional Integration 254
           7.5.2   The Strategy of Regional Joint Representation 255
           7.5.3   The Rules of Fusion.......................... 256
           7.5.4   Wavelet Fusion of Regional Variance 256
    7.6   Image Fusion Using Fuzzy Dempster-Shafer Evidence
           Theory 260
    7.7   Image Quality and Fusion Evaluations 263
           7.7.1   Subjective Evaluation of Image Fusion 264
           7.7.2   Objective Evaluation of Image Fusion 264
    References 268

8   Image Stitching 271
      8.1   Introduction 271
      8.2   Image Stitching Based on Region 272
             8.2.1   Image Stitching Based on Ratio Matching 273
             8.2.2   Image Stitching Based on Line and Plane Feature 276
             8.2.3   Image Stitching Based on FFT 283
      8.3   Images Stitching Based on Feature Points 290
             8.3.1   SIFT Feature Points Detection. 290
             8.3.2   Image Stitching Based on Harris Feature Points 297
             8.3.3   Auto-Sorting for Image Sequence 304
             8.3.4   Harris Point Registration Based on RANSAC
                      Algorithm 307
      8.4   Panoramic Image Stitching 320
      References 327

9   Image Watermarking 329
      9.1   Introduction 329
      9.2   Fragile Watermarking Based on Spatial Domain 334
      9.3   Robust Watermarking Based on DCT 336
      9.4   Semi-fragile Watermarking Based on DWT 344
      References 349

10   Visual Object Recognition 351
      10.1   Face Recognition Based on Locality Preserving Projections . . . 351
      10.2   Facial Expression Recognition Using PCA 375
      10.3   Extraction and Recognition of Characters in Pictures 380
      References 387

Part III   Advances in Video Processing and then Associated Chapters

11   Visual Object Tracking 391
      11.1   Adaptive Background Modeling by Using a Mixture
             of Gaussians. 391
      11.2   Object Tracking Based on Ransac 396
      11.3   Object Tracking Based on MeanShift. 401
             11.3.1   Description of the Object Model 402
             11.3.2   A Description of the Candidate Model. 402
             11.3.3   Similarity Function. 403
             11.3.4   Object Location 403
      11.4   Object Tracking Based on Particle Filter 409
             11.4.1   Prior Knowledge of the Goal 410
             11.4.2   System State Transition 410
             11.4.3   System Observation 411
             11.4.4   Posterior Probability Calculation 412

        11.4.5 Particle Resampling 412
        11.4.6 Implementation Steps 413
    11.5 Multiple Object Tracking 418
    References 427

12 **Dynamic Scene Classification Based on Topic Models** 429
    12.1 Overview 429
    12.2 Introduction to the Topic Models...................... 430
        12.2.1 LDA Model 430
        12.2.2 TMBP Model Based on Factor Graph 433
        12.2.3 TMBP Model Fusing Prior Knowledge 436
    12.3 Dynamic Scene Classification Based on TMBP 439
    12.4 Behavior Recognition Based on LDA Topic Model 451

13 **Image Understanding-Person Re-identification** 475
    13.1 Introduction 475
    13.2 Person Re-ID Scenarios.............................. 477
    13.3 Methodology 478
    13.4 Public Datasets and Evaluation Metrics in Person
          Re-identification 480
        13.4.1 Public Datasets............................ 480
        13.4.2 Evaluation Metrics 483
    13.5 Classic Feature Representations for Person
          Re-identification 484
        13.5.1 Salient Color Names........................ 484
        13.5.2 Local Maximal Occurrence Representation 487
    13.6 An Example of Metric Learning Based Person
          Re-identification Method-XQDA 501
    References 511

14 **Image and Video Understanding Based on Deep Learning** 513
    14.1 Introduction 513
    14.2 Model Analysis of CNN 515
        14.2.1 Basic Modules of CNN 515
        14.2.2 Convolution and Pooling 515
        14.2.3 Activation Function 516
        14.2.4 Softmax Classifier and Cost Function 517
        14.2.5 Learning Algorithm 519
        14.2.6 Dropout.................................. 521
        14.2.7 Batch Normalization........................ 522
    14.3 Typical CNN Models 522
        14.3.1 LeNet 522

14.3.2 AlexNet..................................... 523
14.3.3 GoogLeNet 524
14.3.4 VGGNet 528
14.3.5 ResNet 530
14.4 Deep Learning Model for Lip Recognition Instance 531
14.4.1 Testing Dataset 531
14.4.2 Deep Network Training 532
14.4.3 Code Analysis 536
14.5 Deep CNN Architecture for Event Recognition Instance 539
14.5.1 Testing Dataset 539
14.5.2 Deep Feature Extraction 540
14.5.3 Spatial-Temporal Feature Fusion 540
14.5.4 Fisher Vector Encoding 541
14.5.5 Code Analysis 542
References 553

Appendix: Common Evaluation Criterion........................ 555

About the Authors

Shengrong Gong received his M.S. from Harbin Institute of Technology in 1993 and his Ph.D. from Beihang University in 2001. He is the Dean of School of Computer Science and Engineering, Changshu Institute of Technology, and also a Professor and Doctoral Supervisor. His research interests are image and video processing, pattern recognition, and computer vision.

Chunping Liu received her Ph.D. in pattern recognition and artificial intelligence from Nanjing University of Science and Technology in 2002. She is now a Professor of School of Computer Science and Technology, Soochow University. Her research interests include computer vision, image analysis and recognition, in particular in the domains of visual saliency detection, object detection and recognition, and scene understanding.

Yi Ji received her M.S. from National University of Singapore, Singapore, and Ph.D. from INSA de Lyon, France. She is now an Associate Professor in School of Computer Science and Technology, Soochow University. Her research areas are 3D action recognition and complex scene understanding.

Baojiang Zhong received his B.S. in mathematics from Nanjing Normal University, China, in 1995, M.S. in mathematics, and Ph.D. in mechanical and electrical engineering from Nanjing University of Aeronautics and Astronautics (NUAA), China, in 1998 and 2006, respectively. From 1998 to 2009, he was on the Faculty of the Department of Mathematics of NUAA and reached the rank of Associate Professor. During 2007–2008, he was also a Research Scientist at the Temasek Laboratories, Nanyang Technological University, Singapore. In 2009, he joined the School of Computer Science and Technology, Soochow University, China, where he is currently a Full Professor. His research interests include computer vision, image processing, and numerical linear algebra.

Yonggang Li received his M.S. from Xi'an Polytechnic University in 2005. He is currently pursuing Ph.D. in School of Computer Science and Technology, Soochow University. He is a Lecturer of College of Mathematics, Physics and Information

Engineering, Jiaxing University. His research interests include computer vision, image and video processing, and pattern recognition.

Husheng Dong received his M.S. from School of Computer Science and Technology, Soochow University, in 2008, and he is pursuing Ph.D. currently. He is also a Lecturer of Suzhou Institute of Trade & Commerce. His research interest includes computer vision, image and video processing, and machine learning.

Part I
The Basic Concepts

Chapter 1
Introduction

Abstract In this chapter we introduce some basic concepts and terminology about digital image and video analysis. Then some example applications are listed.

1.1 Basic Concepts and Terminology

1.1.1 Digital Image and Digital Video

The digital image can be understood as a matrix obtained by a two-dimensional function sampling f and quantization of the sampled results. The resulting digital image is usually represented by a two-dimensional matrix.

Sampling is the discretization of an image according to the spatial coordinate of each pixel, which determines the final spatial resolution. As shown in Fig. 1.1, it first uses some grids to cover the analog image, and then averages the brightness of each cell. Or it directly takes the value at each intersection as the value of one grid. In this way, an analog image is discretized by representing the value of each grid as a digital number. This grid is called sampling grid, and it defines the width and height of the final image after sampling and quantization. Each element of the obtained digital image is a discrete value which is usually called a pixel.

Let the numbers of a row and column be M and N, then the size of an image is $M \times N$. The pixel values constitute a real matrix with the size of $M \times N$ which can be represented as

$$f(x, y) = \begin{bmatrix} f(0, 0) & f(0, 1) \cdots & f(0, N-1) \\ f(1, 0) & f(1, 1) \ldots & f(1, N-1) \\ \vdots & & \\ f(M-1, 0) & & \ldots f(M-1, N-1) \end{bmatrix}. \quad (1.1)$$

© Springer International Publishing AG, part of Springer Nature 2019
S. Gong et al., *Advanced Image and Video Processing Using MATLAB*,
Modeling and Optimization in Science and Technologies 12,
https://doi.org/10.1007/978-3-319-77223-3_1

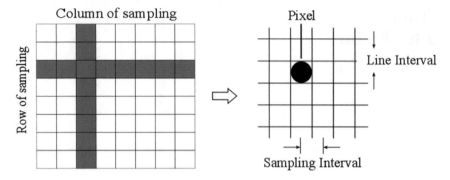

Fig. 1.1 Structure of grid and sampling method

The conversion of pixel value from analog to the discrete amount is called the quantization, which determines the final amplitude resolution of the image. There are two ways of quantizing grayscale values, one is equidistant quantification, and the other is non-equidistant quantification. Non-equidistant quantization is usually performed based on the probability density function of pixel value distribution and the principle of minimum quantification error. Specifically, this means we need to set small quantization interval for the grayscale values appear frequently in the image, while set larger intervals for pixels rarely appear. Due to the probability distribution functions of grayscale values differs on different images, it is impossible to find an optimal non-equidistant quantification scheme for all images. Therefore, equidistant quantification is more widely used in practice.

Figure 1.2 shows an example of an image scaled to 256 gradations evenly. Figure 1.2a is the whole image quantized with 256 gray-scales, Fig. 1.2b is a subgraph of 16×16 pixels cropped from Fig. 1.2a, c is the corresponding quantized data.

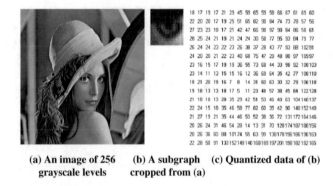

(a) An image of 256   **(b) A subgraph**   **(c) Quantized data of (b)**
grayscale levels   **cropped from (a)**

Fig. 1.2 An example of image quantization

When the number of quantization levels is a constant, the more sampling points of the image, the better quality it has. When the number of sampling points decreases, the block effect on the graph becomes more obvious. Similarly, when the number of sampling points is constant, the more quantization levels, the better image quality it has. When the number of quantization stages decreases, the image quality will be worse.

Video is the dynamic form of static images. In other words, the video is composed of a series of static images in a certain order, and each image is called a 'frame'. These frames are continuously projected onto the screen at a constant speed, resulting in a dynamic effect due to the presence of visual persistence. Similar to the image, video can also be categorized into analog and digital.

The standard of video signals in the process of generation, transmission, and display is called system. Common systems include NTSC, PAL, and SECAM. In the PAL color TV system, YUV color space is adopted where Y represents the brightness signal, U and V represent the color difference signals. The computer displays images in the RGB color space, and it requires the color components of YUV to be converted to RGB values first. The conversion formulas are as follows [1, 2]:

$$Y = 0.299R + 0.587G + 0.14B$$
$$U = -0.169R - 0.331G + 0.5B$$
$$V = 0.5R - 0.419G - 0.81B \tag{1.2}$$

Since the human eyes are less sensitive to color than to brightness, the sampling frequency of the color difference signals can be lower than the luminance signal to reduce the data size. Let Y:U:V represent the sampling ratio of Y, U, and V, in practice the digital video sampling ratios include 4:1:1, 4:2:2, 4:4:4, and the format of 4:2:2 is recommended by ITU-R.[1] In this case, the color difference signal takes half of the sampling frequency of the luminance signal.

Similar to the quantization of digital images, video quantization also requires discretizing the continuous pixel values, and the quantization rate determines the dynamic range of the system. With higher bit rate the digital video will be of high quality, but it needs more storage space and streaming bandwidth in turn.

The bit rate of digital video is determined by the frame rate f_p, frame height M, frame width N, and the total bits of a pixel. For example, if we use (R, G, B) color space and 8 bits per channel to represent PAL standard color digital video (the frame rate is $f_p = 25$ frames per second), the frame size is $0 = 576$ and $N = 720$, then the bit rate is $720 \times 576 \times 25 \times 24 = 249$ Mbit/s.

---

[1]ITU-R: International Telecommunication Union—Radio communications sector.

1.1.2 Image Processing

Image processing refers to the technique of performing a series of operations on an image to achieve some desired purposes [1–3]. It can be divided into two types: analog image processing and digital image processing.

Analog image processing is to process analog images by using optical, photographic and electronic methods. The earliest image processing is the work regarding of the light, such as using magnifier and microscope to magnify objects. Due to the fast processing performance of the optical imaging, many military and astronautics processing still use optical analog processing nowadays. In addition to its solid theoretical foundations, the optical image processing has the advantages of high speed, large capacity, and high resolution. There are also some disadvantages, such as the low processing precision, poor stability, bulky equipment, and inconvenient operation. Because the processing is achieved by optical devices such as lens and prisms, we have to afford for a long time to design and manufacture them, and the accuracy cannot be guaranteed [1, 4, 5].

Different from analog image processing, the digital image processing utilizes computer technologies to obtain some expected results. Generally speaking, computer image processing and digital image processing can be regarded as synonyms. Sometimes digital image processing is also referred as image processing. In this book, "image processing" means "digital image processing" unless specified.

The image processing can be roughly divided into three categories, narrow sense of image processing, image analysis, and image understanding. In its narrow sense, image processing emphasizes the transformation of images, which is a process from image to image on a lower level.

Let $f(x, y)$ represent the source image, $g(x, y)$ be processed image, and T be the processing operation, then the narrow sense image processing can be described as:

$$g(x, y) = T[f(x, y)]. \tag{1.3}$$

1.1.3 Image Analysis

The image analysis mainly aims to detect and measure the objects of interested in the digital image, in order to create specific descriptions. It is also known as scene analysis or image understanding [6]. Its content can be divided into several parts, such as feature extraction [7], object description [8], target detection [9], scene matching and recognition [10]. It is a process from an image to values or symbols, which generates some non-image descriptions or representations by extracting useful data or information first. The data here may be the result of the target feature measurement, or the symbolic representation based on the measurement. They describe the characteristics and features of the target in the image. Thus, image analysis is also referred to middle-level processing.

The image analysis is not only for image region classification but also for complex image description in the variant and unseen scenes. To "understand" the circumstantial fact of an image, the description needs to be more intelligent as a human being in logical inference, thinking and associating with the cognition in the objective world instead of simply represented in symbols. Therefore, image analysis relies on some algorithms to identify the relationships among objects and the background in the scene of an image.

Image analysis and image processing are closely related, although there may be some conceptual overlaps between them, they are different in essence. Image processing mainly focuses on the research of signal processing, such as image transmission, storage, enhancement and restoration. While image analysis emphasizes describing images with symbolic representations, and it also uses a variety of background knowledge for reasoning,analysis, interpretation, and recognition. Therefore, image analysis is more related to pattern recognition and computer vision, and it is generally used to analyze the underlying features or the relationships between objects, by some mathematical models. Image understanding includes three levels: low-level image primitives such as edges, texture elements, or regions; intermediate-level includes boundaries, surfaces and volumes; and high-level includes objects, scenes, or events. For example, in image-to-sentence conversion, it first extracts the features; then uses the classification functions to label the features; use the labels as the words of a sentence. Finally, the system re-orders the words by high-level semantics. It is very difficult to implement this function from the features to the sentence expression. However, it is easier to implement from words to sentence by using the classification. We call it middle semantics.

Image analysis basically has the following four stages:

(1) Pre-processing. The actual scene is converted into a suitable form for computer processing. Sometimes, the three-dimensional scene is converted into a two-dimensional image.

(2) Segmentation. The objects in the scene are recognized and decomposed in this phase, and this requires the application of knowledge of the objective world. In general, image segmentation can be considered as a decision-making process, and its algorithms can be divided into two categories of pixel technology and regional technology. The pixel technique uses the threshold method to classify each pixel. For example, we can obtain the strokes of a character image by comparing the gray level with the threshold of each pixel. The regional technology determines various components of one image by the texture, local area gray-scale contrast, and other characteristics such as boundaries, lines, regions, etc.

(3) Recognition. To name or label the objects which have been segmented from the image, such as pedestrians, cars, buildings, and so on in natural scenes. Generally, they are classified into different categories with decision-making theory and structural methods. We can also construct a series of templates of known objects, and then match the unknown objects with them for identification.

(4) Explanation. To create a hierarchical structure using some heuristic methods or human-computer interaction technologies for objects identification in the scene and relationship description of them. In the case of a three-dimensional scene, the knowledge about the constraints of the objects in the real world can be utilized. For example, we can infer the three-dimensional surface of objects in an image from the shadow, texture changes and the contour. According to the distance, angle, and depth of field information, we can obtain the description and interpretation of 3-dimensional objects in the scene.

1.1.4   Video Analysis

Similar to image analysis, video analysis is also a broad concept which includes the tasks of visual object tracking, human action analysis, abnormal behavior detection, and so on [11, 12]. Intelligent video surveillance is an application of video analysis. In intelligent video surveillance, the users can analyze the monitored video by presetting some alarm rules in different scenarios. Once the rules are violated, the system will automatically send an alarm. Due to various noises in surveillance video, video analysis must have the ability to filter the wind, rain, snow, fallen leaves, birds, floating flags, etc. Usually, we can achieve this by establishing human activity models, excluding non-human interference factors and modeling backgrounds.

In real-world environments, due to the changes in illumination, the target movement complexity, occlusion, color similarity between targets and background, and background clutter, it is difficult to design a robust algorithm for target detection and tracking. The challenges mainly reside in the following aspects:

(1) Background complexity. The changes in illumination may lead to large variations in target color and background color, which usually results in false detection and error tracking. Although using different color spaces can reduce the impact of light changes, it cannot be eliminated completely. When the target color is close to the background, the detection and tracking will be seriously affected. When the target shadow is different from the background color, it may be classified as the foreground incorrectly, which may bring difficulties to segmentation and feature extraction of the moving target.

(2) Target feature selection. The video contains a large number of information that can be used for target tracking, such as motion, color, edge, and texture. However, the features of the target are generally time-varying. Thus it is difficult to select the most appropriate features to ensure the effectiveness of tracking.

(3) Occlusion. When the moving target is partially or completely occluded, or multiple targets occlude each other. The occlusion will affect the stability of tracking due to the missing information of the occluded part of the target.. In order to reduce the ambiguity caused by occlusion, it is necessary to deal with the correspondence between features and target correctly.

(4) The balance between real-time processing and robustness. As video contains a lot of information, we have to choose the algorithms that are less time-consuming such that the target tracking can meet the real-time requirements. Robustness is another aspect that should be considered in target tracking, which means the algorithm should be applicable under complex background, light changes and occlusion, etc. However, this will in turn recur high computational cost. Therefore, it is a non-trivial task to balance the computational cost and robustness.

1.2 Image and Video Analysis

1.2.1 Image and Video Scene Segmentation

Scene segmentation is the key step in the image and video analysis, it refers to divide image or video sequence into some specific parts or subsets with unique characteristics, and then extract the interested target [13]. The purpose is to isolate a meaningful entity from image or video sequence. This meaningful entity is also called an object, which is the basis for the extraction, identification and tracking of the interested target.

In the research of image and video analysis, people tend to be interested in only some special parts which are often called targets or foregrounds (other parts are called backgrounds). These targets typically correspond to some specific, unique areas of images or video frames. The uniqueness here can be the grayscale value of the pixel, object contour curve, color, texture, movement information, etc. Such uniqueness can be used to represent an object as the characteristics between regions of different objects usually change dramatically. The target can correspond to a single region or multiple regions. To identify and analyze the targets, it is necessary to isolate and extract them, such that further identification and understanding can be carried out.

The segmentation of images or video frames can be implemented in a pixel-wise way, or by using some information in the specified field. The basis of image segmentation includes two important concepts called "similarity" and "discontinuity" in the digital image. The so-called pixel similarity means that the pixels in one region have similar characteristics, such as pixel gray level or texture formed by pixel arrangement. The "discontinuity" refers to the discontinuity of the pixel grayscale, which forms a jump step in values and the mutation of texture structure.

Image segmentation is generally achieved by considering the image color, grayscale, edge, texture, and other spatial information. Currently, image segmentation algorithms can be divided into two categories: structural segmentation and non-structural segmentation. Structure segmentation methods are based on the characteristics of the local area of image pixels, including threshold segmentation, region growing, edge detection, texture analysis, etc. These methods assume that the features of these areas are known in advance and they are obtained during processing. Non-structural segmentation methods include statistical pattern recognition, neural network methods, and the methods using the prior knowledge of relationships between objects. For example, the snakes [14] method which uses active contour model to segment objects, is a framework in computer vision for delineating an object outline from a possibly noisy 2D image. A snake is an energy minimizing, deformable spline influenced by constraint and image forces that pull it towards object contours and internal forces that resist deformation. It may be understood as a special case of the general technique of matching a deformable model [9, 15] to an image by means of energy minimization. In two dimensions, the active shape model represents a discrete version of this approach, taking advantage of the point distribution model to restrict the shape range to an explicit domain learnt from a training set. The snakes model is popular, and snakes are greatly used in applications like object tracking, shape recognition, segmentation, edge detection and stereo matching.

Because there is no temporal information in image segmentation, it cannot be used to get satisfactory segmentation results on video sequences. The efficiency of segmentation algorithms can be improved by considering the time correlation of video frames. Therefore, video segmentation jointly uses the spatial and temporal information to achieve this goal.

Besides classical segmentation methods based on edge, threshold, entropy, region, there are also some methods using graph theory, clustering, random models, fuzzy sets, partial differentiation, image fusion, etc.

1.2.2 Image and Video Feature Description

When an image or video sequence is segmented into objects and backgrounds, a further step is to describe the characteristics of the scene with a series of symbols or rules, and then identify, analyze, and categorize the descriptions. This inspires the work of feature engineering [16]. Image features refer to its original properties or attributes. Some are natural and can be perceived by the vision, such as the brightness of the region, edges, texture or color, etc. There are also some artificial ones that need to be transformed or measured, such as transform spectrum, histogram, moment, etc. In general, descriptors refer to a series of symbols that are used for describing the characteristics of an image or video object. A good descriptor should be insensitive to the target's scale, rotation, translation, etc. Feature descriptors generally fall into two main types: global and local.

The global feature is calculated from all the pixels of an image, and it describes the image as a whole. Commonly used features include color, texture and shape features.

Color features reflect the overall characteristics of a color image or video frames. An image or video frame can be approximately represented by its color properties. Compared with other type features, the color feature is less dependent on the scale, rotation angle and viewpoint, thus has stronger stability. In addition, the calculation of color features is generally simple and fast. According to the relationship between color and spatial attributes, we can describe color features by color moments, color histograms, color correlations, and so on. Color moments are generally calculated in the RGB space. As most information is associated with low moments only, in practice we only extract color moment of level-1 to level-3. The color histogram describes the statistical properties of image color distribution. The histograms of a color image can be computed from different color spaces, such as RGB, HSV, Lab, and so on. The color correlation feature is similar to the color histogram, but it also considers spatial information. The color features are essentially global properties. Thus they cannot well capture the local characteristics in image.

The texture features describe the surface properties of an image or local regions. Unlike the color features, texture features are not based on pixel; they are obtained by statistical calculations in local regions that contain multiple pixels. As a statistical feature, texture feature is robust to rotation and noise. However, it also has some drawbacks. An obvious disadvantage is that when the resolution changes, the calculated textures may have large deviations. Besides, it is difficult to accurately describe the difference between different textures perceived by human visual system. The texture features can be divided into two categories: statistical methods and structural methods. The statistical method is based on the statistical analysis of some related properties; The structural method is to find texture primitives, and then explore the rules that they compose the texture structures. For example, the forests, mountains and grasslands in remote sensing images have small texture and no regular rules. Therefore the statistical methods are generally used. For more regular textures, structural methods are generally applied. Among existing statistical methods, there are some works that study the statistical properties of the texture regions, some works that study the first-order statistical properties of grayscale or other second-order or higher-order statistical properties of multiple pixels. There are also some works utilize models (e.g., Markov model, Fractal model) to describe textures. The most classical and commonly used methods of describing the global texture feature mainly include the texture co-occurrence matrix representation, the texture feature set, and the Gabor filter features.

The shape feature describes the shape characteristics of the objects in an image or video frames, in which the edges and regions are mainly described. The commonly used methods of extraction and analysis of shape features include the spatial domain analysis of internal regions, the transformation analysis, and the shape characterization of regional boundaries. The spatial domain analysis extracts the shape features from spatial domain in local regions directly, such as Euler number, concave convexity, distance and region measurement, etc.

Compared with global features, local features describe the local regions in an image with better uniqueness, invariability and robustness, and they have the better robustness to background clutter, local occlusion, and illumination changes. Local features may be points, edges, or blobs in an image, which have the advantages of describing the characteristics of pixels or colors in local regions. Due to its excellent performance, local features have attracted more and more research attention. Local features have been widely used in computer vision tasks, such as image retrieval, image registration, image recognition, image classification, etc. In particular, some local features with strong robustness to illumination and occlusion have been proposed in recent years, such as Moravec corner detector [17], Harris corner detector

[18], Smallest Univalue Segment Assimilating Necleus corner detector (SUSAN) [19], Scale Invariant Feature Transform (SIFT) [20], Difference of Gaussian (DOG) operator and Gradient Location and Orientation Histogram (GLOH), Speeded Up Robust Features (SURF) [21], Maximally Stable Extreme Regions (MSER) [22], Local Binary Pattern(LBP) [23], etc.

1.2.3 Object Recognition in Images/Videos

The recognition ability of humankind is rather powerful, even with dramatic scale changes, large displacement, and heavy occlusion, people can still identify the objects.

In computer vision, image recognition mainly refers to the task of recognizing objects in an image or video sequence [24]. We employs some computational models to extract features from a two-dimensional image to form the digital description, and then establish a classifier for classification and recognition.

Classifier designing is the process of optimizing models using the training samples, which is also a machine learning procedure of minimizing the classification error for all the training samples. The purpose is to train a classification model which can automatically classify unknown data into specified classes. The classifiers can be divided into three categories: Generative Model (including probability density model), Discriminative Model (decision boundary learning model). Recently, the deep learning [25] based models have been widely applied in object recognition task, which can be viewed as another category.

Generative Model is also called Productive Model, which tries to estimate the joint probability distribution of training samples (observations) and their labels. Generative Model has a flexible and clear hierarchy, and the model is interpretive. The input and output variables (and implicit variables) of Generative Model are represented by the joint probability distribution. These variables can be discrete or continuous, or multi-dimensional. Since the generative model is a distribution model for all variables, it can be used for classification or regression through standard marginalization and restricted operations. Popular generative methods include: Gaussian Mixture Model (GMM), Naive Bayes Model (NBM), Mixtures of Multinomial Model (MMM), Mixtures of Experts System (MES), Hidden Markov Models (HMM), Latent Dirichlet Allocation (LDA), Sigmoidal Belief Networks (SBN), Bayesian Networks (BN), Markov Random Fields (MRF), etc. For example, we can learn the attributes from a large number of training samples via the topic model (e.g., LDA) and then apply them to recognize different types of human actions [26], or classify different types of scenes by considering their latent topics [27]. With the foreground targets detected by the GMM [28], we can further conduct motion analysis or object tracking task.

Discriminative Model is also called Conditional Model, or Conditional Probability Model. Compared with the generative model, it is much more straightforward. During the training phase, it tunes the parameters by the samples and their classification labels. Discriminative Model mainly calculates the edge distribution, and its objective function is directly related to classification accuracy. The objective is to look for the optimal classification surface between different categories, which well reflects the difference between heterogeneous data. The discriminative method does not model the basic distribution of variables and labels, it is only interested in the optimization of the mapping between input and output. As there are no intermediate objectives for modeling variables, much higher classification accuracy can be obtained. The commonly used discriminative methods include Linear Discriminant Analysis, Logistic Regression, Artificial Neural Networks, Support Vector Machine, Nearest Neighbor, Boosting trees, Conditional Random Fields, etc. These classification algorithms have been widely applied to face recognition [29], handwritten digits recognition [30], object detection [9, 15], pedestrian detection [31], and so on.

With the substantial increase in the amount of available training data, and continuous improvement of the computing power of hardware devices (especially the rapid development of GPU), deep learning has achieved great success in a number of applications. Different from traditional object recognition pipeline of "feature extraction—classification", the object recognition can be achieved in an "end-to-end" way through deep learning. Recently, deep learning based models have been widely applied to face recognition [32], fine-grained classification [33], pedestrian detection [34] and re-identification [35], visual tracking [36], and so forth. Almost all object recognition tasks has been shined by deep learning in nowadays, and the performances are greatly improved.

Aiming at various specific problems in image recognition, the employed models are different from each other. For example, in the case of multi-objective recognition, we should not only consider the interference of complex background, but also take the situations of mutual occlusion, merger and separation between targets into account. Sometimes, we also need to guide the selection and integration of information via prior knowledge and conduct repeated hypothesis testing or complex feedback processing.

1.2.4 Scene Description and Understanding

Scene description and understanding are the high-level tasks of image understanding [37]. The main objective is to automatically assign labels to the image scene via a set of semantic categories, in order to provide contextual information for other jobs like object recognition.

It is the task of finding out some specific regions in an image based on the organization principle of visual perception, and then automatically labelling them based on a given set of semantic categories. These regions may be the whole image, or some local patches, which may be coastal, mountain, street, city, forest, etc. Scene classification provides an effective contextual semantic information for higher-level image understanding (e.g., object recognition). Scene description and understanding is a hot topic in recent years, most of the existing works focus on the following two aspects:

(1) Modeling low-level scene method directly. These works extract color, texture and shape features from the image first, and then they employ some supervised learning methods to divide the image into several semantic categories, such as indoor, outdoor, urban, and landscape, etc.
(2) Modeling the scene through middle-level semantic description. In this way, the "semantic gap" between the low-level features and the high-level semantic expression is reduced as much as possible, so as to establish a model consistent with human perception process.

Scene description and understanding are closely related to object recognition and low-level visual features. The latest works have tried to generate some sentences for describing a given image. Usually, this is also termed "image caption".

1.3 Examples of Advanced Applications

1.3.1 Image Correction

During the process of image formation, transmission, and recording, the quality may decrease due to various reasons, which will lead to the degradation of the digital image. To compensate the degradation and restore the distorted images, we can resort to the image correction technology [38]. The causes of image distortion may be partial color, blur, geometric distortion, geometric inclination, etc. Therefore, image rectification is actually a process of establishing a reverse mathematical model based on the image distortion procedure, such that the contaminated or distorted image signals can be corrected. To achieve this goal, we need to design a filter to evaluate the predicted image from the distorted image which is most close to the original image according to the specified criterion.

Image correction methods can be divided into two categories: geometric rectification and grayscale rectification. The aim of geometric rectification is to obtain parameters of a mapping from distorted image to the original image, which is the basis of restoring pixel values. The geometric rectification needs to establish a geometric model to describe the degradation at first, and then determine the model parameters with some known conditions. Finally, we could rectify the image according to the estimated model.

Grayscale rectification aims to fix the degraded pixel values in the image formation due to the inhomogeneity of illumination, sensor sensitivity and optical system, so as to obtain the satisfactory visual effect. Gray level rectification is used to rectify the image in a point-wise way by averaging the entire image pixel values. As the imaging is not uniform, there would be the case that one part is dark while the other part is bright due to the inhomogeneous exposure. In this case, gray level Correction is capable of enhancing the image gray-scale contrast for the part lacking exposure. The histogram Correction is also commonly used to improve the pixel value distribution, such that the image's visual quality can meet people's needs.

1.3.2 Image Fusion

Image fusion [39] is to obtain relevant information from two or more channels and then integrate them into a single image, where the information is collected from multiple channels of the same object, or images of the same object obtained at different times in the same channel. The fused image can be used for observation or further processing. The general model of image fusion is shown in Figs. 1.3 and 1.4 gives an example of fusing the MR image with CT image by extreme fusion.

1.3.3 Digital Image Inpainting

Inpainting is the process of reconstructing lost or deteriorated parts of images. The purpose is to further enhance the visual quality of the image in order to make it

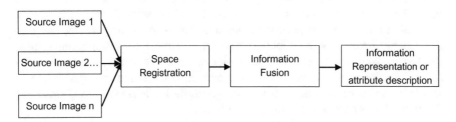

Fig. 1.3 The general model of image fusion

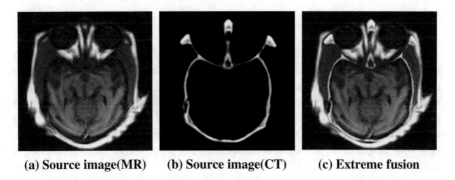

(a) Source image(MR)   **(b) Source image(CT)**   **(c) Extreme fusion**

Fig. 1.4 The results of image fusion

Fig. 1.5 The restoration of
artworks

(a) Damaged image   (b) Restored image

invisible for the observer where the image was defective or has been repaired. Image inpainting [40] is a core technology of image restoration, its applications include the restoration of old photographs and historical relics, movie special effects production, virtual reality, removal of redundant objects (such as delete some characters, text, headings and so on in the video image), data compression, network data transmission, etc. Figure 1.5 shows an example of the restoration of the cracks and scratches in the precious artwork. Due to its great application prospect, image inpainting has attracted extensive attention in recent years.

At present, there exist two major types of inpainting technologies. One is the digital image inpainting technique for repairing small scale defects. During inpainting, it first utilizes the edge information of the area that needs to be patched, and then estimate the direction of the isophote from coarse to fine. Lastly, the communication mechanism is used to spread the information to the whole patch, so as to get better results. The other one is the completion technique for filling in large chunks of lost information in the image, which are generally implemented based on image decomposition or block-based texture synthesis.

1.3.4   Image Stitching

We often need to get a panoramic image with a large field of view and high resolution in our daily life and work. However, due to the limitations of a hardware device, we can only get local images instead. Generally, as the hardware devices (e.g., panoramic cameras, wide-angle lens) for creating panoramic images are expensive, they are not suitable for widely use. This inspires the technology of image stitching [41] which tries to assemble several overlapping images (possibly obtained from different times, different perspectives, or different sensors) into a large, seamless, high-resolution image.

The quality of the image can be always improved by overlapping multiple images of the same scene, Complementary information between multiple images can also help to improve the field-of-view. Besides, by utilizing the information from multiple sources, the obtained image can also reduce some of the uncertainty taken each alone.

Generally speaking, the image stitching consists five steps:

(1) Image preprocessing. This step includes the basic operations of digital image processing (such as denoising, edge extraction, histogram processing, etc.), the establishment of the image matching template, the transformation of the image (such as Fourier transform, wavelet transform, etc.), as well as other operations.

(2) Image registration. In this step, a certain matching strategy is first used to find the corresponding position of the template or feature point in both spliced and reference images, and then the transformation between the two images is determined.

(3) Establishment of the transformation model. According to the relationship between the template or image features, the parameters of the mathematical model are calculated, and then they are employed to establish a mathematical transformation model between two images.

(4) Unified coordinate transformation. In this step, the image that needs to be spliced into the coordinate system of the reference image is transformed according to the established mathematical model.

(5) Fusion and reconstruction, which fuse the overlapped areas of the spliced image and the reference image to obtain a seamless panoramic image.

The image stitching technology has been widely used in the fields of computer graphics, photogrammetry, video communication, image processing, and computer vision.

1.3.5 Digital Watermarking

Digital watermarking [42] is a technology of embedding some digital data into multimedia content, but it cannot affect the visual quality of the original content. In other words, it is invisible by human perception system. The embedded data can be extracted only through a dedicated detector or reader. The embedded data is usually called "digital watermark" which can be the author's serial number, company logo, meaningful text, and other digital data that can be used to identify the file, image or music products source, version, the original author, owner, issuer, legitimate users, etc. An example of digital watermarking is shown in Fig. 1.6. Noting that the Fig. 1.6a, c look the same though the digits "Copyright" has been embedded.

Unlike encryption, digital watermarking does not prevent the occurrence of piracy, but it can determine whether the object is protected. Therefore, digital watermarking technology can be used to identify the authenticity and illegal copy of the media content, resolve the copyright dispute and provide evidence to the court.

For example, the owner of a digital work can use a key to generate a watermark and embed it into the original data, and then publish its watermarked works publicly. When the work is pirated or there is a copyright dispute, the owner can use the

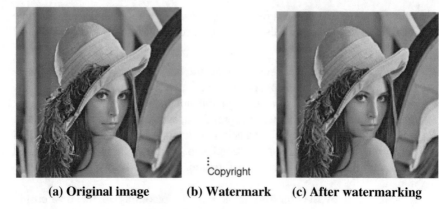

(a) Original image   (b) Watermark   (c) After watermarking

Fig. 1.6 The image is embedded with a digital watermark

watermark as a basis to protect his own interests. The owner of the digital work can also add a unique watermark to each copy of the work to protect the author's legitimate rights and interests. Once an unauthorized copy is present, the source of the copy can be determined through the watermark recovered from this copy. In addition, the watermark detector can be copied and controlled in the photocopying device. Once the detector finds that the copied work contains watermarks, it will stop copying, thus helping to suppress illegal copies and protect the copyright. Digital works are also commonly used in court, medical, news and business, and digital watermarking techniques can be used to determine whether their content has been modified, falsified or specially treated. Therefore, digital watermarking technology has a very wide range of application prospects.

1.3.6 Visual Object Recognition

Visual object recognition [43] is one of the central tasks in computer vision, image processing, and pattern recognition, it plays an important role in video surveillance, military equipment, traffic management, and other fields. Many desired applications demand the ability to recognize objects, such as terrain recognition of cruise missile, terrain reconnaissance of side-view radar, RPV (Remote Piloted Vehicle) guidance, vigilance system and automatic artillery control, anti-camouflage reconnaissance, fingerprint automatic identification, iris recognition, face recognition, and so on.

(1) Iris Recognition

The external view of human eye is composed of three parts: sclera, iris and pupil. The iris is located between the sclera and the pupil, and the boundary connected to them is approximately circular. The iris is unique for each person and it provides important geometric information for matching. The iris recognition system uses a camera about 0.9 m away from the human eye to capture images of iris. Then the captured image is matched with the databases. The similarity between images determines whether the image is from the same object as well as determines rejection or acceptance of the individual.

(2) Fingerprint Identification

Fingerprint identification technology has been widely used in civil fields such as contract since it was discovered. Due to the two important characteristics of human fingerprints: (1) human fingerprints are unique in the entire life and (2) the probability of a pair exactly matched fingerprint from different people is extremely low, it can be assumed that two different people in the world cannot have the same fingerprints. As a result, fingerprint identification technology has been used in many fields, and it has achieved excellent performance. The applications of fingerprint identification include data communication, information security, financial security, and so on.

(3) Face Recognition

Face recognition refers to the technology that uses a computer to identify a person via the human facial characteristic information. Face recognition consists of image preprocessing, face detection, location and recognition. It plays an important role on many occasions. For example, it can be used to find a specific person from a large face database according to the user's needs, thus greatly improving the work efficiency. In daily life, we can use face recognition to assist credit card payment and prevent non-credit card owners from using these cards. Face recognition has also been applied in the field of leisure and entertainment. For example, the autofocus of digital camera can greatly enhance the photograph quality, and the smile shutter technique can judge whether a human is smiling or not.

(4) Optical Character Recognition

Optical character recognition (OCR) is the process of scanning text data and then analyzing the image files to get the contained text. With the development of information technology, OCR has been deployed in mobile devices to extract text captured by the device's camera. OCR has been widely used in the fields like text input in office automation, automatic mail processing, and other jobs associated with automatic access to text processes.

(5) Emotion Recognition

The emotion recognition utilizes computer to implement the intelligent interaction between human and machine by inferring human mental state according to the facial expression. Facial expression recognition technology can be applied in many fields, such as the management of the nuclear power plant and the long-distance bus driver inspection where the safety needs to put more emphasis. Once the fatigue and drowsiness sign occur, the warning system will be triggered to alarm in time to avoid danger. It can also be used in robot operation and electronic nursing, which could detect physical changes according to the changes of patient's facial expressions. In distance education, the teacher can learn from the students' expression to determine what degree they have understood. These needs an expression identifier to map the students' expressions to the level of mastering the course, so that the teachers can make a corresponding response.

(6)   Autonomous Cars
An autonomous car [44] and an unmanned ground vehicle is a vehicle that is capable of sensing its environment and navigating without human input [45].

Autonomous cars use a variety of techniques to detect their surroundings, such as radar, laser light, GPS, odometry and computer vision. Advanced control systems interpret sensory information to identify appropriate navigation paths, as well as obstacles and relevant signage [46]. Autonomous cars must have control systems that are capable of analyzing sensory data to distinguish between different cars on the road.

1.3.7   Object Tracking

Moving object tracking [47] refers to find moving objects of interest (for example, vehicles, pedestrians, animals, etc.) in a continuous video sequence. It is an important branch of computer vision, and it has a wide range of applications in the military guidance, visual navigation, robotics, intelligent transportation, public safety, and other fields.

The key of object tracking is to extract the robust feature of moving object and identify it accurately. Sometimes, we also need to consider the time cost of tracking algorithm in implementing the real-time system. Moving object tracking methods can be divided into two types of moving-analysis based methods and image matching based methods.

The methods based on moving-analysis are generally implemented by inter-frame differencing and optical flow segmentation. The inter-frame differencing subtracts the adjacent frames first, and then use a specified threshold value to extract the moving object. The optical flow segmentation method detects the moving object by the different speed between the object and the background.

The methods based on image matching can identify the moving object and determine its relative position. One important performance index of image matching based method is the positioning accuracy. Based on the principle of matching, this type methods can be divided into the types of region matching, feature matching, model matching, and frequency domain matching.

Moving object tracking system generally contains the following steps:

(1)   Extracting effective descriptors of the object. The object tracking system depends on the effectiveness of the descriptors that are used to capture the object characteristics. The most used features include image edge, contour, shape, texture, moments, transform coefficients, etc.

(2)   Similarity metrics. In moving object tracking, the similarity of moving target between adjacent frames is usually measured by some similarity metrics like

European distance, Mahalanobis distance, chessboard distance, weighted distance, similarity coefficient, and correlation coefficient.
(3) Object region matching. In moving object tracking, estimating the region of the moving object can greatly reduce exhaustive searching and speed up the tracking system. The commonly employed object region searching algorithms contains Kalman filter, particle filter, mean shift and so on.

1.3.8 Dynamic Scene Classification

With the increasing use of digital video surveillance systems, the content of dynamic scenes has becoming more and more complex and this makes it a great challenge to manage video data manually. While by automatic classification of video scenes, people will feel free to find their interested content quickly and accurately. For example, if we want to find a video of a forest fire, it would be much better if the computer can automatically locate the forest fire scene of the video and find a specific object. Therefore, it has become an urgent task of making computer to classify videos into different scene categories such as tsunamis, waterfalls, volcanic eruptions, streams, beaches, etc. This job can assist manual labeling and the management of digital images. It can also provide supports for a deeper analysis of digital video content.

The so-called scene refers to a series of video frames with the same or similar semantics, which is a high-level semantic category. The scene classification [48] not only needs to understand the content of an image, but also have to rely on some context information. Dynamic scene classification can be divided into two categories of tracking based methods and feature based ones. The former first track the moving objects in the dynamic scene and obtains its trajectory, and then performs classification by analyzing the trajectory. The latter is based on the feature extraction of the dynamic scenes, which employs not only the low-level visual features but also some intermediate semantics for classification. The low-level visual features extracted from the dynamic scenes are generally about color, texture and shape. To achieve high classification accuracy, these features are usually combined together to feed to some supervised training models. However, when the scene is too complex the classification accuracy will not be ideal. In this case, the middle semantics can greatly help to fill the gap between the low-level features and high-level semantics, so as to improve the classification performance. A typical dynamic scene classification procedure may be generating a visual dictionary by hierarchical clustering of low-level features first, and then using the probability latent semantic analysis (PLSA) model or topic model to train a generative model.

1.3.9   Pedestrian Re-identification

Pedestrian re-identification [49] refers to the task of retrieving a particular pedestrian captured by camera monitoring network with non-overlapping field-of-views. Pedestrian re-identification is the basis of many applications in video surveillance, such as criminal investigation and human retrieval. In the multi-camera pedestrian tracking, when the tracked object disappears from one camera and appears in another camera, we need to assign the same label to him/her so as to ensure the unity of the identity label. In the sing-camera tracking, sometimes the tracked pedestrian may be occluded for a while in single-camera tracking, when he/she appears again on the screen, we also need to perform re-identification.

In order to obtain a wider range of monitoring, the cameras are generally placed in a relatively high position. Due to the uncontrollable imaging environment of the cameras, the pedestrian images in the re-identification task are usually of low quality and resolutions. As a result, the pedestrian re-identification task can only rely on the pedestrian's clothing appearance information to implement cross-camera matching. Therefore, the clothing texture, color and contour information plays a vital role for re-identification.

Since the pedestrian re-identification is entirely dependent on the pedestrian's appearance information, when the pedestrian's clothes are changed, the basis of pedestrian re-identification will be lost, and the matching results will be no longer reliable. Thus, existing pedestrian re-identification works generally assume that pedestrians do not change clothes in different camera views.

Pedestrian re-identification can be grouped into different categories with different criteria. For example, according to the provided media types, the re-identification can be divided into image-based and video-based scenarios. According to the implementation method types, it can be divided into descriptor-based and matching model-based scenarios. We can also categorize it into "open set" or "closed set" re-identification according to whether the training and testing are carried on the same dataset.

1.3.10   Lip Recognition in Video

Lip recognition [41] aims to create a lip model library by self-learning technology using a large number of lip images and text labels. To achieve this goal, we need to train a matching model library with lip images and the corresponding text results. From the lip images, we extract the features of the lip first, and then use lip model library to "read" or "read a part of" what is saying. This job is a useful supplement to the visual understanding of human-computer interaction.

In order to obtain the contour of the lip, the images are usually transformed into gray space first. Then some deformable templates or active contour models are employed to extract the lip contour. However, the recognition accuracy may be not high enough in this way since only the gray information is explored. In latest years, some methods have tried to use the rich color information to locate lip quickly, accurately, and robustly. Then the lip contour can be extracted in two different ways. The first one is a pixel-based approach which uses the gray image containing the mouth to extract the contour directly. However, in this way the result is sensitive to translation, rotation, scaling, and illumination. Besides, the obtained feature vector may have high redundancy. The second way is based on model learning which has the advantage of low-dimensional feature and robustness to translation, rotation, shrinkage, and illumination changes. However, there is still a drawback in this way, that is, the established model may be incapable of including all relevant lips information.

The basic steps of the lip recognition include: (1) positioning the lips, the commonly employed positioning methods are the active shape model (ASM) or active appearance model (AAM); (2) collecting training samples for training, in lip recognition the mostly used samples are some special point features; (3) training supervised models like Support vector machine (SVM) and neural networks; and (4) deploy the model on new samples.

References

1. Gonzalez RC, Woods RE (2008) Digital image processing, 3rd edn. Prentice Hall
2. Sonka M, Hlavac V, Boyle R (2008) Image processing, analysis, and machine vision. Cengage
3. Jahne B (2002) Digital image processing. Springer (2002)
4. Gibson AP, Hebden JC, Arridge SR (2005) Recent advances in diffuse optical imaging. Phys Med Biol 50(4):R1–R43
5. Shack RV (1970) Image processing by an optical analog device. Pattern Recogn 2(2):123, IN13, 125–124, IN20, 126
6. Russ JC, Russ JC (2007) Introduction to image processing and analysis. CRC Press
7. Bouthemy P, Garcia C, Ronfard R, Tziritas G, Veneau E, Zugaj D (1999) Scene segmentation and image feature extraction for video indexing and retrieval. Springer, Berlin, Heidelberg
8. Kim ZW, Nevatia R (2003) Expandable bayesian networks for 3d object description from multiple views and multiple mode inputs. IEEE Trans Pattern Anal Mach Intell 25(6):769–774
9. Felzenszwalb PF, Girshick RB, McAllester D, Ramanan D (2003) Object detection with discriminatively trained part based models. IEEE Trans Pattern Anal Mach Intell 25(6):769–774
10. Logothetis NK, Sheinberg, DL (1996) Visual object recognition. Wiley, Inc.
11. Snoek CGM, Worring M, Smeulders AWM (2005) Early versus late fusion in semantic video analysis. In: ACM international conference on multimedia, pp 399–402
12. Wang JL, Singh S (2003) Video analysis of human dynamics—a survey. Real-Time Imag 9(5):321–346
13. Murray DW, Buxton BF (1987) Scene segmentation from visual motion using global optimization. IEEE Trans Pattern Anal Mach Intell 9(2):220–228
14. Kass M, Witkin A, Terzopoulos D (1988) Snakes: active contour models. Int J Comput Vis 1(4):321–331
15. Felzenszwalb PF, Girshick RB, Mcallester D (2013) Cascade object detection with deformable part models. Commun ACM 56(9):97–105
16. Gupta R, Patil H, Mittal A (2010) Robust order-based methods for feature description. In: Computer vision and pattern recognition, pp 334–341

17. Morevec HP (1977) Towards automatic visual obstacle avoidance. In: International joint conference on artificial intelligence, pp 584–584
18. Chen J, Zou L, Zhang J, Dou L (2009) The comparison and application of corner detection algorithms. J Multimed 4(6):435–441
19. Smith SM, Brady JM (2015) Susan—a new approach to low level image processing. Int J Comput Vis 45–78
20. Lowe DG (2004) Distinctive image features from scale-invariant keypoints. Int J Comput Vision 60(2):91–110
21. Bay H, Ess A, Tuytelaars T, Van Gool L (2008) Speeded-up robust features. Comput Vis Image Underst 110(3):404–417
22. Nistér D, Stewénius H (2008) Linear time maximally stable extremal regions. In: Proceedings of computer vision—ECCV 2008, European conference on computer vision, Marseille, France, 12–18 October, 2008, pp 183–196
23. Ahonen T, Hadid A, Pietikainen M (2006) Face description with local binary patterns: application to face recognition. IEEE Trans Pattern Anal Mach Intell 28(12):2037–2041
24. Lowe DG (2002) Object recognition from local scale-invariant features. In: IEEE international conference on computer vision, pp 1150
25. Lecun Y, Bengio Y, Hinton G (2015) Deep learning. Nature 521(7553):436
26. Fu Y, Hospedales TM, Xiang T, Gong S (2014) Learning multimodal latent attributes. IEEE Trans Pattern Anal Mach Intell 36(2):303–316
27. Li X, Ouyang J, Zhou X (2015) Supervised topic models for multi-label classification. Neurocomputing 149(PB):811–819
28. Li WT, Chang HS, Lien KC, Chang HT, Wang YC (2013) Exploring visual and motion saliency for automatic video object extraction. IEEE Trans Image Process 22(7):2600–2610
29. Belhumeur PN, Hespanha JP, Kriegman DJ (2002) Eigenfaces vs. fisherfaces: recognition using class specific linear projection. IEEE Trans Pattern Anal Mach Intell 19(7):711–720
30. Liu CL (2007) Normalization-cooperated gradient feature extraction for handwritten character recognition. IEEE Trans Pattern Anal Mach Intell 29(8):1465
31. Gool LV, Mathias M, Timofte R, Benenson R (2012) Pedestrian detection at 100 frames per second. In: Computer vision and pattern recognition, pp 2903–2910
32. Ouyang W, Zeng X, Wang X, Qiu S, Luo P, Tian Y, Li H, Yang S, Wang Z, Li H (2014) Deepid-net: deformable deep convolutional neural networks for object detection. IEEE Trans Pattern Anal Mach Intell PP(99):1–1
33. Wang J, Song Y, Leung T, Rosenberg C, Wang J, Philbin J, Chen B, Wu Y (2014) Learning fine-grained image similarity with deep ranking. In: IEEE conference on computer vision and pattern recognition, pp 1386–1393
34. Ouyang W, Wang X (2014) Joint deep learning for pedestrian detection. In: IEEE international conference on computer vision, pp 2056–2063
35. Zhao H, Tian M, Sun S, Shao J, Yan J, Yi S, Wang X, Tang X (2017) Spindle net: Person re-identification with human body region guided feature decomposition and fusion. In: IEEE conference on computer vision and pattern recognition, pp 907–915
36. Wang N, Yeung DY (2013) Learning a deep compact image representation for visual tracking. Adv Neural Inf Process Syst 809–817
37. Sturgess P, Alahari K, Ladicky L, Torr PHS (2009) Combining appearance and structure from motion features for road scene understanding. In: British machine vision conference
38. Liang J, Dementhon D, Doermann D (2008) Geometric rectification of camera-captured document images. IEEE Trans Pattern Anal Mach Intell 30(4):591
39. Li H, Manjunath BS, Mitra SK (1995) Multi-sensor image fusion using the wavelet transform. Gr Models Image Process 57(3):235–245
40. Bertalmio M, Vese L, Sapiro G, Osher S (2003) Simultaneous structure and texture image inpainting. IEEE Trans Image Process 12(8):882–889
41. Brown M, Lowe DG (2007) Automatic panoramic image stitching using invariant features. Int J Comput Vis 74(1):59–73

42. Cox IJ, Miller, ML, Bloom JA, Fridrich J, Kalker T (2008) Digital watermarking and steganography, 2nd edn. Morgan Kaufmann Publishers
43. Marszalek M, Schmid C (2010) Semantic hierarchies for visual object recognition. In: IEEE conference on computer vision and pattern recognition, pp 1–7 (2010)
44. Thrun S (2010) Toward robotic cars. Commun ACM 53(4):99–106
45. Gehrig SK, Stein FJ (1999) Dead reckoning and cartography using stereo vision for an autonomous car. Int Conf Intell Robots Syst 3:1507–1512
46. Zhu W, Miao J, Jiangbi H, Qing L (2014) Vehicle detection in driving simulation using extreme learning machine. Neurocomputing 128(5):160–165
47. Yilmaz A (2006) Object tracking: a survey. ACM Comput Surv 38(4):13
48. Bosch A, Zisserman A, Muñoz X (2008) Scene classification using a hybrid generative/discriminative approach. IEEE Trans Pattern Anal Mach Intell 30(4):712
49. Gong S, Cristani M, Yan S, Chen CL (2014) Person re-identification. Springer Publishing Company, Incorporated

Chapter 2
Matlab Functions of Image and Video

In this chapter, we begin introducing the basic usage of MATLAB. Then, some important tools for image and video processing are introduced, such as the graphics and visualization, the image processing toolbox, and the functions for processing video.

2.1 Introduction to MATLAB for Image and Video

MATLAB is the abbreviation of Matrix Laboratory, which is a commercial mathematical software produced by The MathWorks company. It integrates data visualization, data analysis and numerical calculation with an easy-to-use environment [1]. MATLAB is an interactive system whose basic data element is a no-defined array. Also, instruction expressions are very similar to those used in mathematics and engineering. Therefore, using MATLAB to solve many numerical problems is much easier than using C, FORTRAN and other languages. MATLAB has a strong openness and applicability, and the corresponding toolkits have been developed for different fields, such as control system design and analysis, image processing, signal processing and communication, financial modeling and analysis, etc.

The basic data element in MATLAB is an array of real or complex numbers [2], and the image is also represented as an array of real values made up of grayscale or color data elements [3, 4]. MATLAB usually uses a two-dimensional array to store images, and each element of an array corresponds to a pixel value of the image. Videos can be viewed as an extension of images in time or perspective. Each frame of a video is a static image. Videos are stored by adding a dimension to the image array, which represents the time and view information [5].

© Springer International Publishing AG, part of Springer Nature 2019
S. Gong et al., *Advanced Image and Video Processing Using MATLAB*,
Modeling and Optimization in Science and Technologies 12,
https://doi.org/10.1007/978-3-319-77223-3_2

2.2   Basic Elements of MATLAB

2.2.1   *Working Environment*

(1)   Software Interface

Figure 2.1 shows a screenshot of the running MATLAB, mainly including Command Window, Workspace, Command History, and Current Folder Window. Users can make different window Settings for MATLAB according to their usage habits. Some Windows are visible and some are not visible. Some can set Windows open, others close.

(2)   Commonly Used Window

In the process of using MATLAB, the commonly used window functions and purposes are as follows:

Command Window: It is the main interactive window for MATLAB to input commands and displays all execution results except graphics. The "≫" in the command window is a command prompt, indicating that MATLAB is in the ready state. Type the command after the command prompt then press enter, MATLAB interprets the commands that are entered and gives the result after the command. When multiple commands need to be executed together, they can be separated by a semicolon.

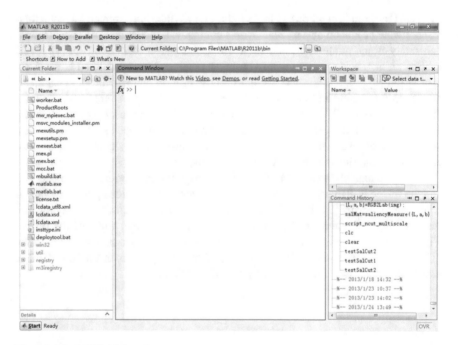

Fig. 2.1   The MATLAB interface

Workspace: It is used to store the various variables and the space of results where the user can easily view, edit, load, and save the various variables of MATLAB.

Current Folder: It refers to the path of the current running file of MATLAB and the files under that path, which can only be run or invoked if the file is in the current directory or the search path.

Command History: It automatically maintains a history of all the used commands from the installation, and also indicates the use of time to facilitate user queries. Double-clicking on these commands can also execute the history command again.

Editor Window: It provides users with windows to create, edit, run, and debug M files.

Launch Pad: Users can easily open and call MATLAB programs, functions and help files in the Launch Pad.

Help Browser: It provides easy and quick online help for users.

2.2.2 Data Types

Each type of data in MATLAB is based on an array and is derived from the array, including logical, char, numeric, cell, structure, Java classes and function handle. The relationship between data types is shown in Fig. 2.2.

The most commonly used data types are double and char. All calculations in MATLAB treat the data as double to process. Other data types are only used in some

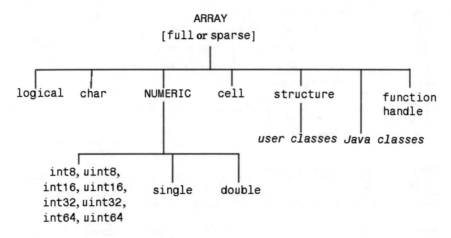

Fig. 2.2 Data structure in MATLAB

special conditions. For example, an unsigned 8-bit integer is generally used to store image data; Cell array and structure arrays are generally used in large programs.

1. Logical Data

The logical data in MATALAB is only "1" and "0", respectively representing the logical true and the logical false. The common logical functions are shown in Table 2.1.

2. Char Data

In MATLAB, the input character is used in single quotation marks, and the string is stored as an array of characters, each of which is an element of the string and each element occupies two bytes. Commonly used string manipulation functions are shown in Table 2.2.

3. Numeric Data

Numerical types in MATLAB include symbols and unsigned integers, single and double precision floating-point numbers. By default, all values in MATLAB are stored and operated on according to the double floating-point type. As shown in Table 2.3.

Table 2.1 Common logic functions

Function	Instruction
all	Whether all the elements are non-zero
any	Whether at least one element is non-zero
isempty	Whether the matrix is empty
isequal	Whether the two matrices are the same
isinf	Whether there is an inf (infinity) element
isnan	Whether there is a nan (not quantitative) element
isnumeric	Whether it is a numeric type
isinteger	Whether it is an integer
isfloat	Whether it is a float
isreal	Whether it is a real number

Table 2.2 String manipulation functions

Function	Instruction	Function	Instruction
isstr	Determine whether or not a character	strcmp	String comparison
blanks	Generate a blank string	strfind	Look up another string in a string
deblank	Delete the space at the end of the string	strcat	Concatenation of string
upper	Capitalize the string	strmatch	Look up the match string
lower	Lowercase the string	strrep	Replace another string with one string

Table 2.3 Numeric data

Data type	Instruction
uint8	8-bit unsigned integer, range $[0, 2^8 - 1]$, occupying 1 byte of memory space
uint16	16-bit unsigned integer, range $[0, 2^{16} - 1]$, occupying 2 bytes of memory space
uint32	32-bit unsigned integer, range $[0, 2^{32} - 1]$, occupying 4 bytes of memory space
uint64	64-bit unsigned integer, range $[0, 2^{64} - 1]$, occupying 8 bytes of memory space
int8	8-bit signed integer, range $[-2^7, 2^7 - 1]$, occupying 1 byte of memory space
int16	16-bit signed integer, range $[-2^{15}, 2^{15} - 1]$, occupying 2 bytes of memory space
int32	32-bit unsigned integer, range $[-2^{31}, 2^{31} - 1]$, occupying 4 bytes of memory space
int64	64-bit unsigned integer, range $[-2^{63}, 2^{63} - 1]$, occupying 8 bytes of memory space
single	Single precision floating-point number, range $[-10^{38}, 10^{38}]$, occupying 1 byte of memory space
double	Double precision floating-point number, range $[-10^{308}, 10^{308}]$, occupying 8 bytes of memory space

4. Cell Array

Each element in the cell array is called a cell, and each cell can contain any data type in MATLAB, defined using curly braces { }. The array content can be accessed through the curly brace form index, and the form of the original parenthesis can only get the description of the variable, as shown below.

```
>> cell={'Lily', 20, [85:90:78]}:
>> cell{1}

ans =

Lily

>> cell(3)

ans =

    [3x1 double]
```

5. Structure

The structure is the same as an array of cells, which is to focus different types of data in a single variable. The difference is that the structure is indexed by the field, which refers to the internal field through the dot operator ".", where the field must have a different name for the distinction, as shown in the graphics below.

```
>> struct.name='Lily':
>> struct.age=20:
>> struct.score=[85:90:78]:
>> struct

struct =

    name: 'Lily'
     age: 20
   score: [3x1 double]

>> struct.score

ans =

    85
    90
    78
```

2.2.3   Array and Matrix Indexing in MATLAB

MATLAB supports a large number of powerful indexes that not only simplify array operations, but also improve the efficiency of the program.

1.   Vector Index

The number of dimensions in MATLAB is $1 \times N$, which is called row vector. The access of elements in a row vector is performed using a one-dimensional index, such as A (1), which is the first element of the vector A. The elements defined by the vector are enclosed in square brackets, separated by spaces or commas. Using the transpose operator ".", the row vectors can be converted to column vectors.

```
>> A=[1 3 5 7 9]:
>> A(1)

ans =

    1

>> A.'

ans =

    1
    3
    5
    7
    9
```

For the row vector A, the method for extracting data blocks of elements is shown below.

```
>> A=[1 3 5 7 9]:
>> A(2:4)%Extract the second to fourth elements of vector A

ans =

    3    5    7

>> A(1:end)%Extract the first one to the last element in vector A

ans =

    1    3    5    7    9

>> A(1:2:end)%Extract the first to last element of vector A in steps of 2

ans =

    1    5    9
```

2. Matrix index

By defining a matrix in MATLAB, use semicolons to separate rows and rows, using commas (or spaces) to separate columns and columns to define directly (the matrix's subscripts start at 1). Selecting an element from a matrix is the same as selecting an element from a vector, only two indexes are needed: one to determine the row location, and the other to determine the corresponding column location.

```
>> B=[1,3:5,7]:
>> B(1,2)%Take the first row of the second column of the elements

ans =

    3

>> B(1,:)%Extract the first line

ans =

    1    3

>> B(1:end,2)%Extract the first row to the last row, the second column of a sub-block

ans =

    3
    7

>> B(:)%The matrix is stored in columns to give a long column vector

ans =

    1
    5
    3
    7
```

2.2.4   Standard Arrays

In the design of image processing program with MATLAB, some simple image arrays will be used to test the algorithm of image processing. Some of the important generation functions of standard arrays are as follows:

(1)   eye(m, n): generate a matrix of m rows and n columns with a main diagonal line of 1, which can be abbreviated as eye(n) when m=n. The matrix is an n-dimensional identity matrix.

(2)   zeros(m, n): generate a zero matrix of m rows and n columns.

(3)   ones(m, n): generate a matrix of m rows and n columns, whose all elements are 1.

(4)   true(m, n): generate a logic matrix of m rows and n columns, whose all elements are ture.

(5)   false(m, n): generate a logic matrix of m rows and n columns, whose all elements are false.

(6)   rand(m, n): generate a random matrix, which is distributed evenly in the interval [0, 1].

(7)   randn(m, n): generate a random matrix with the standard normal distribution, the mean of which is 0 and the variance of which is 1.

(8)   randperm(n): generate random permutations of integers between 1 and n.

(9)   magic(n): generate an n-order magic matrix, in which the sum of each row, the sum of each column element, and the sum of the main diagonal elements are equal.

(10)   blkdiag(a, b, c, d, ...): generate a matrix, whose diagonal elements are a, b, c, d, ...

(11)   hilb(n): produce an n-order Hilbert matrix, whose elements are $H(I, j) = 1/(I + j - 1)$.

(12)   invhilb(n): generate an n-order matrix, which is the inverse of the Hilbert matix.

2.2.5   Command-Line Operations

After the launch of MATLAB, the execution result or operation result will be given immediately after the input command or MATLAB statement after the command window prompt "≫." If the semicolon ";" is entered at the end of each line of command, the command window does not display the results of the execution immediately, and the results are saved in the workspace.

MATLAB allows to type multiple statements in the same line, separated by a semicolon; split the same statement in multiple lines to facilitate reading, as long as you connect with three dots "..." at the end of the line. In MATLAB, "%" represents an annotation, which is similar to the "//" comment in C/C++. Common command line operations are shown in the following Table 2.4.

Table 2.4 Common command line operations

Command	Instruction	Command	Instruction
cd	Set the current working directory	exit	Close/exit MATLAB
dir	List files and subdirectories in the specified directory	quit	Close/exit MATLAB
clc	Clear the contents of Command Window	type	Display the contents of the specified M file
clear	Clear the saved variables in Workspace	more	Make the subsequent content to display in the form of paging
Help	Display help information in Command Window	which	Indicate the directory where the subsequent file is located
Doc	Display help information in Help Browser	who/whos	Check memory variables
lookfor	Look up a function or command that contains the following contents	save	Save variables in memory using files
diary	Record the input of Command Window as a file	load	Read variables to memory from files

2.3 Programming Tools: Scripts and Functions

2.3.1 M-Files

MATLAB provides an extremely rich internal function, and the user can resolve a lot of works by calling them through the command line. However, to use MATLAB more efficiently, it cannot be separated from MATLAB programming. Users can complete an independent function (script file programming) by organizing a sequence of MATLAB commands. Or abstract the M-file to form a functional block (function file programming) that can be reused. The M-file is a text file that can be created and edited with any editing program, and the text editor provided by MATLAB is generally used and most convenient. The extension of M-files is ".m," which can be divided into script files and function files according to its contents and functions.

When dealing with some simple problems in MATLAB, you can enter processing commands directly in Command Window. When the problem is more complex, you can enter a series of commands into a text file. As long as the file name is entered in Command Window, all the commands in the file are executed according to the design process, resulting in the desired result, which is called the script file.

The function file consists of the function definition line, "H1" line, function help information, function body and annotation, where the function definition line and function body are required.

(1) Function Definition Line: function [outputs] = name(inputs)
     MATLAB allows multiple input parameters (inputs) and returns parameters
     (outputs), separated by commas, and the square brackets can be omitted if only
     one parameter is returned.
(2) "H1" Line: The first comment line in the M-file, starting with a percent, must
     be followed by the function definition line, with no rows in the middle, or blank
     characters or indentation. The contents of this line will appear in the first line
     when using the help command. It will only be searched in the line when using
     lookfor function to search a function associated with a word.
(3) Function Help Information: explain the function which the function file imple-
     ment, the value of variables, parameters, and copyright information.
(4) Function Body: the MATLAB code part of the file implementation function.
(5) Annotation: mainly used to annotate the specific operation process of the func-
     tion file's function body for easy reading and modification.

The difference between a function file and a script file is shown in Table 2.5.

2.3.2   Operators

Operators in MATLAB can be divided into three categories: arithmetic operators,
relational operators, and logical operators. The following Tables 2.6, 2.7 and 2.8
shows.

The order of the calculation in MATLAB is the same order as the general mathematical evaluation: the expression is executed from left to right, and if there are
parentheses, the expression in parentheses is calculated. The priorities of operators
are shown in the following Table 2.9.

Table 2.5   The difference between a function file and a script file

	Script file	Function file
Input and output	No input parameters, no return output	Bring input parameters as well as output
Variable operating	Operate only workspace variables (global variables)	Operate workspace variables (global variables need to be specified by global) and local variables
Run mode	Run directly	The method of calling a function

Table 2.6 Arithmetic operators

Operator	Instruction	Operator	Instruction
+	Plus	–	Minus
*	Multiplication of matrixes	.*	Multiplication of arrays
^	Power of matrixes	.^	Power of arrays
\	Left-division of matrixes	.\	Left-division of arrays
/	Right-division of matrixes	./	Right-division of arrays
.'	Transpose of matrixes and vectors	'	Transpose of plural matrixes

Table 2.7 Relational operators

Operator	Instruction	Operator	Instruction
>	Greater than	<	Less than
>=	Greater than or equal to	<=	Less than or equal to
==	Equal to	~=	Unequal to

Table 2.8 Logical operators

Operator	Instruction	Operator	Instruction
&	And	|	Or
~	Non	xor	Xor
&&	Prerequisite and	||	Prerequisite or

Table 2.9 Operators priorities in MATLAB

Priority	Operator
High	matrix transposition (.'), conjugate transpose ('), Power of matrixes (^), Power of arrays (.^)
	Logic non (~)
	Multiplication of arrays (.*), Multiplication of matrixes (*), Left and right division of arrays(.\, ./), Left and right division of matrixes(\, /)
	Plus and minus (+, −)
Low	The colon operation (:)
	The class of equal (<, <=, >, >=, ==, ~=)
	Logic and (&)
	Logic or (|)
	Prerequisite and (&&)
	Prerequisite or (||)

Table 2.10 List of internal variables in MATLAB

Special variables	Instruction
ans	Output variable
pi	Circumference
Inf or inf	Infinity, like 1/0
NaN or nan	Not quantitatively, like 0/0
I and j	Unit virtual value
eps	The relative precision of floating point operations
realmax	The largest positive floating-point number
realmin	The smallest positive floating-point number
nargin	The number of input parameters of functions
nargout	The number of output parameters of functions
lasterr	Recent error information
lastwarning	Recent warning information
computer	Computer type
version	MATLAB version

2.3.3   Important Variables and Constants

Variables can save the intermediate results and numerical information, such as the output variable naming rules and some common in MATLAB programming language like (begin with a letter and may contain numbers, underscores, and letters, but it cannot contain spaces), and the variable names are case sensitive. In MATLAB, there is no need to specify the type of variable, and the system can determine the data type of the variable automatically based on the value of the expression or the value of the input. However, if use the same name as the previously defined variable, the original variable will be automatically overridden and the system will not give an error message. When using variables, consciously avoid repetition, and do not have the same name as some internal variables and reserved words. Table 2.10 presents some important internal variables and constants.

2.3.4   Number Representation

In MATLAB, the values are used in the decimal system and they are expressed in double precision by default. This is consistent with the general mathematical representation. When define other type variables, it needs to specify the data type of the variable first, or convert the double-precision floating-point number to the specified data type though a conversion function. Commonly used numerical conversion functions are shown in Table 2.11.

Table 2.11 Common numerical conversion functions

Function name	Instruction
double	Convert to data of double precision type
single	Convert to data of single precision type
int8, int16, int32, int64	Convert to signed integer data
uint8, uint16, uint32, uint64	Convert to unsigned integer data
dec2hex	Convert decimal number to hexadecimal number
hex2dec	Convert hexadecimal number to decimal number
hex2num	Convert hexadecimal number to double—precision floating-point number
int2str	Convert integer to string
num2str	Convert number to string
mat2str	Convert matrix to string

Table 2.12 Common complex functions

Function name	Instruction
real	Give a real part of a complex number
image	Give the imaginary part of a complex number
abs	Give the module of a complex number
angle	Give the amplitude of a complex number in radians
conj	Give the conjugate of a complex Number
complex	Give a complex number that is created using real and imaginary numbers

The complex is an extension of the real number. In MATLAB, the complex numbers are expressed in the same form of mathematics, with the characters i and j representing the imaginary part. We can also use the complex function to define them. Commonly used complex functions are presented in Table 2.12.

2.3.5 Flow Control

As with other programming languages, MATLAB provides a statement for process control, that is, flow control. As shown in Table 2.13.

Table 2.13 Methods of process control

Statement	Standard format	Instruction

if	if expression1
	
	   commands1
	
	elseif expression2
	
	   commands2
	
	else expression3
	
	   commands3
	
	end
		The sequence of commands must be conditional on the test of the relationship. Statements can be used to implement multi-branch structures, and if elseif uses too many layers, consider switching to switch statements instead

for	for index=start:increment:end commands end	Allow a set of commands to be repeated at fixed and scheduled times. Increment is the specified step length and the default is 1. It can be nested
while	while expression commands end	It repeats a set of commands with an unfixed number of times. Commands are executed only if expression is true. It can be nested

switch	switch expression1
	
	case test_expression1
	
	   commands1
	
	case test_expression2
	
	   commands2
	
	otherwise
	
	   commands3
	
	end
		The branching structure of multiple choices, according to the different values of expression1, respectively executes different statements. When the statement of any branch is executed, the next sentence of the switch statement is executed directly

try-catch	try commands1 catch commands2 end	The try statement first tries to perform the commands1, and if commands1 is wrong in the execution, it goes to execute commands2

(continued)

Table 2.13 (continued)

Statement	Standard format	Instruction
break	–	Terminate the execution of the loop and jump out of the loop to continue the next statement of the loop statement
continue	–	Skip the statement after continue in the loop body and continue the next cycle
return	–	Cause the function to exit normally, and return the function that called it to continue running

2.3.6 Input and Output

MATLAB provides some functions for the input and output of data during a program operation, mainly including data input function and data display function.

(1) input function: to input a parameter from the keyboard to the computer. Call format: A = input('prompt message') or A = input('prompt message', 's')
(2) disp function: to output variables in Command Window, which can be a string or in the format of the matrix call: disp(output)

```
>> n=input('please input name:','s');
please input name:Lily
>> disp(n);
Lily
>> a=input('please input age:');
please input age:22
>> disp(a);
    22
```

2.4 Graphics and Visualization

Visualization is the theory, method and technique of using computer graphics and image processing to transform data into graphics or images on the screen and interact with them. MATLAB provides a powerful graphical processing and editing function that allows the data to be graphically represented so that users can visually observe the relationship between data.

1. Graphics Window

(1) Graphics Window creation: figure function:

figure: create a new graphics window with default attribute values.

figure('PropertyName', PropertyValue, ...): create a new graphics window with the specified property value (PropertyValue) for the specified attribute (Property-Value), and default values for the properties that are not specified.

figure(h): If h is an existing graphics handle, the graphics window is the active window, which is the output of the image. If h is an entire parameter, a new graphics window is created, and the handle (the title of the graphics window) is shown as h and the window is active.

H = figure: create a new graphics window and return the handle H.

(2)   Graphics Window division: subplot function:

subplot(m, n, p): The current graphics window is divided into $m \times n$ plots, the area numbers are numbered according to row priority, and the p area is selected as the current activity area.

(3)   Graphics Window retaining: hold on function:

hold on: In the existing graphics window, keep the original graphics (not refresh) and add new drawing graphics.
hold off: In the existing graphics window, overwrite the original graphics (refresh) to add new drawing graphics.
hold: Toggle the status of the current graphics window refresh.

(4)   Other Functions:

set function: set the properties of the graphics object.
reset function: reset the property of the graphics object to their default values.
delete function: delete a graph object.
gcf function: gets the handle to the current graphics window.
clf function: clears the current graphics window, and when it executes from a callback, the command simply deletes the graphics object whose HandleVisibility property is on.
close function: delete the specified graphics window.

2.   Two-dimensional Curve Graphics

(1)   Basic Plane Figure Function: plot

plot(Y): If Y is a real vector, the vector subscript is the horizontal axis and the element value is the horizontal axis; If Y is a real matrix ($m \times n$), then the Y is decomposed into n column vectors in the direction of the column, drawn by column elements, and a total of n curves; If Y is a complex matrix, the real part is the horizontal axis, the imaginary part is the vertical axis, to draw many curves.

plot(X, Y): If X, Y is the same dimensional real vector, then X is the horizontal axis, Y is the vertical axis, and the corresponding points are traced in the plane. If X, Y is equal to the same dimensional same type real matrix ($m \times n$), then each column element is drawn with a curve, a total of n curves; If the X, Y is a vector, another as the matrix, and the dimensions of the vector is equal to the number of rows or columns of the matrix, the matrix according to the direction of the vector

Table 2.14 The instruction of "point", "line" and "point-line color"

Point symbol	Instruction	Line symbol	Instruction	Point-line color symbol	Instruction
d	Rhombus	–	Thin lines	b	Blue
h	Hexagon	:	Imaginary point line	g	Green
o	Hollow circle	-.	Point line	r	Red
p	Pentagon	--	Dash line	c	Cyan
s	Rectangle			m	Magenta
x	Fork operator			y	Yellow
.	Solid black spots			k	Black
+	Cross character			w	White
*	Asterisk character				
^	Upper triangle				
<	Left triangle				
>	Right triangle				
v	Lower triangle				

is decomposed into several vectors, paired with the vector and draw respectively, matrix can be decomposed into several vectors corresponding to the lines.

plot(X1, Y1, X2, Y2, …): take the data Xi and Yi in order to draw.

plot(X1, Y1, 'S1', …): Respectively draw the curve in order which is defined by three parameters Xi, Yi, 'S1'. The "S1" is a string of characters from the "point," "line" and "point-line color," which are used to name the curve (Table 2.14).

plot(X1, Y1, 'S1', 'PropertyName', PropertyValue, …): the first three parameters Xi, Yi and 'S1' are the same as the above definitions, PropertyName and PropertyValue indicate the property name and attribute value, the most commonly used attribute name/attribute value is shown in Table 2.15.

(2) Graphic Identification

title('s'): add the graphics title.

xlabel('s'): add the name of the horizontal axis.

ylabel('s'): add the name of the vertical axis.

text(Xi, Yi, 's'): annotate character in the specified position (Xi, Yi) of the graphic.

legend('s1', 's2', …): identify legends in the upper right corner of graphics. The number of strings is equal to the number of curve graphics identifies by the legends, and the sequence of the strings is the same as the sequence of the different curves.

Table 2.15 Common attribute names and attribute values of the line object

Implication	Attribute name	Attribute value
Point-line color	Color	$[v_r, v_g, v_b]$, each element in the RGB tritple can take any value in [0, 1]
Line style	LineStyle	4 types of lines are shown in the table above
Line width	LineWidth	Positive real number, the default line width is 0.5
Point style	Marker	14 types of points are shown in the table above
Point size	MarkerSize	Positive real number, the default size is 6.0
Point boundary color	MarkerEdgeColor	$[v_r, v_g, v_b]$, each element in the RGB tritple can take any value in [0, 1]
Point domain color	MarkerFaceColor	$[v_r, v_g, v_b]$, each element in the RGB tritple can take any value in [0, 1]

(3) Axis Setting

   axis([Xmin, Xmax, Ymin, Ymax]): set the maximum and minimum values of the axis
   axis auto: return the axis system to the natural default state
   axis equal: let the horizontal axis, the vertical axis set to equal length scale
   axis normal: the rectangular axis system (default)
   axis square: generate a square axis system
   axis on: display the axis system
   axis off: cancel the axis system.

(4) Grid and Axis Border

   grid: Switch the state of the current grid line
   grid on: draw the grid line of the frame
   grid off: not draw the grid line of the frame (default)
   box: switch the state of the current axis border
   box on: give the axis a border line (default)
   box off: not give the axis a border line.

(5) Other Axis Systems Drawing

   polar function: polar axis system drawing
   semilogx function: single logarithmic axis system drawing
   loglog function: double logarithmic axis system drawing
   plotyy function: double y-coordinate axis system drawing.

3. Three-dimensional Curve Graphics

(1) Three-dimensional Curve Drawing: plot3 function

   plot3(X1, Y1, Z1, 'S1', ...): When X1, Y1, and Z1 are the same dimensional vectors, draw the three-dimensional curves of X1, Y1 and Z1 corresponding to the elements x, y and z. When X1, Y1, and Z1 are the same dimensional matrix, then

X1, Y1, and Z1 correspond to the column elements x, y, and z, respectively. The number of curves is equal to the number of columns in the matrix. The 'S1' meaning is the same as in two dimensions, and is used to specify a string of points, lines, and point-line colors.

(2) Three-dimensional Grid: mesh function

mesh(Z): generate the three-dimensional grid determined by the matrix Z, we can get x = 1:n and y = 1:m from [m, n] = size(Z). The axis grid generated by the points of the horizontal axis and vertical axis is that [X, Y] = meshgrid(x, y), which means that Z is a single-valued function defined in the grid partition area.

mesh(X, Y, Z): If X and Y are vectors, we can get length(X) = n and length(Y) = m from [m, n] = size(Z). The point $(X(j), Y(i), Z(i, j))$ is the focus of the grid lines, which mean that X corresponds the column of Z and Y corresponds to the row of Z. If X and Y are both matrixes, the point $(X(i, j), Y(i, j), Z(i, j))$ is the focus of the grid lines.

mesh(X, Y, Z, C): generate the grid determined by X, Y and Z, in which X controls the x coordinate and Y controls the y coordinate. (X, Y) determines the z coordinate and (X, Y, Z) forms the grid point of three-dimensional space; C is used to specify the color of the grid. When there is no need to draw a fine three-dimensional surface structure diagram, we can show the three-dimensional surface by drawing the three-dimensional grid.

(3) Three-dimensional Surface: surf function

surf(Z): generate the three-dimensional surface determined by the matrix Z.

surf(X, Y, Z): the data Z is the height of the surface and is the matrix of color. If X and Y are both vectors, we can get length(X) = n and length(Y) = m from [m, n] = size(Z). The point $(X(j), Y(i), Z(i, j))$ is the node on the surface. If X and Y are both matrixes, the point $(X(i, j), Y(i, j), Z(i, j))$ is the focus of the surface.

surf(X, Y, Z, C): draw a three-dimensional surface using the specified color C.

4. Drawing of Special Graphics

In addition to the drawing of ordinary graphics, MATLAB also provides a series of functions for drawing special graphics, as shown in Table 2.16.

5. Animation Production

Animation production can carry on physical simulation, digital simulation and so on, is very meaningful. MATLAB provides two animation methods:

(1) Movie mode: preserve a set of images in the image buffer, and then play the group by frame. Because the human vision has a short stay, then produce an animation effect. This method computes a large amount of memory and is suitable for complex objects. Basic steps to make a movie animation:

   ① Call the getframe function capture graphs, which is stored in an array of frames.

Table 2.16 Drawing functions of special graphics

Function	Instruction	Function	Instruction
stem	Two-dimensional discrete data graph	stem3	Three-dimensional discrete data graph
bar/barh	Two-dimentional vertical histogram/horizontal histogram	bar3/bar3h	Three-dimentional vertical histogram/horizontal histogram
pie	Pie chart	pie3	Three-dimentional pie chart
comet	Two-dimensional comet map	comet3	Three-dimensional comet map
quiver	Two-dimensional gradient field	quiver3	Three-dimensional gradient field
contour	Two-dimensional contour line	contour3	Three-dimensional contour line
fill	Fill the graphic	fill3	Fill the three-dimensional graphic
area	Regional figure	sphere	Draw the sphere
hist	Probability distribution map	cylinder	Draw the cylinder
stairs	Ladder graphic	waterfall	Waterfall figure
errorbar	Error figure	feather	The vector diagram that diverges from the point at which the horizontal line is evenly spaced
rose	Probability distribution in polar coordinates	compass	The vector diagram that diverges from the pole in polar coordinates

    ② Call the movie function to play the movie animation at the specified speed and number of times.

(2) Object mode: to keep most of the pixel color in the graphics window, but only change the color of some pixels to make up the motion image. This method can be applied to situations where there is less variation and less graphic accuracy, which means no complex animation can be produced. Basic steps of making object animation:

    ① Draw the motion trajectory graph of the active object.
    ② Calculate the new location of the active object and display it in the new location and set EraseMode to be in the mode of xor.
    ③ Erase the original object and refresh the screen.

2.5   The Image Processing Toolbox

2.5.1   The Image Processing Toolbox: An Overview

MATLAB is an advanced programming language based on an array rather than a scalar, so it essentially provides support for the image. The digital image is actually a set of discrete and ordered data, and MATLAB can be used to deal with the matrix of discrete data. The relevant toolkits of image processing include:

(1) Image Acquisition Toolbox
(2) Image Processing Toolbox
(3) Signal Processing Toolbox
(4) Wavelet Toolbox
(5) Statistics Toolbox
(6) Bioinformatics Toolbox

The image processing functions in MATLAB are mainly included in the Image Processing Toolbox. The IPT is constituted of functions which support a series of image processing and operations, these operations mainly include image geometry transform, neighborhood and block operation, linear filtering, filtering design, image transformation, image analysis, image enhancement, mathematical morphology processing, image smoothing and region of interest (ROI) operation.

2.5.2 Essential Functions and Features

All the functions in the image processing toolbox are M-files, which can be checked by typing function_name. Also, we can extend the toolbox through coding MATLAB functions by ourselves. Some common image processing functions are shown in Tables 2.17, 2.18, 2.19, 2.20, 2.21, 2.22, 2.23, 2.24, 2.25, 2.26, 2.27, 2.28, 2.29, 2.30, 2.31 and 2.32.

Table 2.17 Image display functions

Function	Instruction	Function	Instruction
colorbar	Display the bar of colors	imcountour	Shows an outline of an image
getimage	Obtain the data of an image from the coordinate axis	immovie	Create a movie animation with multiple frames
montage	Multiple images are displayed simultaneously	imshow	Display all kinds of images
image	display an image	truesize	Resize the size of the display of images
imagesc	Display a brightness image	zoom	The image area is shrunk or enlarged
subimage	Display multiple images in a graphics window	warp	Display the image to the surface of the texture map

Table 2.18 The input/output functions of image files

Function	Instruction
imread	Read the image file
imwrite	Output the image file
imfinfo	View the information of the image file

Table 2.19 Image geometry operation functions

Function	Instruction	Function	Instruction
imcrop	Image clipping	imrotate	Image rotation
imresize	Image resizer	interp2	Two-dimentional data interpolation

Table 2.20 Image pixel values and statistical functions

Function	Instruction	Function	Instruction
corr2	Calculate Two-dimensional correlation coefficients of two image matrixes	improfile	Calculate the pixel value of a path in the image
std2	Calculate the standard deviation of the image matrix	impixel	Display the pixel color value of the selected image
mean2	Calculate the mean of the image matrix	imcontour	Show the outline of the image
imfeature	Calculate the feature size of the image area	imhist	Display the histogram of the image

Table 2.21 Image analytic functions

Function	Instruction	Function	Instruction
edge	Image edge detection	qtgetblk	Get the value of the quadtree decomposition block
qtdecomp	Image quadtree decomposition	qtsetblk	Set the value of the quadtree decomposition block

Table 2.22 Image enhancement functions

Function	Instruction	Function	Instruction
histeq	Histogram equalization	medfilt2	Two-dimensional median filtering
imadjust	Contrast adjustment	ordfilt2	Two-dimensional sequential statistical filtering
imnoise	Add image noise	wiener2	Two-dimensional adaptive de-noising filtering

Table 2.23 Linear filtering functions

Function	Instruction	Function	Instruction
conv2	Two-dimensional convolution	convn	Multidimensional convolution
convmtx2	Calculate the two-dimensional convolution matrix	filter2	Two-dimensional linear filtering

Table 2.24 Two-dimensional linear filtering design functions

Function	Instruction	Function	Instruction
fspecial	Produce a predefined filter	ftrans2	The two-dimensional FIR filter designed by frequency conversion
freqz2	Calculate the two-dimensional frequency response	fwind1	The two-dimentional FIR filter designed by a one-dimensional window
fsamp2	The two-dimensional FIR filter designed by frequency sampling	fwind2	The two-dimentional FIR filter designed by a two-dimensional window
freqspace	Determine the interval of the two-dimensional frequency response	fsample	Generate the filter

Table 2.25 Image transformation functions

Function	Instruction	Function	Instruction
dct	Calculate the discrete cosine transform	fft2	Calculate the two-dimensional fast Fourier transform
dct2	Calculate the two-dimensional discrete cosine transform	fftn	Calculate the multidimensional fast Fourier transform
dctmtx	Calculate the discrete cosine transform matrix	fftshift	Move the dc component to the center of the spectrum
dctmtx2	Calculate the two-dimensional discrete cosine transform matrix	idct	Calculate the inverse transformation of the discrete cosine transform
radon	Calculate the Radon transform of the image at the specified angle	idct2	Calculate the inverse transformation of the two-dimensional discrete cosine transform
iradon	The inverse transformation of Radom	ifftn	Calculate the inverse transformation of multidimensional fast Fourier transform

Table 2.26 Image neighborhood and block operation functions

Function	Instruction	Function	Instruction
blkproc	Blockoing of images	col2im	The matrix columns do the rearrangement of image blocks
bestblk	Determine the size of the block operation	colfilt	Use colunmn functions to do operations of domains
nlfilter	Do operations of general fields	im2col	Images are rearranged by matrix columns

Table 2.27 Operation functions of binary images

Function	Instruction	Function	Instruction
makelut	Create lookup tables	Bwmorph	Morphological operation of binary images
applylut	Use lookup tables for domain operations	Bwperim	Extract the target boundary of binary images
bwarea	Calculate the area of the target region of the binary image	Bwselect	Determine the target of the binary image
bweuler	Calculate the euler number of the binary image	imdilate	Expansion operation of binary images
bwlabel	Mark different targets in the image	imerode	Erosion operation of binary images

Table 2.28 Image processing functions based on the region

Function	Instruction	Function	Instruction
rolpoly	Select the polygonal region to be processed	roifill	Fast interpolation of the target region
roifilt2	Filter the image target region	roicolor	Select the target region according to the color

Table 2.29 Operation functions of color images

Function	Instruction	Function	Instruction
brighten	Increase or decrease the brightness of the color image	imapprox	Approximate the index image with the less color image
cmpermute	Rearrange colors of the color image	rgbplot	Draw the RGB color image
colomap	Get the current color image	comunique	Find the specific color and the corresponding image in the color image

Table 2.30 Conversion functions of color space

Function	Instruction	Function	Instruction
hsv2rgb	Convert the HSV value to RGB color space	rgb2ntsc	Convert the RGB value to NTSC color space
ntsc2rgb	Convert the NTSC value to RGB color space	rgb2ycbcr	Convert the RGB value to YCBCR color space
rgb2hsv	Convert the RGB value to HSV color space	ycbcr2rgb	Convert the YCBCR value to RGB color space

Table 2.31 Image types and conversion functions of types

Function	Instruction	Function	Instruction
dither	Transform the image with the dithering method	im2bw	Convert the image to a binary image
gray2ind	Convert the grayscale image to an index image	im2double	Convert the image matrix to double
grayslice	Convert the grayscale image to an index image	im2unit8	Convert the image matrix to uint8
isbw	Judge if it is a binary image	im2unit16	Convert the image matrix to uint16
isgray	Judge if it is a grayscale image	ind2gray	Convert the index image to a grayscale image
isind	Judge if it is an index image	ind2rgb	Convert the index image to a RGB image
isrgb	Judge if it is a RGB image	rgb2ind	Convert the RGB image to an index image
mat2gray	Convert the matrix to a grayscale image	rgb2gray	Convert the RGB image to a grayscale image

Table 2.32 Demonstraction functions of image processing

Function	Instruction	Function	Instruction
dctdemo	Image compression demonstration of two-dimensional DCT	Landsatdemo	Demonstration of Terrestrial satellite color synthesis
edgedemo	Edge detection demonstration	nrfiltdemo	Demonstration of noise elimination filtering
firdemo	Two-dimensional FIR filters and filter demonstration	qtdemo	Quadtree decomposition demonstration
imadjdemo	Demonstration of grayscale and adjustment and histogram equalization	roidemo	Demonstration of specific region handling

Table 2.33 File formats

Format	Instruction	Format	Instruction
'bmp'	Windows bitmap	'pgm'	Portable grayscale image
'cur'	Windows cursor resources	'png'	Portable grid image
'gif'	Graphics interchange format	'ppm'	Portable pixel map
'ico'	Windows chart resources	'ras'	Sun grating
'jpg'/'jpeg'	Static image compression standard	'tif'/'tiff'	The format of marked image files
'pbm'	Portable bitmap	'hdf'	The format of hierarchical data
'pcx'	windows paintbrush	'xwd'	X Windows heap

2.5.3 Displaying Information About an Image File

In image processing, the imfinfo function is used to obtain the details of the image file, and the file information may be different according to the different types of files. But no matter what type of the image file, the file information must contain the file name (path), the file format, the version number of the file format, the modified time, the size of the file, the width of the image (pixels), the length of the image (pixels), the number of bits of each pixel, the type of the image, and so on. The specific invoking format is as follows:

info = imfinfo(filename, fmt);
info = imfinfo(filename).

Info is the returned structure, which includes the specific information of the image file. The parameter filename is the string which assigns the name of the image file, and the parameter fmt is the string that specifies the file format. The file must be in the current directory or the path of MATLAB, and if imfinfo cannot find a file named filename, then it will look for the file named filename.fmt (Table 2.33).

If the parameter filename is a TIFF, HDF, ICO, GIF or CUR file, and be contained in more than one image, info is an element of an array of structures (a single structure) for each image.

2.5.4 Reading an Image File

In image processing, the imread function is used to read image data. In brief, the data of the image file is a two-dimensional array, which stores the color index or color value of each pixel in images.

A = imread(filename, fmt) read a grayscale or color image named filename into A. If the file contains a grayscale image, it is a two-dimensional array; if the file contains a true color (RGB) image, it is a three-dimensional array ($m \times n \times 3$). Filename is a string that specifies the name of the image file and the string fmt specifies the format of the image file. If the image file is not in the current directory or the path of MATLAB, we need to specify the name of the full path of the image file on the system.

[X, map] = imread(filename, fmt) reads the index image in filename to X and reads the associated color map to map, whose value will be rezoomed in the interval [0, 1].

[…] = imread(filename) trys to infer its format from the contents of the file.

[…] = imread(URL, …) reads the image from an Internet URL.

[…] = imread(…, Param1, Val1, Param2, Val2, …) uses parameter values to control the read operation.

2.5.5 Data Classes and Data Conversions

In MATLAB, the image is represented by one or more matrix, and MATALAB's powerful matrix operation can be applied to the image completely, and the syntax applicable to matrix operations is also applicable to the image. The default image data type supported in the image processing toolbox is the unsigned 8-bit integer (uint8), that is, each data in the image matrix occupies one byte. However, many matrix operations do not support types other than double precision type (double). In this case, the built-in image data type conversion function in the image processing toolbox can be used, and functions for conversion of data types are:

(1) im2uint8: convert the input image data (logical, uint16, double) into the type of unit8.
(2) im2uint16: convert the input image data (logical, uint8, double) into the type of unit16.
(3) im2double: convert the input image data (logical, uint8, uint16) into the type of double.
(4) im2bw: convert input image data (uint8, uint16, double) into the type of logical (binary image).
(5) mat2gray: convert input image data (double) into the type of normalized double (range [0, 1]).

The image types supported by the image processing toolbox include binary image, grayscale image, true color image, index image and multi-frame image (video).

1. Binary Image: The binary image is also called black and white image, and each pixel in the image has only two gray values (black or white), or the pixel value of the binary image is 0 or 1.
2. Grayscale Image: further add many color depths between black and white into the binary image, which constitute the grayscale image. Each pixel of the grayscale image is a quantified gray value. If the pixel of the grayscale image is the type of unit8, the range of the pixel is [0, 255]; if the pixel of the grayscale image is the type of unit16, the range of the pixel is [0, 65,535].

3. True Color Image: the RGB image. The colors of each pixel are composed of red (R), green (G), blue (B), so a RGB image whose size is $m \times n$ requires a three-dimensional matrix ($m \times n \times 3$) to store. The true color image can be stored in double precision, and the range of the brightness value is [0, 1]. The common method of storage is unsigned integer, and the range of the brightness value is [0, 255].

4. Index Image: The index image is the image whose pixel value is as the subscript of the RGB palette. The index image contains two matrices, the image data matrix and the palette (also known as the color map) matrix. Palette is the color image matrix which has three columns and a number of lines. Each row of the matrix represents a kind of color, with three columns represent the intensity of red, green, and blue colors of doubles, forming a particular color. The color intensity of the palette in MATLAB is [0, 1]. 0 is the darkest, and 1 is the brightest. The image data is uint8 or double.

5. Multi-frame Image Sequence (Video): The image processing toolbox supports connecting multi-frame images to image sequences. The image sequence is a 4-dimensional array, and the sequence number of the image frame constitutes the fourth dimension after the height, width and color depth of the image. You can use the cat function to merge the scattered images into the image sequence, provided that the image size must be the same, and if it is an index color image, the palette must be the same.

By default, MATLAB will store most of the data in the type of double to ensure the accuracy of operation. There is a large space overhead for the data type of image. Sometimes we have to convert the image storage format when using some image processing functions. MATLAB provides many functions of image type conversion:

(1) dither function: enhance the color resolution of the output image by color jitter. This function can convert the RGB image to an index image or convert the grayscale image to a binary image.

X = dither(RGB, map): The true color image RGB is dithered to an index image X by the specified color map.

X = dither(RGB, map, Qm, Qe): An index image is generated from an RGB image. The parameter Qm represents the quantized number of color graphs along each color axis, and Qe represents the quantification of the error in the color space calculation. If Qe is < Qm, then no jitter is performed, returning an index image without jitter. The default value is Qm = 5, and Qe = 8.

BW = dither(I): jitter the grayscale image I into a binary image BW.

(2) gray2ind function: convert the grayscale image to an index image.

[X, map] = gray2ind(I, n): convert the grayscale image I to an index image X in accordance with the specified grayscale, the color map is gray(n), and the default value of n is 64.

(3) grayslice function: convert the grayscale image to an index color image by setting the threshold value.

X = grayslice(I, n): The grayscale image I is quantized uniformly into n grades and converts to an index image.

X = grayslice(I, v): According to the specified threshold vector v (each element is between 0 and 1), the range of the image I is divided into the index image X.

(4) im2bw function: The grayscale image, index image and true color image are converted to binary images by setting threshold values.

BW = im2bw(I, level): convert the grayscale image I to a binary image BW based on the normalized threshold level.

BW = im2bw(X, map, level): convert the index image X to a binary image BW based on the normalized threshold level.

BW = im2bw(RGB, level): convert the RGB image to a binary image BW based on the normalized threshold level.

(5) ind2gray function: convert the index image to a grayscale image.

I = ind2gray(X, map): convert the index image X with the color map to a grayscale image I and return the same image as the storage type of the original image.

(6) ind2rgb function: convert the index image to a true color image.

RGB = ind2rgb(X, map): convert the index image with the color map to a true color image RGB and return the RGB image to the type of double.

(7) mat2gray function: convert the data matrix into a grayscale image using the normalized method.

I = mat2gray(A, [amin amax]): convert the data matrix A to a grayscale image I according to the specified value interval [amin amax], in which amin corresponds to 0 (darkest), and amax corresponds to 1 (brightest).

I = mat2gray(A): The smallest element in A is amin, and the largest element is amax, which converts A to a grayscale image I.

(8) rgb2gray function: convert the true color image to a grayscale image.

I = rgb2gray(RGB): convert the RGB image to a grayscale image I, and return the image to the storage type of the original image.

newmap = rgb2gray(map): convert the color map to the grayscale, and the types of input and output are double.

(9) rgb2ind function: convert the true color image to an index image.

[X, map] = rgb2ind(RGB, n): Using the minimum variance quantization method, the RGB image is converted to an index image, and n is the number of colors in the color map.

X = rgb2ind(RGB, map): The RGB is converted to an index image with a color map by matching the color of the RGB with the most similar color in the color map.

2.5.6 Displaying the Contents of an Image

The image processing toolbox in MATLAB provides special functions for the display of image files, common functions include:

1. colorbar function: diplay the color bar

colorbar updates the color bar to the right of the current coordinate axis. If there is no color bar, the right side of the current axis will display a vertical color bar.

colorbar('off'), colorbar('hide'), colorbar('delete'): delete all the color bars of the current coordinate axis.

colorbar('vert'): add a vertical color bar to the current axis.

colorbar('horiz'): add a horizontal color bar to the current axis.

colorbar(…,'peer', axes_handle): generate the color bar associated with the axes-handle to substitute the current axis.

2. image function: display the image

image(C): show the matrix C as the image. Each element in C defines the color of the image. C can be an $m \times n$ order matrix, or an array of $m \times n \times 3$.

image(x, y, C): where x and y are vectors of two elements that specify C (1, 1) and C (m, n).

image(x, y, C, 'PropertyName', PropertyValue, …): draw image C in the specified position x and y with the specified attribute name and attribute value pair.

3. imagesc function: zoom out the image data to make it in the interval of the current color image and display the image.

imagesc(C): show the matrix C as an image. Each element in C corresponds to a rectangular area in the image, and each element in C is a pointer to the current color map, and the current color map determines the color of each block.

imagesc(x, y, C): display the matrix C as an image and specify the range of the coordinate axis through vector x and y.

imagesc(…, clims): normalize the value in C to be in the range specified by clims, and C is displayed as an image. The parameter clims is a vector of one or two elements, limiting the range of data values in C. These values are mapped to the full range of the current color map.

4. imshow function: display the image.

imshow(I): use a grayscale system palette (R=G=B) to display the grayscale image I. In most cases, grayscale images are rarely saved with the color image table. However, when displaying grayscale images, MATLAB uses the system's predefined default grayscale color image table in the backstage.

imshow(I, [low high]): show the grayscale image I with a specified grayscale range [low high]. The grayscale value in the image is equal to or lower than the low is shown as black. The grayscale value greater than or equal to high is shown as white. The grayscale value between low and high is shown as the default value of the gray level. If an empty matrix ([]) is used to replace [low high], the function will use [min(I(:)) max(I(:))] to replace, which means that the minimum value in I is shown as black and the maximum is shown as white.

imshow(RGB): display the true color image RGB.

imshow(BW): display the binary image BW, where pixel 0 is black and pixel 1 is white.

Imshow(X, map): display the index image X with the specified color map, in which the map is a data matrix of $p \times 3$, and each row represents the color value of red, green and blue, whose brightness is in the range of [0, 1].

imshow(filename): display an image stored in the image file named filename. This file must contain images that can be read using imread or dicomread functions. The imshow function calls the imread or dicomread function to read the image from the

```
>> imtool
>>
```

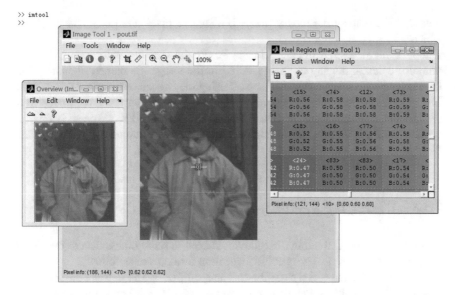

Fig. 2.3 Display an image and use the pixel region tool to explore the content of the image

file, but does not store the image data in the MATLAB workspace. If this file contains multiple images, the imshow function displays the first image of the file. This file must be in the current directory or the MATLAB path (Fig. 2.3).

2.5.7   *Exploring the Contents of an Image*

MATLAB provides the imtool function for more careful observation of the image. Not only the function of displaying images such as the imshow function, but also the Pixel Region tool, Image Information tool, Adjust Contrast tool and some other tools. These tools can also be used by direct access library functions impixelinfo, imageinfo, and imcontrast.

imtool opens a new image tool using the menu option "File" to open or import an image selected from Workspace Window to display the image.

imtool(I): display the grayscale image I.

imtool(I, [low, high]): diplay the grayscale image I with a specified grayscale range [low, high]. The grayscale value in the image which is equal to or lower than low will be shown as black. The grayscale value which is greater than or equal to high will be shown as white. The grayscale value between low and high is shown as the default value of the gray level.

imtool(RGB): display the true color image RGB.

imtool(BW): display the binary image BW, where pixel 0 is black and pixel 1 is white.

imtool(X, map): display the index image X with the specified color map.

imtool(filename): display an image stored in the image file named filename. This file must contain images that can be read using the imread or dicomread functions. The imtool function calls the imread or dicomread function to read the image from the file, but does not store the image data in the MATLAB workspace. If this file contains multiple images, the imtool function displays the first image of the file. This file must be in the current directory or the MATALAB path.

2.5.8   Writing the Resulting Image onto a File

In image processing, the imwrite function is used to write image data to the image file and store it on disk. The specific invocation formats are as follows:

imwrite(A, filename, fmt): use the format specified by fmt to write the images in A to the file filename. A can be a grayscale image ($M \times N$) or a true color image. The parameter filename is a string that specifies the output filename and does not allow empty image data. The possible value of the parameter FMT is determined by the MATLAB file format.

imwrite(X, map, filename, fmt): write the index image of X and their associated color map to the file filename. The file format is specified by fmt. If X is a uint8 or uint 16 format, the function imwrite writes the actual value of the array to the file. If X is a double precision type, then the function imwrite makes the value in the array on the bias before using uint8(X − 1) to write the data. The parameter map must be a MATLAB color map. Note that most image file formats do not support color maps with more than 256 elements.

imwrite(…, filename): write the image to the filename and derives the used format from the extension of the file which must be the valid value of fmt.

imwrite(…, Param1, Val1, Param2, Val2 …): specify the different characteristic parameters of the control output file.

2.6   Video Processing in MATLAB

MATLAB provides basic functions of video processing, and it mainly supports video files with the AVI format. Starting with MATLAB7.5 (R2007b), the new library functions (mmreader) can support AVI, MPEG, and WMV formats. Additionally, Simulink provides video and image processing blockset to support some video processing applications.

2.6.1 Reading Video Files

MATLAB provides some functions to read video files:

1 aviinfo function: return a structure of AVI file information, including file name, file size, file generated date, file frames, the number of frames per second, the width of each frame, the height of each frame, file type, file compression ratio, image quality and the number of colors of color images.

fileinfo = aviinfo(filename): return the AVI file information of filename. Among them, the filename is the name of AVI file which in the form of a string, and if there is no expanded-name, the system will choose '.avi' as default file expanded-name. And this file must be in the current working directory or MATLAB search path.

2. aviread function: read AVI video files.

mov = aviread(filename): read an AVI file named filename into the video structure of MATLAB. Then, it returns value mov which own two fields: cdata and colormap, and the contents are changed according to the type of images.

mov = aviread(filename, index): read the frame marked by the index in the AVI file named filename, which can be a single index or an index array. In AVI files, the index value of the first frame is 1, and the second frame is 2, and so on.

3. mmreader function: create the object to read video files. The supported video formats vary from different platforms. Windows supports AVI (.avi), mpeg-1 (.mpg), Windows Media videos (.wmv, .asf, .asx), and all formats supported by Microsoft DirectShow.

obj = mmreader(filename): create the object obj to read a video file named filename. If we cannot create an object for any reason, mmreader will generate an error.

obj = mmreader(filename, 'PropertyName', PropertyValue): read a video file named filename for the created object obj according to the properties.

4. videoreader class: MATLAB has a special video reading class VideoReader to complete the function of reading. obj = VideoReader(fileName).

2.6.2 Processing Video Files

Processing video files usually includes the following steps (if we do the same for all frames in a video, we can nest them in the for loop):

1. Use the frame2im function to transform a movie frame to an image.

[X, Map] = frame2im(F): convert a single movie frame F to the index image X and the corresponding color map Map. If the data contained in this frame is true color data, the value of the matrix Map ($m \times n \times 3$) is empty. We can create a movie frame using the getframe and im2frame functions.

2. The functions of the image processing toolbox are used to deal with images.

3. The im2frame function is used to transform the result of image processing to a movie frame.

f = im2frame(X, map): convert the index image X to a movie frame f according to the corresponding color map Map. If X is a true color image, then X is optional and does not produce any effect.

f = im2frame(X): If X contains the index image, convert the index image X to the movie frame f according to the current color map.

2.6.3   Playing Video Files

Functions provided by MATLAB to play video files are:

1. movie function: play the recorded movie frames.

movie(M): play the movie in the matrix M once, where M is the matrix constituted by movie frames.

movie(M, n): play the movie in the matrix M n times. If n is negative, the loop will play forward first and then play back. If n is a vector, then the first element of the vector is the number of play times, and the rest of the elements form a sequence of movie frames.

movie(M, n, fps): play the movie at the speed of FPS frames per second, by default of 12 frames per second. Computers that cannot reach the specified speed will play at maximum possible speed.

movie(h, ...): play movies in the image window or axis specified by the handle h.

movie(h, M, n, fps, loc): specifie the position vector of a 4 element [x, y, 0, 0] to hold the lower left corner of the movie frame (only the first two elements of the vector). The above position is relative to the lower left corner of the image window or axis specified by the handle.

2. implay function: play movies, videos or image sequences.

implay opens a movie player to play MATLAB movies, videos or image sequences. Use the menu bar "File" to select the movie or image sequence to play. Use toolbar buttons or menu bar options to play movies, jump to a specific frame in the sequence, change the speed of the frame playing, or perform other exploration operations (Fig. 2.4).

implay(filename): open the movie player and plays the contents of the file specified by filename.

implay(I): open the movie player and plays the first frame of the multi-frame image matrix specified by I. I can be MATLAB movie structure, or binary sequence, grayscale image sequence or true color image sequence. A binary or grayscale image sequence can be a $m \times n \times 1 \times k$ matrix or a $m \times n \times k$ matrix. The true color image sequence is a $m \times n \times 3 \times k$ matrix.

implay(..., FPS): play a movie or image sequence at a specified speed. The frame rate is one frame per second. If not, the default value is 20.

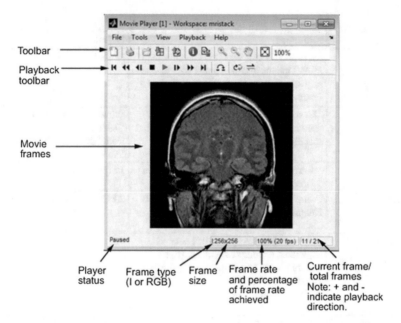

Fig. 2.4 The implay movie player

2.6.4   *Writing Video Files*

Functions provided by MATLAB to write video files are:

1.   avifile function: create a new AVI file.

aviobj = avifile(filename): create a AVI file named filename. All attributes of the AVI file object are taken by default. If the extension is not included in the filename, the extension .avi is automatically added. aviobj is a handle to the AVI file object, allowing other functions to reference the object.

avifile(filename, ParameterName, ParameterValue): use the specified parameter to create a AVI file named filename. Possible parameters are shown in the following Table 2.34.

2.   movie2avi function: AVI video files generated by MATLAB movie files.

movie2avi(mov, filename): convert the MATLAB movie mov to an AVI video named filename. mov is an array of structures ($1 \times n$), representing the number of video frames, and each frame is a structure containing two fields: cdata and colormap.

movie2avi(mov, filename, ParameterName, ParameterValue): convert mov to an AVI video.

3.   VideoWriter(filename, profile): create video files, static images or MATLAB videos from graphics with open, writeVideo and close methods.

Table 2.34 Parameters of avifile function

Parameter name	Instruction	Default
'colormap'	An $m \times 3$ matrix, which is the color map of AVI videos to define index colors, where $m \leq 256$ (if it is the compression format of Indeo, it can be 236)	None
'compression'	The string is used to specify the used decoder. Create a non-compressed file that specifies the value 'None'	Windows operating system: 'Indeo5' UNIX operating system: 'None'
'fps'	Specify the number of frames per second for AVI movies	15 fps
'keyframe'	Compressors that support time and space compression, the number of keyframes per second	2.1429 keyframe/s
'quality'	The value between [0, 100] is only applicable to compress videos. High quality parameters output high quality videos, and large files. Low quality parameters output low quality videos, and small file	75
'videoname'	Descriptive name of video stream. No more than 64 characters	The default is the filename

2.6.5   Basic Digital Video Manipulation in MATLAB

Other basic functions of video operations:

1. getframe function: get the movie frames.

    getframe returns a movie frame. The frame is a snapshot of the current coordinate axis or image window (pixel map).
    F = getframe: get the frame from the current axis.
    F = getframe(h): get the frame from the image window or coordinate axis marked by the handle h.
    F = getframe(h, rect): specify a rectangular area to copy the pixel map image. The parameter rect is a four-element vector [left, bottom, width, height], where width and height define the rectangle size. The parameter rect is defined relative to the lower left corner of the image window or the coordinate axis h, and the unit is the pixel.

2. addframe function: add a movie frame to an AVI file.

    aviobj = addframe(aviobj, frame): add data from the frame to AVI files labeled by aviobj, which has previously been created by calling avifile. If the frame is not the first frame of the AVI file, its size must be the same as the previous frame. aviobj is a handle to AVI file objects to update AVI file objects.
    aviobj = addframe(aviobj, frame1, frame2, frame3, …): add multi-frame images frame1, frame2, frame3, … to the AVI file object aviobj.

aviobj = addframe(aviobj, mov): add the mov image sequence in the MATLAB movie to the AVI file object aviobj. If the color image is not defined previously, the index image sequence frames stored in the MATLAB movie uses the color image of the first frame image as the color image of the AVI file.

aviobj = addframe(aviobj, h): add the image obtained from the image window and the coordinate axis handle h to the AVI file object aviobj.

3. close function: close AVI files.

aviobj = close(aviobj): close the write of the AVI file object aviobj and closes the file.

4. immovie function: create movie snippets from multi-frame images.

mov = immovie(X, map): create the movie snippet mov based on the multi-frame index image X. X contains multi-frame index images, and index images have the same size and the same color map. X is a $m \times n \times 1 \times k$ matrix, and k represents the number of images.

mov = immovie(RGB): create the movie snippet mov according to the multi-frame true color image X. X contains multi-frame true color images, and true color images have the same size. X is a $m \times n \times 3 \times k$ matrix, and k represents the number of images.

References

1. Gilat A (2016) MATLAB: an introduction with applications, 6th edn, Wiley
2. Attaway S (2016) A practical introduction to programming and problem solving, 4th edn, Elsevier
3. Gonzalez RC, Woods RE, Eddins SL (2009) Digital image processing using MATLAB, 2nd edn, Gatesmark Publishing
4. Solomon C, Gibson S (2011) Fundamentals of digital image processing: a practical approach with examples in Matlab, Wiley
5. Marques O (2011) Practical image and video processing using MATLAB, Wiley-IEEE Press

Chapter 3
Image and Video Segmentation

Abstract This chapter is devoted to some segmentation method of image and video. For image segmentation, five types of methods are detailed, including threshold segmentation, region-based segmentation, partial differential equation based segmentation, clustering based segmentation, and the graph theory based segmentation. For video segmentation, we shall introduce the motion region extraction method based on cumulative difference.

3.1 Introduction

People are only interested in regions of inputs which called target or foreground, and it generally corresponds to certain specific or unique properties regions in the image or video frame. Relatively, a region in which people are not interested is called background. Generally, internal features of one object are similar or the same while it will cause a sharp change tendency between the region of the different objects.

Segmentation [1] means dividing an image or a video frame into a number of specific and unique property parts or subsets according to the principle and extracting the target of interest for a higher level of analysis and understanding. Therefore, it is the basis of feature extraction, recognition and tracking. Image and video segmentation is one of the most basic and important fields of low-level vision in computer vision. It is a classic problem and it is also the basic premise of visual analysis and pattern recognition. There is neither a general method for image segmentation so far, nor an objective criterion for judging whether the segmentation is successful.

Image segmentation can be built on the two basic concepts: similarity and discontinuity. The similarity of pixels means that pixels of the image in a certain region have some similar characteristics, such as pixel grayscale, or texture formed by the arrangement of pixels. The discontinuity refers to the discontinuity of some features of pixels, such as the mutation of grayscale value, the mutation of color and the mutation of texture structure. A more formal definition of image segmentation is as follows.

© Springer International Publishing AG, part of Springer Nature 2019   65
S. Gong et al., *Advanced Image and Video Processing Using MATLAB*,
Modeling and Optimization in Science and Technologies 12,
https://doi.org/10.1007/978-3-319-77223-3_3

The set R represents the whole image, and the image segmentation of R can be viewed as that divide the image into N non-empty subsets $R_1, R_2, \ldots, R_N$ which satisfy the following conditions:

(1) $\bigcup_{i=1}^{N} R_i = R$;
(2)   For $i = 1, 2, \ldots, N, P(R_i) = TRUE$;
(3)   For $\forall i, j, i \neq j, R_i \cap R_j = \emptyset$;
(4)   For $\forall i, j, i \neq j, P(R_i \cup R_j) = FALSE$;
(5)   For $i = 1, 2, \ldots, N, R_i$ is the connected region.

Here condition (1) point out that the sum of all the segmentation regions should include all the pixels in the image, or segmentation should divide each pixel in the image into some sub regions. It is important that $\bigcup_{i=1}^{N} R_i = R$ represents the union of all the sub-regions of the segmentation, which is a sufficient condition for ensuring that each pixel in the image is processed. In condition (2), $P(R_i)$ indicates that pixels obtained in the same region after segmentation are supposed to have the same characteristics. Condition (3) indicates that each sub-region is non-overlapping, or that one pixel cannot belong to both regions at the same time. Condition (4) indicates that the pixels in different regions obtained after segmentation should have some different characteristics, and they do not have common features. Condition (5) requires that pixels in the same sub-region should be connected.

Video segmentation technology is developed with the technology based on static image segmentation, normally, the image or video sequence is divided into region of interest (ROI) in order to extract meaningful features or video objects (VO) refer to the Human vision. These objects always have consistent properties in some respects, such as brightness, color, shape, and motion. The video scene can be accessed and manipulated by VO. That is to say, the video object is a certain high-level "semantic" area, and it is more in line with the abstract expression of people on the visual perception of things in real life.

In addition to edge-based segmentation, threshold-based segmentation, entropy-based segmentation, morphological watershed segmentation, region-based segmentation and other classical segmentation methods, the image and video segmentation methods should also include graph-based segmentation, cluster bases segmentation, stochastic model based segmentation, fuzzy set based segmentation, partial differentiation based segmentation, fusion based segmentation and so on.

3.2   Threshold Segmentation

Image threshold segmentation is a widely used segmentation technique. Firstly, it compares the gray level differences between the target regions and the background regions, then selects a more reasonable threshold to determine the object pixels and background pixels. Finally, it generates a corresponding binary image. The key in using a threshold to segment image is to find the appropriate threshold to distinguish image object and background. Threshold method has become the most basic and widely used segmentation technology in image segmentation because of the simple realization, small amount of calculation and stable performance. It is applied to many fields, such as target segmentation in the infrared imaging tracking system, target segmentation in the synthetic aperture radar image, the magnetic resonance image segmentation, and so on.

Threshold segmentation is a region-based image segmentation technique. Basic principle is that the image pixels are divided into several classes by setting different feature thresholds. Regular features include the grayscale or color features that come directly from the original image, or features obtained from the original grayscale or color values. Assuming that the original image is $f(x, y)$, according to certain criteria, the eigenvalue t is found in $f(x, y)$ to divide the image into two parts. Segmentation results is that:

$$g(x, y) = \begin{cases} b_0 & f(x, y) < t \\ b_1 & f(x, y) \geq t \end{cases} \tag{3.1}$$

If you take $b_0 = 0$ (black), $b_1 = 1$ (white), that is, the image binarization. Figure 3.1 shows an example of image binarization.

According to the different constraints on the threshold, we can get three different types of threshold like the point correlation global threshold which only related to the grayscale value of the point; the region correlation global threshold that related to the grayscale value of points and local neighborhood characteristics of this point; local threshold or dynamic threshold that related to the location of the point, and the grayscale value of the point and the neighborhood characteristics of the point. All thresholding methods can be divided into non-contextual methods (also called point-dependent methods) and contextual methods (also known as region-dependent methods), according to the use of local information or global information. It can be divided into global thresholding and local thresholding, also known as adaptive

Fig. 3.1 An example of image binarization

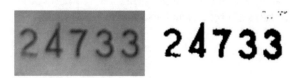

(a) The original image   **(b) The binary image**

thresholding, according to the use of uniform thresholds for the whole graph or the use of different thresholds for different regions. In addition, it also can be divided into bi-thresholding and multi-thresholding.

3.2.1   Global Threshold Image Segmentation

When the background of the image is relatively simple, and the image grayscale histogram is obviously bimodal distribution, the global threshold is used to segment the image and satisfactory results can be obtained. There are many methods for global threshold image segmentation, such as histogram threshold image segmentation, interclass difference threshold image segmentation, maximum entropy threshold image segmentation and so on. This section only introduces the basic principle and implementation process of interclass difference threshold image segmentation briefly. The processing of the method is as follows:

Select an initial estimate (the average grayscale for the image) for the global threshold;
Split the image with T. Generating two sets of pixels: $G1$ is composed of pixels whose grayscale value is greater than T, $G2$ is composed of less than or equal to T pixels;
Calculate the average grayscale values $m1$ and $m2$ of $G1$ and $G2$ pixels;
Calculate a new threshold: $T = (m1 + m2)/2$;
Repeat steps 2–4 until the ΔT value in successive iterations reaches the desired set of parameters;
Split image by $im2bw$ function.

The parameter ΔT is used to control the number of iterations when speed plays an important role. Usually, the larger the ΔT, the fewer the number of iterations performed by the algorithm. This can be proved that if the initial threshold is chosen between the maximum and minimum grayscale values in the image, and then the algorithm converges within a finite number of steps. According to the segmentation, the algorithm works well when there is a fairly clear valley between the histogram patterns of objects and background.

(a) The source image being processed   (b) Global threshold processed image

Fig. 3.2 Experimental results of a fingerprint, **a** original image; **b** processed image using global threshold method

The specific MATLAB code, as shown in PROGRAMME 3.1:

PROGRAMME 3.1: Global threshold segmentation

```
count=0;
T=mean2(f);
done=false;
while ~done
    count=count+1;
    g=f>T;
    Tnext=0.5*(mean(f(g))+mean(f(~g)));
    done=abs(T-Tnext)<0.5;
end
g=im2bw(f,T/255);
imshow(f);
figure,imhist(f);
figure,imshow(g);
```

Where f refers to the original image to be segmented, and Fig. 3.2 is the experimental result obtained by using the global threshold segmentation algorithm.

3.2.2 Local Dynamic Threshold Segmentation

The basic idea of the local threshold segmentation algorithm is dividing the image into several sub-images and select the corresponding segmentation threshold for each sub-image to complete the segmentation of the image. It is used for normalizing the non-uniform brightness intensity of images. The concrete steps are:

The image is divided into m blocks, where the size of each block image may not be equal;

Calculating the segmentation threshold for each sub-image;

Each sub-image is segmented by thresholds, and finally the blocks are merged to complete the segmentation of the entire image.

1. Image segmentation

In the process of image segmentation, there is still a large degree of gray-scale difference between the target and the background with different regions in the sub-image and the segmentation effectiveness is low when the sub-image is too large. There will be more serious block effect if the sub-image is too small when object dividing into different sub-blocks. The following is an adaptive image segmentation method, and the specific steps are:

Read the original image and divide it at the first time, and then get the same size n sub-images $o_1, o_2, \ldots, o_n$;

Histogram analysis is performed for each sub-image. Calculate the gray scale mean $L1$, the maximum gray scale value $g1$, and the minimum gray scale value $g2$ of the sub-image. When the ratio of target and background in the image is too wide, the average gray scale of the image is close to the highest gray level or the lowest gray level. If it is less than a set threshold, the algorithm failed. At this point, the segmentation condition is judged to be unsatisfied, while it is considered to satisfy the segmentation condition in other cases;

If the sub-image satisfies the segmentation condition, the continuing segmentation is stopped, and these sub images are successively recorded as $p1, p2, \ldots, pr, 1 \leq r \leq n$;

If the sub-image does not satisfy the segmentation condition, the secondary segmentation will perform, and the image will be divided into four sub-images of equal size. The divided sub-images are successively recorded as $q1, q2, \ldots, pk, 1 \leq k \leq 4n$.

2. Selection of local thresholds

First, two sets of sub-image sets $\{p1, p2, \ldots, pr\}$ and $\{q1, q2, \ldots, qk\}$ are analyzed, and then two sets of sub-images are respectively used in the appropriate way to obtain the split threshold and stored in the threshold matrix. Its specific steps are as follows:

For each sub-image in the sub-image set $\{p1, p2, \ldots, pr\}$, the global Otsu algorithm is used to sequentially compute the division threshold T and the four copies are stored in the corresponding positions of the threshold matrix;

The histogram of each sub-image in the sub-image set $\{q1, q2, \ldots, qk\}$ is analyzed, and the grayscale mean value $L2$, the maximum gray scale value $g3$ and the minimum gray scale value $g4$ of each sub-image are calculated. At the same time, the method proposed in Step 2 of Sect. 3.2 is used to judge whether the segmentation condition satisfies or not;

If the sub-image satisfies the segmentation condition, the global Otsu algorithm will be used to calculate the segmentation threshold and stored in the threshold matrix.

If the sub-image does not satisfy the segmentation condition, the sub-image segmentation threshold is determined by judging the relationship between $L2$ and $g3$ and $g4$. If $L2$ is close to $g3$, then a small threshold close to $g3$ is specified; if $L2$ is close to $g4$, then a large threshold close to $g4$ is specified;

The segmentation is terminated, and the threshold matrix $M1$ of $N \times N$ is obtained. The number of elements is $4n$.

3. Elimination of block effects

When the local threshold algorithm is used to segment an image, the segmented image is often accompanied by varying degrees of block effect. It is shown that the target object in the different sub-image will produce a pixel gray-scale mutation at the junction of the neighboring sub-image, and the target object in the different sub images may also be given different grayscales. Through the following method can eliminate the block effect: Firstly, smoothing the threshold matrix so that the threshold value of each sub-image merges the threshold information of the surrounding sub images to reduce the mutation among the neighboring threshold elements. After completion, the threshold matrix is interpolated by the bilinear interpolation, so that the transition of the threshold can be uniform and smooth. Finally, getting the threshold matrix for image segmentation. The specific steps are as follows:

(1) Smoothing the matrix $M1$. That is, each threshold element in $M1$ is added to each threshold elements in the 8 connected neighborhoods and then using the mean value to replace the original threshold. After the smoothing work is completed, we obtain the smoothing matrix $M2$ of $N \times N$;

(2) Interpolating the matrix $M2$. Considering the optimal matching of processing effect and time complexity, this algorithm adopts bilinear interpolation as the interpolation method of threshold matrix. After the interpolation is complete, a new threshold matrix $M3$ is obtained which is equal to the number of pixels of the original image, where each element forms a one-to-one relationship with the pixels at the corresponding position in the original image.

4. Binarization of the image

The original image is binarized by using the threshold matrix $M3$ obtained by the above steps. In processing, all the pixels in the original image are sequentially compared with each element in the threshold matrix pixel by pixel. Get the target and background of the binarized image.

The MATLAB code is shown in PROGRAMME 3.2.

PROGRAMME 3.2: Local dynamic threshold segmentation

```matlab
original_image=imread('scarlett.jpg');              % read the image
gray_image=rgb2gray(original_image);
subplot(1,2,1);
imshow(original_image);
subplot(1,2,2);
gray_image=double(gray_image);
[m,n]=size(gray_image);
result=zeros(m,n);
block_size=input('Block size=');                    % enter the block size
for i=1:block_size:m
    for j=1:block_size:n
        if ((i+block_size)>m)&&((j+block_size)>n)        % block
            block=gray_image(i:end,j:end);
        elseif ((i+block_size)>m)&&((j+block_size)<=n)
            block=gray_image(i:end,j:j+block_size-1);
        elseif ((i+block_size)<=m)&&((j+block_size)>n)
            block=gray_image(i:i+block_size-1,j:end);
        else
            block=gray_image(i:i+block_size-1,j:j+block_size-1);
        end
        t=mean(block(:));
        t_org=t;
        is_done=false;
        count=0;
        while ~is_done                                   % iterative threshold
            r1=find(block<=t);
            r2=find(block>t);
            temp1=mean(block(r1));
            if isnan(temp1);
                temp1=0;
            end
            temp2=mean(block(r2));
            if isnan(temp2)
                temp2=0;
            end
            t_new=(temp1+temp2)/2;
            is_done=abs(t_new-t)<1;
```

```
                t=t_new;
                count=count+1;
                if count>=1000
                    Error = 'Error:Cannot find the ideal threshold.';
                    return
                end
            end
            block(r1)=255;
            block(r2)=0;
            if ((i+block_size)>m)&&((j+block_size)>n)        % merge results
                result(i:end,j:end)=block;
            elseif ((i+block_size)>m)&&((j+block_size)<=n)
                result(i:end,j:j+block_size-1)=block;
            elseif ((i+block_size)<=m)&&((j+block_size)>n)
                result(i:i+block_size-1,j:end)=block;
            else
                result(i:i+block_size-1,j:j+block_size-1)=block;
            end
        end
    end
resule=uint8(result);
figure;
imshow(result);
```

The experimental results are shown in the Fig. 3.3.

(a) The original image   (b) The processed image

Fig. 3.3 Experimental results

3.3   Region-Based Segmentation

In addition to the histogram threshold method, the image-based random field model
method and the relaxation labeling region segmentation method for the segmentation method is based on region in image segmentation. There are two representative
methods: the regional growth method and the split merge method, which are charac-
terized by the segmentation process is decomposed into a sequence of steps, where
in the subsequent steps they are determined based on the results of the preceding
steps.

3.3.1   Region Growing

The basic idea of region growing is to combine the pixels with similar properties
to form a region. First, finding a seed pixel as the starting point for each growth
area. And then combine the pixels in the neighborhood of the seed pixel with the
same or similar properties of the seed (determined by a predetermined growth or
similar criterion) into the region where the seed pixel is located. A region grows
while the new pixels continue to seed to grow around them whilst the pixels meet
the conditions.

Now we give an example of region growth. Given a known matrix A:

$$\begin{bmatrix} 1 & 0 & 4 & 7 & 5 \\ 1 & 0 & 4 & 7 & 7 \\ 0 & 1 & 5 & 5 & 5 \\ 2 & 0 & 5 & 6 & 5 \\ 2 & 2 & 5 & 6 & 4 \end{bmatrix}$$

We choose "5" in the middle of the matrix as a seed. From the beginning of the
seed to around each pixel value and the seed value take the absolute difference with
the grayscale value. When the absolute value is less than a certain threshold T, the
pixel will grow into a new seed, and to the surrounding of each pixel growth: If the
threshold $T = 1$, the result of the region growth is:

$$\begin{bmatrix} 1 & 0 & 5 & 7 & 5 \\ 1 & 0 & 5 & 7 & 7 \\ 0 & 1 & 5 & 5 & 5 \\ 2 & 0 & 5 & 5 & 5 \\ 2 & 2 & 5 & 5 & 5 \end{bmatrix}$$

Visible pixels around the grayscale value of 4, 5, 6 pixels are well wrapped into the
growth area, and to the border pixels at the grayscale value of 0, 1, 2, 7 have become
boundaries. The upper right corner of the 5 can also be a seed, it is also located

outside the growth area because it does not contain a seed around the seed. Now take the threshold $T = 3$, the new region growth result is:

$$\begin{bmatrix} 5\,5\,5\,5\,5 \\ 5\,5\,5\,5\,5 \\ 5\,5\,5\,5\,5 \\ 5\,5\,5\,5\,5 \\ 5\,5\,5\,5\,5 \end{bmatrix}$$

The entire matrix is divided into a region. Thus, threshold selection is very important.

The selection of growth criteria depends not only on the specific problem itself, but also on the type of image data used, such as RGB and grayscale map. The general growth process ceases when it continues to meet the pixels of the growth conditions, and to increase the ability of the region growth often takes into account the criteria relating to the global nature of images and targets such as size and shape.

The implementation steps for the region growth method are as follows:

(1) The image is scanned by line to find no attributable pixels;
(2) Take the pixel as the center and check its neighboring pixels. That is, the pixels in the neighborhood are compared with it one by one. If the gray level difference is less than the predetermined threshold, merge them;
(3) With the newly merged pixels as the center, step 2 is detected until the region cannot be further expansion;
(4) Returning to step 1, continue scanning until you cannot find the pixels that are not attributed, and the entire growth process is over.

The above method is to be scanned first, which has a relatively large dependence on the choice of region growth starting point. To overcome this problem, the method can be improved as follows:

(1) Set the threshold of the gray scale difference to zero. Expand the region with the above method, and merge the pixels with the same gray scale;
(2) Obtain an average gray-scale difference between all neighboring regions, and merge a neighboring region with a minimum gray-scale difference;
(3) Set the termination criteria and merge the regions in turn by repeating the operations in step 2 until the termination criteria are met. The growth process is completed.

When there is a region with slowly varying gray scales in the image, the above method may cause the different regions to merge and produce errors. In order to overcome this problem, instead of using the grayscale value of the new pixel to compare with the grayscale value of the neighborhood pixel, the grayscale value area of the new pixel will be compared with the grayscale value of each neighborhood pixel. For an image area with N pixels, the gray scale is:

$$m = \frac{1}{N} \sum_{R} f(x, y) \tag{3.2}$$

The comparison of pixels is:

$$\max_{R}|f(x, y) - m| < T \tag{3.3}$$

where T is the threshold.

If the region is uniform, the gray scale change in the region should be as small as possible. If the region is non-mean (general condition) and consists of two parts, the proportion of these two pixels in R is $q1$ and $q2$ respectively, and the grayscale values are $m1$ and $m2$. Then the region mean is $q1m1 + q2m2$ and for the pixel with a grayscale value of $m1$, and the difference between the two values is $Sm = m1 - (q1m1 + q2m2)$.

We can see that the probability of correct judgment is:

$$P(T) = \frac{1}{2}[P(|T - s_m|) + P(|T + s_m|)] \tag{3.4}$$

This suggests that the gray scale difference between different pixels should be as large as possible when considering the gray scale mean.

The MATLAB code for the region growing method is shown in PROGRAMME 3.3.

PROGRAMME 3.3: Region growing

```
image=imread('mri1.jpg');
I=rgb2gray(image);
figure,imshow(I),title(' original image ');
I=double(I)/255;
[M,N]=size(I);            % Get the number of rows and columns of the image
[y,x]=getpts;             % Obtain the starting point of region growing
x1=round(x);              % Abscissa rounding
y1=round(y);              % Ordinate rounding
seed=I(x1,y1);            % The gray value of the growing start point is stored in seed
Y=zeros(M,N);             % Do a image matrix Y with all-zero and as large as the original image as
     the output image matrix
Y(x1,y1)=1;               % Set the point in Y to the corresponding position of the point to the white
     field
sum=seed;                 % Store the sum of the gray values of the points that meet the region growing
     conditions
suit=1;                   % Store the number of the points that meet the region growing conditions
```

```
count=1;                  % Record the number of new points that meet the criteria at eight points
    around a point each time
threshold=0.05555;        % Threshold
while count>0
s=0;                      % Record the sum of the gray values of the new points meeting the criteria at
    eight points around a point
  count=0;
  for i=1:M
    for j=1:N
      if Y(i,j)==1
        if (i-1)>0 & (i+1)<(M+1) & (j-1)>0 & (j+1)<(N+1)      % Determine if this point is on the
boundary
          for u= -1:1              % Determine whether the surrounding points meet the field conditions
            for v= -1:1            %u,v is offset
% It is judged whether or not it is present in the output matrix Y and is a point that satisfies the field
    condition
              if   Y(i+u,j+v)==0 & abs(I(i+u,j+v)-seed)<=threshold
% In accordance with the above two conditions, the point corresponding to the position in Y is set to
    white field
                Y(i+u,j+v)=1; count=count+1;
                s=s+I(i+u,j+v);                        % The gray level of this point is added to s
            end
          end
        end
      end
    end
  end
end
suit=suit+count;                          % Add n to the count counter
sum=sum+s;                                 % Add s to the sum of the point grayscale
seed=sum/suit;                             % Calculate the new gray average
end
figure,imshow(Y),title(' Segmented image ')
```

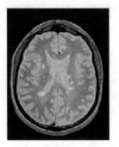
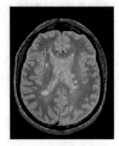

| (a) The original image of region growing method | (b) The image after the selecting seed point | (c) The image after region growing method |

Fig. 3.4 Region growing results

In the region growing method, the initial seed points need to be selected, and then extended around this point, to pick out the eligible points (Fig. 3.4).

3.3.2   Region Splitting and Merging

The basic idea of the region splitting and merging algorithm is to determine the criterion of a split and merge, that is, the measure of the regional feature consistency. When the characteristics of a region in the image are inconsistent, the region is divided into four equal sub-regions. When the neighboring sub-regions meet the consistency characteristics, they are combined into a large area until all regions no longer meet the conditions of the split merge.

The region splitting and merging steps are as follows:

(1) Split the image into $R1, R2, R3, \ldots, Rm$ by threshold set;
(2) Generate a region adjacency graph (RAG) from the segmentation description of the image;
(3) For each $Rj, i = 1, 2, \ldots, m$, determine all $Rj, j \neq i$ from RAG, such as Ri and Rj adjacency;
(4) For all i and j, calculate the appropriate similarity measurement Sij between Ri and Rj;
(5) If $Sij > T$, then merge Ri and Rj;
(6) According to the similarity criteria, (1)–(5) are repeated until there is no merged area.

Splitting and merging may be combined for the complex scenes segmentation, which can guide the application of split and merge operations based on rules.

The MATLAB code for the region splitting and merging method is shown in PROGRAMME 3.4.

PROGRAMME 3.4: Split and merge segment procedures

```
img=imread('lanhua.jpg');
subplot(1,2,1);              % The image display is divided into one row and two columns, and a
    coordinate system is created on the first graph on the left
imshow(img);                 % Show the original image
[M N]=size(img);    % The image size is stored in the array
divide_range=70;
merge_range=40;
Seg_matrix(1,12)=0;
Seg_matrix(1,2:6)=1;
Seg_matrix(1,4)=N;
Seg_matrix(1,6)=M;
Seg_matrix(1,9)=0;
%*********** Segment Algorihm ************
seg=1;
seg_tot=1;
while seg<=seg_tot;
    left=Seg_matrix(seg,3);   % =1
    right=Seg_matrix(seg,4);  % =N
    up=Seg_matrix(seg,5);          % =1
    down=Seg_matrix(seg,6);         % =M
    if abs(left-right)>1 && abs(up-down)>1
        tot_gray_sum=0;
        single_segment=1;
        for p=up:down
            for q=left:right
                if p==up && q ==left;
%The gray value of the pixel is assigned to tot_gray_sum
                    tot_gray_sum=double(img(p,q));pixel_count=1;
                else
% The average value of the gray level of pixel in the region is obtained
                    average=tot_gray_sum/pixel_count;
                    if abs(average-double(img(p,q)))>divide_range
                        single_segment=0;
                        break;
                    end
                    tot_gray_sum=tot_gray_sum+double(img(p,q));
```

```
                pixel_count=pixel_count+1;        % The number of pixels
            end
        end
        if single_segment==0
            break;
        end
    end
end
%********************* divide segment ****************
if single_segment==0
    seg_tot=seg_tot+1;
    Seg_matrix(seg,2)=0;                    % Not using
    Seg_matrix(seg_tot:seg_tot+3,2)=1;   % using == yes
    Seg_matrix(seg_tot:seg_tot+3,9)=2;   % All will connection to two segments
    mid_x=left+floor((right-left)/2)-1;
    mid_y=up+floor((down-up)/2)-1;
    %                          seg1 LU     Split the upper left part
    Seg_matrix(seg_tot,3:6)=[left mid_x up mid_y];
    cell_array(seg_tot,1)={seg_tot+1};%      2
    cell_array(seg_tot,2)={seg_tot+2};%      3
    %                          seg2   RU     Split the upper right part
    seg_tot=seg_tot+1;
    Seg_matrix(seg_tot,3:6)=[mid_x+1 right up mid_y];
    cell_array(seg_tot,1)={seg_tot-1};%      1
    cell_array(seg_tot,2)={seg_tot+2};%      4
    %                          seg3   LD     Split left lower part
    seg_tot=seg_tot+1;
    Seg_matrix(seg_tot,3:6)=[left mid_x mid_y+1 down];
    cell_array(seg_tot,1)={seg_tot-2};%      1
    cell_array(seg_tot,2)={seg_tot+1};%      4
    %                          seg4   RD     Split right lower part
    seg_tot=seg_tot+1;
    Seg_matrix(seg_tot,3:6)=[mid_x+1 right mid_y+1 down];
    cell_array(seg_tot,1)={seg_tot-2};%      2
    cell_array(seg_tot,2)={seg_tot-1};%      3
    %************** checking the old connection exist ****************
    old_conection=Seg_matrix(seg,9);
    for o_c=1:old_conection
        con_seg=cell_array{seg,o_c};
        if Seg_matrix(con_seg,2)~=0
            B1(1:4)=Seg_matrix(con_seg,3:6);
            for child_seg=seg_tot:-1:seg_tot-3;
```

```
                        B2(1:4)=Seg_matrix(child_seg,3:6);
    if B2(2)+1<B1(1) || B1(2)+1<B2(1) || B2(4)+1<B1(3) || B1(4)+1<B2(3)
    else
    if (B2(2)+1==B1(1) || B1(2)+1==B2(1)) && (B2(4)+1==B1(3) || B1(4)+1==B2(3))
    else
    Seg_matrix(child_seg,9)=Seg_matrix(child_seg,9)+1;% Updating child connection
    cell_array(child_seg,Seg_matrix(child_seg,9))={con_seg};
    Seg_matrix(con_seg,9)=Seg_matrix(con_seg,9)+1;% Updating connected segments
    cell_array(con_seg,Seg_matrix(con_seg,9))={child_seg};
                            end
                        end
                    end
                end
            end
        else
            Seg_matrix(seg,7)=tot_gray_sum;
            Seg_matrix(seg,8)=pixel_count;
        end
else
    tot_gray_sum=0;
    pixel_count=1;
    for p=up:down
        for q=left:right
            tot_gray_sum=tot_gray_sum+double(img(p,q));
            pixel_count=pixel_count+1;
        end
        end
        Seg_matrix(seg,7)=tot_gray_sum;
        Seg_matrix(seg,8)=pixel_count-1;
    end
    seg=seg+1;
end
%*************************** mearge algorithm ***************************
cnt=0;
for seg=1:seg_tot
    if Seg_matrix(seg,1:2)==[0 1];
        curr_gry_sum=Seg_matrix(seg,7);
        curr_pxl_cnt=Seg_matrix(seg,8);
        cnt=cnt+1;
        Seg_matrix(seg,10)=cnt;
```

```
loop_flag=1;
curr_seg=seg;
Seg_matrix(curr_seg,11)=0;
Seg_matrix(seg,1)=1;
while(loop_flag==1)
    connection_found=0;
    while Seg_matrix(curr_seg,12)<Seg_matrix(curr_seg,9)
        Seg_matrix(curr_seg,12)=Seg_matrix(curr_seg,12)+1;
        next_seg=cell_array{curr_seg,Seg_matrix(curr_seg,12)};
        if Seg_matrix(next_seg,1:2)== [0 1];
            curr_average=curr_gry_sum/curr_pxl_cnt;
            next_average=Seg_matrix(next_seg,7)/(Seg_matrix(next_seg,8));
            if abs(next_average-curr_average)<merge_range
              connection_found=1;
              break;
            else
                f=1;
            end
        end
    end
    if connection_found==1                    % Merging
        curr_gry_sum=curr_gry_sum+Seg_matrix(next_seg,7);
        curr_pxl_cnt=curr_pxl_cnt+Seg_matrix(next_seg,8);
        Seg_matrix(next_seg,10)=cnt;
        Seg_matrix(next_seg,1)=1;
        Seg_matrix(next_seg,11)=curr_seg;
                    curr_seg=next_seg;
            else
                curr_seg=Seg_matrix(curr_seg,11);
            end
            if curr_seg==0
                loop_flag=0;
            end
        end
        if curr_pxl_cnt>100
                valid_seg_index(cnt)=1;
        end
    end
end
```

```
max_cnt=cnt;
valid_seg_index(cnt+1)=0;
for seg=1:seg_tot
    if Seg_matrix(seg,1)==1;
        cnt=Seg_matrix(seg,10);
        if valid_seg_index(cnt)==1;
            L=Seg_matrix(seg,3);
            R=Seg_matrix(seg,4);
            U=Seg_matrix(seg,5);
            D=Seg_matrix(seg,6);
            result(U:D,L:R,cnt)=img(U:D,L:R);
            img_seg(U:D,L:R)=cnt;
        end
    end
end
k=1;
for cnt=1:max_cnt;
    if valid_seg_index(cnt)==1;
        res(:,:,k)=result(:,:,cnt);
        k=k+1;
    end
end
Lrgb = label2rgb(img_seg);
subplot(1,2,2);
imshow(Lrgb)
```

The splitting and merging method uses the traditional quadtree method. First, image segmentation, then the merger, the results shown in Fig. 3.5.

Figure 3.5 shows the original grayscale image and the processed images using split and merge algorithm. The objects in processed image are clearly separated into different colours, the programme marked the background in dark blue, leaves and solid in light blue and the petals in white.

The watershed algorithm is a kind of mathematical morphology image segmentation method which has been developed. The basic idea is considering the image as a natural terrain covered by water. The grayscale value of each pixel in the image represents the altitude of the point, and each of its minimal local value and its affected region is called the catchment basin, while the boundary of the basin is a watershed. There are usually two ways to describe watershed transformation: one way is "raindrops," that is when a drop of rain from the different position of terrain surface began to decline, which will eventually flow to the different local minimum altitude regions

Fig. 3.5 The original image of the splitting and merging method (left) and the processed image (right)

(called minimal regions). Those converged to the same minimal region raindrop trajectory to form a connected region, known as the catchment basin; another way is to simulate the "overflow" process, namely, first on the surface of each minimal region make a small hole, while water gushed out from the hole, and slowly submerged the regions around the minimal regions. Thus, the range of the minimal regions spread is the corresponding water catchment basin. Either way, the boundaries of water flow in different regions are the expected watershed.

Applying to image segmentation, watershed transformation refers to the original image is converted into a label image, which all belong to the same catchment basin are assigned the same label, and used a special label to identify the point on the watershed.

The calculation process of the watershed is an iterative annotation process, and the classical calculation method is proposed by L. Vincent. In the algorithm, watershed computation is divided into two steps, one is the sorting process, and the other is the submerged process. Firstly, the gray level of each pixel is sorted from low to high. Then in the process of low to high submergence, the FIFO structure is used to judge and annotate the regions that effected by any minimal local value on the h-order height.

The watershed transform is obtained by the input image of the catchment basin, and the boundary point between the basins is watershed. Obviously, the watershed represents the maximal point of the input image. Therefore, in order to obtain the edge information of the image, the gradient image is usually used as an input image, that is:

$$g(x, y) = grad(f(x, y)) = \{[f(x, y) - f(x - 1, y)]^2[f(x, y) - f(x, y - 1)]^2\}0.5$$
$$(3.5)$$

In the formula, $f(x, y)$ represents the original image, $grad\{\}$ represents gradient operation.

The watershed algorithm has a good response to the weak edge. The noise in the image and the slight change of the gray level of the object surface both cause the phenomenon of over segmentation. At the same time, it should be noted that the watershed algorithm has a good response to the weak edges and is guaranteed by the closed continuous edge. In addition, the closed catchment basin obtained by watershed algorithm provides the possibility of analyzing the regional features of images.

In order to eliminate the over segmentation of watershed algorithm, two methods of processing are usually used. One is to use the prior knowledge to remove irrelevant edge information. The second is to modify the gradient function so that the catchment basin only responds to the target you want to detect.

In order to reduce the over segmentation of watershed algorithm, it is common to modify the gradient function. A simple method is to threshold the gradient image to eliminate the over segmentation of the small changes in gray scale. That is

$$g(x, y) = \max(grad(f(x, y)), g\theta) \qquad (3.6)$$

In the formula, $g\theta$ represents the threshold value.

If the target object in the image is connected together, it is more difficult to segment it, and the watershed algorithm is often used to deal with such problems, and usually achieve better results. The watershed segmentation algorithm regards the image as a "topographic map." Among them, regions with strong brightness have larger pixel values, while pixels in dark regions are smaller. And the images are segmented by looking for "catchment basin" and "watershed boundaries."
Algorithm steps:

1. Reading the image;
2. Obtaining the boundary of the image, on this basis it can be directly applied to the watershed segmentation algorithm, but the effect is not good;
3. Labeling the foreground and background of the image, in which the foreground pixels inside each object are connected, and each pixel value in the background does not belong to any target object;
4. Calculating the segmentation function, and applying the watershed segmentation algorithm.

Its implementation code as shown in PROGRAMME 3.5:

PROGRAMME 3.5: Reading the image and finding the boundary of the image

```
rgb = imread('E:\rui\0.jpg');              % read the original image
I = rgb2gray(rgb);                         %convert to gray scale image
figure;subplot(121);                       %show the gray scale image
imshow(I);
hy = fspecial('sobel');                    %sobel operator
hx = hy';
Iy = imfilter(double(I),hy,'replicate');   %filter the y-direction edge
Ix = imfilter(double(I),hx,'replicate');   %filter the x-direction edge
gradmag = sqrt(Ix.^2 + Iy.^2);        %model
subplot(122);imshow(gradmag,[]);           %display gradient
title(' Gradient map ');
%directly use gradient model values for watershed algorithm（often exist over segmentation,not work
    well）
L = watershed(gradmag);            %wastershe algorithm is applied directly
   Lrgb = label2rgb(L);            %convert to RGB image
figure;imshow(Lrgb);               %show the segmented images
   title(' Gradient map after watershed transformation ');
% The foreground and background are marked separately
se = strel('disk',20);             % circular structure element
Io = imopen(I,se);                      % morphological opening operation
figure;subplot(121);
imshow(Io);                             %show the image after opening operation
title('Opening operation processed image');
Ie = imerode(I,se);                %image eroding
Iobr = imreconstruct(Ie,I);        % morphological reconstruction
subplot(122);imshow(Iobr);         %show the reconstructed image
title(' Reconstructed image ')
Ioc= imclose(Io,se);               %morphological closing operation
figure;subplot(121);
imshow(Ioc);                            %show the image after closing operation
title('Closing operation processed image');
```

```
      Iobrd = imdilate(Iobr,se);          %dilating image
      Iobrcbr = imreconstruct(imcomplement(Iobrd);
imcomplement(Iobr));Iobrcbr = imcomplement(Iobrcbr); % image inversing
      subplot(122);imshow(Iobrcbr);                    %show the image after recionstructing and
      inversing
      title(' Reconstructed and inversed image ');
fgm = imregionalmax(Iobrcbr);         %local maximal value
      figure;imshow(fgm);                          % show reconstructed local maximal value image
      title(' Reconstructed local maximal value image ');
I2 = I;
I2(fgm) = 255;                        % the pixel value at the local maximum is set to 255
      figure;imshow(I2);                       %show the maximum value region on the orginal image
      title(' the maximum value region on the orginal image ');
se2 = strel(ones(5,5));               % structure element
      fgm2 = imclose(fgm,se2);        %close operation
      fgm3 = imerode(fgm2,se2);       %eroding image
      fgm4 = bwareaopen(fgm3,20);        %open operation
      I3 = I;I3(fgm4) = 255;            %foreground set value 255
      figure;subplot(121);
      imshow(I3);                              %show the modified maximal value region
      title('the modified maximal value region');
      bw = im2bw(Iobrcbr,graythresh(Iobrcbr));  %convert to binary image
      subplot(122);imshow(bw);                      %show the binary image
      title('Binary image');
% watershed transform and display
D = bwdist(bw);                       %calculate the distance
      DL = watershed(D);               %watershed transformation
      bgm = DL == 0;                   %obtain the segement boundary
      figure;imshow(bgm);              %show the segemented boundary
      title('The segemented boundary');
gradmag2 = imimposemin(gradmag,bgm | fgm4);    %set minimum value
      L = watershed(gradmag2);          % watershed transformation
      I4 = I;I4(imdilate(L == 0,ones(3,3)) | bgm | fgm4) = 255;      % set 255 to foreground and boundary
      value
      figure;subplot(121);
      imshow(I4);                              % highlight the foreground and boundary
      title(' Highlight the foreground and boundary ');
        Lrgb = label2rgb(L,'jet','w','shuffle');   %convert to a pseudo RGB image
        subplot(122);imshow(Lrgb);                     % show the pseudo RGB image
      title(' The pseudo RGB image ')
```

3.4   Segmentation Based on Partial Differential Equation

Image segmentation based on a partial differential equation [2, 3] as a relatively new and effective image segmentation method, has gradually become research hotspots. This section mainly introduces the C-V image segmentation model based on partial differential equation (PDEs) and its MATLAB implementation. The idea is that if you can find a closed curve c divides the image into internal Ω_1 and external Ω_2. So that the average grayscale value of Ω_1 and Ω_2 just reflects the difference between the object and the background. Then this closed curve c can be seen as the outline of the object. The energy function is as follows:

$$E(C_1, C_2, C) = \mu \oint_c ds + \lambda_1 \iint_{\Omega_1} (I - C_1)^2 dxdy + \lambda_2 \iint_{\Omega_2} (I - C_2)^2 dxdy \quad (3.7)$$

Where the mentioned equation is called active contours without edges model (or C-V model). In this equation, I is the image intensity, the first item is the arc length of c. The second and third terms are the square errors of the internal and external greyscale values with scalar C_1 and C_2. Only when c reaches the correct position, the values of these two items can be minimized at the same time. Using the horizontal diversity method. Introduce the Heaviside function in the above formula firstly and modify it to the functional of the embedded function u:

$$E(C_1, C_2, U) = \mu \iint_{\Omega} \delta(u)|\nabla u| dxdy + \lambda_1 \iint_{\Omega} (I - C_1)^2 H(U) dxdy + \lambda_2 \iint_{\Omega} (I - C_2)^2 [1 - H(U)] dxdy$$

$$E(C_1, C_2, u) = \mu \iint_{\Omega} \delta(u)|\nabla u| dxdy + \lambda_1 \iint_{\Omega} (I - C_1)^2 H(u) dxdy + \lambda_2 \iint_{\Omega} (I - C_2)^2 [1 - H(u)] dxdy$$

$$(3.8)$$

So that the function u in the fixed conditions relative to C_1, C_2 to minimize the formula. It can be obtained that:

$$C_i = \frac{\iint_{\Omega_i} I dxdy}{\iint_{\Omega_i} dxdy}, \quad i = 1, 2 \quad (3.9)$$

C_1, C_2 are the average values of the input image (internal curve) and (external curve). In the C_1, C_2 fixed conditions, minimize the formula relative to u. It can be obtained that:

$$\frac{\partial u}{\partial x} = \partial \varepsilon [\mu div(\frac{\nabla u}{|\nabla u|}) - \lambda_1 (I - C_1)^2 + \lambda_2 (I - C_2)^2] \quad (3.10)$$

Thus, the stable solution through the equation to obtain the segmentation result. The MATLAB code based on the above idea is shown in PROGRAMME 3.6:

PROGRAMME 3.6: Segmentation based on partial differential equations

test.m Main function

```
%achieve Active Contours Without Edges
Img=imread('picture1.bmp');          % A well-behaved CV model
%Img=imread('vessel.bmp');           % A bad-behaved CV model
%Img=imread('twoCells.bmp');         % A bad-behaved CV model
U=Img(:,:,1);
% get size
[nrow,ncol] =size(U);
ic=nrow/2;
jc=ncol/2;
r=20;
phi_0 = sdf2circle(nrow,ncol,ic,jc,r);
delta_t = 0.1;
lambda_1=1;
lambda_2=1;
nu=0;
h = 1;
epsilon=1;
mu = 0.01*255*255;
I=double(U);
phi=phi_0;
figure;
imagesc(uint8(I));colormap(gray)
hold on;
plotLevelSet(phi,0,'r');
numIter = 1;
for k=1:50,
phi=EVOLUTION_CV(I,phi,mu,nu,lambda_1,lambda_2,delta_t,epsilon,numIter);
% Update the level set function
    if mod(k,10)==0
        pause(.5);
        imagesc(uint8(I));colormap(gray)
        hold on;
        plotLevelSet(phi,0,'r');
    end
end;
% backward_gradient.m
function [bdy,bdx]=backward_gradient(f)
[nr,nc]=size(f);
bdx=zeros(nr,nc);
```

```
bdy=zeros(nr,nc);
bdx(2:nr,:)=f(2:nr,:)-f(1:nr-1,:);
bdy(:,2:nc)=f(:,2:nc)-f(:,1:nc-1);
% binaryfit.m
function [C1,C2]= binaryfit(phi,U,epsilon)
H = Heaviside(phi,epsilon);a= H.*U;
numer_1=sum(a(:));
denom_1=sum(H(:));
C1 = numer_1/denom_1;
b=(1-H).*U;
numer_2=sum(b(:));
c=1-H;
denom_2=sum(c(:));
C2 = numer_2/denom_2;
% BoundMirrorEnsure.m
function B = BoundMirrorEnsure(A)
  [m,n] = size(A);
  if (m<3 || n<3)
      error('either the number of rows or columns is smaller than 3');
end
  yi = 2:m-1;
xi = 2:n-1;
B = A;
B([1 m],[1 n]) = B([3 m-2],[3 n-2]);B([1 m],xi) = B([3 m-2],xi);
B(yi,[1 n]) = B(yi,[3 n-2]);
% BoundMirrorExpand.m
function B = BoundMirrorExpand(A)
[m,n] = size(A);
yi = 2:m+1;
xi = 2:n+1;
B = zeros(m+2,n+2);
B(yi,xi) = A;
B([1 m+2],[1 n+2]) = B([3 m],[3 n]);B([1 m+2],xi) = B([3 m],xi);
B(yi,[1 n+2]) = B(yi,[3 n]);
% BoundMirrorShrink.m
function B = BoundMirrorShrink(A)
[m,n] = size(A);
yi = 2:m-1;
xi = 2:n-1;
B = A(yi,xi);
% curvature.m
function K=curvature(f)
[f_fx,f_fy]=forward_gradient(f);
```

```
[f_bx,f_by]=backward_gradient(f);
 mag1=sqrt(f_fx.^2+f_fy.^2+1e-10);
n1x=f_fx./mag1;
n1y=f_fy./mag1;
mag2=sqrt(f_bx.^2+f_fy.^2+1e-10);
n2x=f_bx./mag2;
n2y=f_fy./mag2;
mag3=sqrt(f_fx.^2+f_by.^2+1e-10);
n3x=f_fx./mag3;
n3y=f_by./mag3;
mag4=sqrt(f_bx.^2+f_by.^2+1e-10);
n4x=f_bx./mag4;
n4y=f_by./mag4;
nx=n1x+n2x+n3x+n4x;
ny=n1y+n2y+n3y+n4y;
magn=sqrt(nx.^2+ny.^2);
nx=nx./(magn+1e-10);
ny=ny./(magn+1e-10);
[nxx,nxy]=gradient(nx);
[nyx,nyy]=gradient(ny);
K=nxx+nyy;
% Delta.m
function Delta_h = Delta(phi,epsilon)
Delta_h=(epsilon/pi)./(epsilon^2+ phi.^2);
% EVOLUTION_CV.m
function phi = EVOLUTION_CV(I,phi0,mu,nu,lambda_1,lambda_2,delta_t,epsilon,numIter);
I=BoundMirrorExpand(I);
phi=BoundMirrorExpand(phi0);
for k=1:numIter
    phi=BoundMirrorEnsure(phi);
    delta_h=Delta(phi,epsilon);
    Curv = curvature(phi);
    [C1,C2]=binaryfit(phi,I,epsilon);
phi=phi+delta_t*delta_h.*(mu*Curv-nu-lambda_1*(I-C1).^2+lambda_2*(I-C2).^2);
end
phi=BoundMirrorShrink(phi);
% forward_gradient.m
function [fdy,fdx]=forward_gradient(f)
[nr,nc]=size(f);
fdx=zeros(nr,nc);
fdy=zeros(nr,nc);
```

```
a=f(2:nr,:)-f(1:nr-1,:);
fdx(1:nr-1,:)=a;
b=f(:,2:nc)-f(:,1:nc-1);
fdy(:,1:nc-1)=b;
% get_phi.m
function [xcontour,ycontour] = get_phi(I,nrow,ncol,margin)
count=1;
x=margin;
for y=margin:nrow-margin+1,
    xcontour(count) = x;
    ycontour(count) = y;
    count=count+1;
end;
y=nrow-margin+1;
for x=margin+1:ncol-margin+1,
    xcontour(count) = x;
    ycontour(count) = y;
    count=count+1;
end;
x=ncol-margin+1;
for y=nrow-margin:-1:margin,
    xcontour(count) = x;
    ycontour(count) = y;
    count=count+1;
end;
y=margin;
for x=ncol-margin:-1:margin+1,
    xcontour(count) = x;
    ycontour(count) = y;
    count=count+1;
end;
%Heaviside.m
function H = Heaviside(phi,epsilon)
H = 0.5*(1+ (2/pi)*atan(phi./epsilon));
% plotLevelSet.m
function [c,h]=plotLevelSet(u,zLevel,style)
  [c,h] = contour(u,[zLevel zLevel],style);
% sdf2circle.m
function f = sdf2circle(nrow,ncol,ic,jc,r)
[X,Y] = meshgrid(1:ncol,1:nrow);
f = sqrt((X-jc).^2+(Y-ic).^2)-r;
% signed_distance.m
```

```
function u = signed_distance(I,xcontour,ycontour,margin)
[nrow,ncol] = size(I);
[temp,contsize] = size(xcontour);
Mark = zeros(nrow,ncol);
for y=1:nrow
    for x=1:ncol
        if   (x > ncol-margin+1) || (x < margin) ||(y < margin) || (y > nrow-margin+1)
            Mark(y,x) = -1;
        end
    end
end
for y = 1:nrow,
    for x =1: ncol,
        u(y,x) = sqrt(min((x-xcontour).^2+(y-ycontour).^2));

        if Mark(y,x) == -1
            u(y,x) = -u(y,x);
        end
    end
end
```

The result of segmentation is as follows (Figs. 3.6 and 3.7).

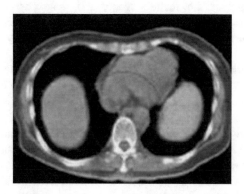

Fig. 3.6 Original test image

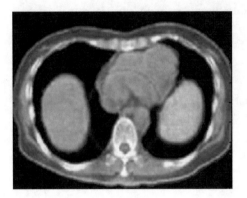

Fig. 3.7 Segmented image

3.5 Image Segmentation Based on Clustering

Clustering [4, 5] is the process of distinguishing and classifying things according to certain requirements and rules. Generally, it is necessary to give the number of clusters and the initial clustering centers. The image segmentation of clustering method is to represent the pixels in the image space with corresponding feature space points. According to their clustering in the feature space to segment the feature space, and then map them back to the original image space, get the segmented results. The classical clustering segmentation algorithm includes K-means clustering and fuzzy C-means clustering.

K-means algorithm first selects K initial class mean, and then each pixel into the average of its nearest class and calculate the new class mean. Iterate the previous steps until the difference between the old and new class mean is less than a threshold.

This method requires that each individual data can only and must belong to a class. By given the initial classification number and clustering center, iterate, and finally converge to the extreme point to achieve the effect of segmentation. The fuzzy C-means algorithm is a generalization of K-means algorithm on the basis of fuzzy mathematics, and it is to achieve clustering by optimizing a fuzzy objective function. It does not like the K-means clustering that each point can only belong to a certain class, but to give each point a class of membership degree. With the membership to better describe the edge of the pixel is also the characteristics of this, and suitable for dealing with things inherent uncertainty. Using the fuzzy C means (FCM) unsupervised fuzzy clustering calibration features for image segmentation, can reduce the artificial intervention, and more suitable for the image of uncertainty and fuzzy characteristics.

The advantage of the clustering segmentation method is that it does not require a priori knowledge and belongs to unsupervised segmentation method, which greatly improves the degree of automation of segmentation and improves the efficiency of segmentation. The disadvantage is that all typical clustering methods are sensitive to the initial value and the segmentation effect is not stable. In addition, no considered of image space between the context of information, so the segmentation effect is not ideal.

The pixel clustering steps of color images based on gray space are as follows:

(1) read the color image and convert the RGB value to grayscale value;
(2) arbitrarily select k objects from n data objects as the initial clustering center;
(3) according to the mean of the objects in the class, each object will be reassigned to the most similar class;
(4) update the mean of the class, that is, calculate the mean of the objects in each class. According to the mean (central image) of each clustering object, the distance between each object and the central objects is calculated. And according to the minimum distance to re-divide the corresponding object;
(5) loop (3)–(4) until each cluster no longer changes.

The MATLAB code based on clustering is shown in PROGRAMME 3.7.

PROGRAMME 3.7: Clustering algorithm for image segmentation

```
function kmeans_demo1()
clear;close all;clc;
%% Read the test image
im = imread('city.jpg');
imshow(im), title('Imput image');
%% Transform the color space of the image to get the sample
cform = makecform('srgb2lab');      % rgb is a matrix of m * n * 3
lab = applycform(im,cform);         % converts the rgb value in m * n * 3 to the value of lab
ab = double(lab(:,:,2:3));          % let the a, b component of the lab be a matrix of m * n * 2
nrows = size(lab,1);                %take m from lab
ncols = size(lab,2);                %take n from lab
X = reshape(ab,nrows*ncols,2)';     % The three-dimensional m*n*2 matrix in ab is transformed into a
     two-dimensional matrix of 2 * (m * n), where the first row is the value of component a and the
```

second row is the value of component b

figure, scatter(X(1,:)',X(2,:)',3,'filled');   box on; % Shows the spatial distribution of the two-dimensional sample space after the conversion of the color space, where the first row of X is the abscissa and the second row of X is the ordinate

%print -dpdf 2D1.pdf

%% Kmeans clustering the sample space

k = 5;   % Number of clusters

max_iter = 100;   % maximum number of iterations

[centroids, labels] = run_kmeans(X, k, max_iter);

%% Kmeans clustering the sample space

figure, scatter(X(1,:)',X(2,:)',3,labels,'filled');   % Display the two-dimensional sample space clustering effect, where labels is the category tag and number is one of the 1-5, which can be used as a color area to sort the good category, all the points by cluster results for color classification

hold on; scatter(centroids(1,:),centroids(2,:),60,'r','filled')   % Mark the center of the cluster where centroids is the center of the coordinate matrix

hold on; scatter(centroids(1,:),centroids(2,:),30,'g','filled')

box on; hold off;

%print -dpdf 2D2.pdf

pixel_labels = reshape(labels,nrows,ncols);% The two-dimensional classification results into rgb space and display the image

rgb_labels = label2rgb(pixel_labels);

figure, imshow(rgb_labels), title('Segmented Image');

%print -dpdf Seg.pdf

end

function [centroids, labels] = run_kmeans(X, k, max_iter)

%% The K-means ++ algorithm is used to initialize the clustering center

    centroids = X(:,1+round(rand*(size(X,2)-1)));% Randomly take a column as the first cluster center

    labels = ones(1,size(X,2));

    for i = 2:k

        D = X-centroids(:,labels);

        D = cumsum(sqrt(dot(D,D,1)));

        if D(end) == 0, centroids(:,i:k) = X(:,ones(1,k-i+1)); return; end

        centroids(:,i) = X(:,find(rand < D/D(end),1));   %why

        [~,labels] = max(bsxfun(@minus,2*real(centroids'*X),dot(centroids,centroids,1).'));

    end

```
%% standard Kmeans algorithm
  for iter = 1:max_iter
        for i = 1:k, l = labels==i; centroids(:,i) = sum(X(:,l),2)/sum(l); end
        [~,labels]  =  max(bsxfun(@minus,2*real(centroids'*X),dot(centroids,centroids,1).'),[],1);%
     Value output labels,~ means ignoring the output
     end
end
```

Figure 3.8 gives the experimental result image.

The K-mean clustering programme classified the objects and marked them in different colours. In Fig. 3.8b, the grass and the leaves are marked in yellow, and the walls of the buildings are marked as dark blue. The K-means algorithm is used to cluster the sample data, whether it is the choice of the initial point or the completion of an iteration of the data adjustment, are based on randomly selected sample data. Thus, it improves the convergence of the algorithm speed. The experimental results show that the image segmentation method based on K-means clustering algorithm has clear contour, and it is an effective segmentation algorithm of grayscale image.

3.6 Image Segmentation Method Based on Graph Theory

3.6.1 Introduction

The segmentation method based on graph theory maps the image as a weighted undirected graph. The pixel is used as the node of the graph, and the optimal segmentation of the image is obtained by using the minimum shear criterion. This method transforms the problem of image segmentation into an undirected graph $G = (V, E)$ optimization problem. The node v_i in the undirected graph represents the pixels

(a) Original image   **(b) Segmentation image**

Fig. 3.8 Experimental result, **a** original image, **b** segmentation image

in the image. The edge e_{ij} between the nodes v_i and v_j represents the relationship between the pixels, and a weight $w(v_i, v_j)$ is assigned to each edge e_{ij} according to a certain rule. By using certain optimization criterion, the edges in the region of the segmentation result have lower weights, and the edges between regions have higher weights.

The cost function of dividing the binary image into two regions A, B can be defined as

$$cut(A, B) = \sum_{i \in A, j \in B} w(i, j) \tag{3.11}$$

The optimal segmentation of the graph is the partition minimized the cost function. The weight function is generally defined as the similarity between two nodes, as shown in (3.12):

$$w_{ij} = \exp(-\frac{\|F_i - F_j\|}{\sigma_I^2}) \times \begin{cases} \exp(-\frac{\|X_i - X_j\|_2^2}{\sigma_x^2}), & \|X_i - X_j\|_2 < r \\ 0, & \text{other} \end{cases} \tag{3.12}$$

F_i is the grayscale value of the pixel, X_i is the space coordinates of the pixel, σ_1 is the standard deviation of the gray-scale Gaussian function, σ_x is the standard deviation of the space distance Gaussian function, r is the effective distance between two pixels. When the distance between the two pixels exceeds r, they are considered to have a similarity of zero. From the similarity function of Eq. (3.12), it is not difficult to find that the closer the effective distance between the two pixels, the greater the similarity between the two pixels, the closer the grayscale value between the two pixels, the greater the similarity between them.

According to the above-mentioned optimal segmentation criteria and similarity function, we can see that the basic principle of graph theory segmentation is to make the internal similarity of two regions (A, B) divided into maximum, and the similarity between regions (A, B) is minimum. And the segmented regions should avoid skew (i.e., biased for small regions). In order to obtain accurate segmentation results, it is important to design the cut set criterion, and the common cut set criteria are Minimum cut, Average cut, Normalize cut, Min-max Cut, Ratio cut, etc.

Table 3.1 lists several common cut sets criteria. Normalize cut is a more normative form among them. The criterion can be transformed into the eigenvector problem of the matrix.

There are two ways to implement the optimal criteria: one is to use the defined criteria to reduce the tree graph directly. Another is to convert the optimal criterion into solving the matrix equation.

3.6.2 GraphCut and Improved Image Segmentation Method

As a combinatorial optimization technique based on graph theory, GraphCut is used by many researchers to compute the minimum energy function, and the maximum flow minimum cut theorem is used to complete the image segmentation problem. It finds the boundary by people's perception of the target, through the appropriate interaction to obtain high-performance segmentation effect. GraphCut divides the image into two regions: "target" and "background." First, the user forces to define some "hard constraints," that is, some of the pixels in the image are definitely targeted or certainly the background of the pixels are manually marked out, respectively, as the target and background seed pixels. The definition of these hard constraints can directly reflect the user's segmentation intention. Then, calculate the optimal global solution in all the segments that satisfy the hard constraint, the other pixels of the image are automatically segmented into the target or background.

As shown in Fig. 3.9a, O is the user labeled foreground, and B is the background. As shown in Fig. 3.9b, the foreground, background and ordinary pixels in the image are constructed into a graph. According to the theory of graph theory, apply the maximum flow minimum cut algorithm on the corresponding image segmentation, and the optimal global segmentation of the graph is obtained. As shown in Fig. 3.9c, the optimal segmentation in Fig. 3.9d is also obtained. Figure 3.9 describes the segmentation process in detail.

General speaking, Graph Cut uses the minimum cut method for cutting, and the Minimum cut is generated by the Max-flow method. According to the maximum flow and minimum cut theorem, the maximum flow and the minimum cut are equivalent, and the maximum flow is the maximum flow of the network constructed by the image, and the minimum cut is the optimal global value of the required energy function. Therefore, GraphCut obtains the minimum cut by the maximum flow of the network, and this minimum cut is the objective function, that is, the minimum value of the

Table 3.1 Some common cut set criteria

Criteria	Form	Achieve
Minimum cut	$cut(A, B) =$ $\sum_{u \in A, v \in B} w(u, v)$	Tree graph reduction
Average cut	$Ncut(A, B) =$ $\frac{cut(A,B)}{assoc(A,V)} + \frac{cut(A,B)}{assoc(B,V)}$ $Avecut(A, B) =$ $\frac{cut(A,B)}{|A|} + \frac{cut(A,B)}{|B|}$	Solving equation $(D - W)x = \lambda x$
Normalize cut	$assoc(A, V) =$ $\sum_{u \in A, v \in V} w(u, v)$	Solving equation $(D - W)x = \lambda x$
Min-max cut	$Mincut(A, B) =$ $\frac{cut(A,B)}{w(A)} + \frac{cut(A,B)}{w(B)}$	Solving equation $(D - W)x = \frac{\lambda}{1-\lambda} Dx$
Ratio cut	$Rcut(A, B) = \frac{cut_1(A,B)}{cut_2(A,B)}$	Tree graph reduction

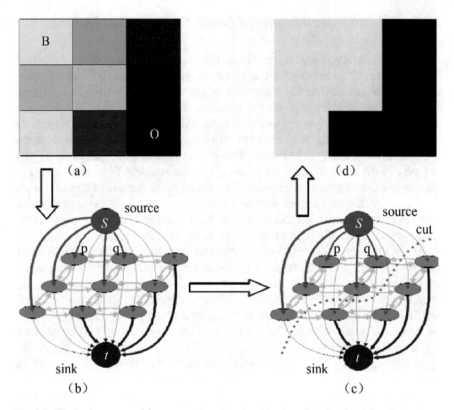

Fig. 3.9 The basic process of image segmentation algorithm based on the GraphCut theory

energy function. Thus, the key to image segmentation with the GraphCut theory is to solve the following two problems.

Constructing network graph. By constructing the network, we can find the maximum flow of the network, that is, the minimum cut value. The minimum value of the energy function can be used to segment the image accurately.

Constructing energy functions. GraphCut is to solve the problem of optimizing the energy function. For image segmentation, the energy function is the sum of the data item and the smoothing term.

Graph Cut is consist of the following steps.

1. Map image to the graph

The first step of the image segmentation method based on Graph Cut theory is to represent the image as a network graph, and the core problem is how to establish the corresponding relationship between image and graph. The basic element of image processing is a pixel, and the basic element of graph processing is a node. The spatial relationship between pixels corresponds to the edges of the graph, and the similarity between the pixels corresponds to the weight of the edges.

Figure 3.10 depicts the relationship between graph and image. The closer the two pixels are physically or the similarity of the gray scale, the higher the similarity is, and the lower the pixel with the similarity should be divided in the same class. And pixels with low similarity should be divided into different classes.

In order to use GraphCut for image segmentation, it is necessary to manually add some foreground and background annotation. So that the original image has 3 parts: foreground, background and ordinary image pixels. In order to make use of these 3 parts of information to construct a weighted graph (x, y), we need to complete the following 3 problems.

(1)   Construct the vertex set V of graph G

The vertex set V consists of two parts.

$$V = P \bigcup \{S, T\} \tag{3.13}$$

p is the set of pixels in the original image and p forms the intermediate nodes of the graph. The symbol S is the manual annotation of the foreground "object" pixel set, which constitutes the source of the graph source. T is a manually annotated background "background" pixel set, which constitutes the map of the sink.

(2)   Construct the edge set E of graph G

E is the edge set of the connected nodes, including N-links and T-links, and the edges between the ordinary pixels in the vertex set V are called N-connections, denoted as $N(\{p, q\}, p, q \in V)$. The definition uses eight neighborhoods, except for the boundary points of the image, and each pixel has eight N-connections. T-connection refers to the vertex of the ordinary pixels are connected with the source

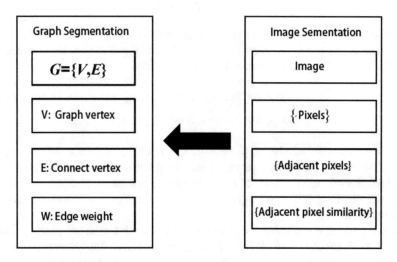

Fig. 3.10   The correspondence between graph and image

Table 3.2 Weight allocation rule of edge

Edge	Weight	Condition
$\{p, q\}$	$B_{\{p,q\}}$	$\{p, q\} \in N$
$\{p, S\}$	$\lambda \cdot R_p(\text{``bkg''})$	$p \in P, p \notin o \cup B$
	K	$p \in o$
	0	$p \in B$
$\{p, T\}$	$\lambda \cdot R_p(\text{``obj''})$	$p \in P, p \notin o \cup B$
	0	$p \in o$
	K	$p \in B$

and sink, each pixel has two such connections, denoted as $\{p, S\}, \{p, T\}, p \in V$. Therefore, the definition of edge set E is shown in Eq. (3.14).

$$E = N \bigcup_{p \in P} \{\{p, S\}, \{p, T\}\} \tag{3.14}$$

(3) Weight E for the corresponding edge W

According to the rules in Table 3.2, all non-negative weights $\cos t(p, q)$ are assigned to all edges in E.
  Among them,

$$K = 1 + \max_{p \in P} \sum_{q:\{p,q\} \in N} B_{\{p,q\}} \tag{3.15}$$

$$B_{\{p,q\}} = \frac{\exp(-\frac{(I_p - I_q)^2}{2\delta^2})}{dist(p, q)} \tag{3.16}$$

Equation (3.16) represents the boundary attribute of the image, I represents the brightness of the pixel, and $dist(p, q)$ represents the physical distance between the pixel p and q. The region attribute R_p is shown below

$$R_p(\text{``obj''}) = -lb(sum_p / sum_{obj}) \tag{3.17}$$

$$R_p(\text{``bkg''}) = -lb(sum_p / sum_{bkg}) \tag{3.18}$$

sum_p represents the number of pixels with the specified brightness value. Respectively, The sum_{obj} and sum_{bkg} represent the total number of pixels of the foreground and background, and the ratio of sum_p to sum_{obj} and sum_{bkg} respectively represents the pixels of different brightness values in the foreground and background distribution histogram.
  δ in Eq. (3.16) is a threshold of luminance difference between two pixels p and q. From Eqs. (3.15) and (3.16), if $|I_p - I_q| > \delta$, the luminance difference of two pixels is larger, the value of $B_{\{p,q\}}$ is smaller. The corresponding N-connection is called the main object of segmentation. Therefore, the larger the parameter A, the greater

the likelihood that the pixel having a large luminance difference is segmented into the same region on the image. According to the above method, the mapping of the image to the graph is completed after the weight is assigned to each edge E.

2. Construct energy function

The artificially interactively labeled pixels constitute some rigid limits of segmentation, which provide clues to segmenting the target. By constructing the region term and the boundary term of the energy function, the energy function is used as the segmentation model, and the remaining pixels in the image are segmented by calculating the optimal value of the energy function.

For the weighted graph $G(V, E, W)$, the vector $A = (A_1, \ldots, A_p, \ldots, A_{|p|})$ is defined as a segmentation result, and A_p is the label of the pixel p in the set P, which can be a foreground or a background. The cost function of the vector A divided into two regions is defined as the sum of the cost of the image boundary property $B(A)$ and the region property $R(A)$, as shown in Eq. (3.19).

$$E(A) = \lambda R(A) + B(A) \tag{3.19}$$

where

$$R(A) = \sum_{p \in P} R_p(A_p) \tag{3.20}$$

$$B(A) = \sum_{\{p,q\} \in N} B_{\{p,q\}} \delta(A_p, A_q) \tag{3.21}$$

$$\delta(A_p, A_q) = \begin{cases} 1 \text{ if } A_p \neq A_q \\ 0 \text{ other} \end{cases} \tag{3.22}$$

where λ is a nonnegative coefficient that describes the relative importance of the region term (data item) $R(A)$ and the boundary term (smoothing term) $B(A)$ in the image segmentation process. The region term can reflect the degree of fit of the pixel p's luminance value for all target histogram models. The boundary term constitutes the boundary property of segment A, and $B_{\{p,q\}}$ is a penalty between pixels p, q for their discontinuity. According to the previous analysis, the $B_{\{p,q\}}$ is larger when the brightness values of pixels p and q are relatively close. When the difference between the brightness values of the pixels p and q is large and larger than the specified threshold, $B_{\{p,q\}}$ is zero.

For an image, the smallest segmentation of the cost function in Eq. (3.19) is the optimal segmentation of the image.

3. Minimize the energy function

In order to minimize the energy function in Eq. (3.19), Boykov et al., designed the Maxflow-mincut algorithm, which is based on the augmented-path. Firstly, two search trees S and T are established. The root of S is the source point V_S, and the root of T is the sink point V_T. The edges of all the father nodes to the child nodes in

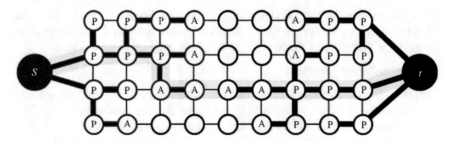

Fig. 3.11 Search tree

the search tree S are unsaturated, and all the edges from the child node to the father node in the T are also unsaturated. The nodes in the search tree S and T are free nodes. The nodes in the search tree S and T are divided into "Active" and "Passive" nodes, the active node is on the outer edge of the tree, and the passive node is inside the tree. The active node captures new subnodes from a set of free nodes via the unsaturated edges, allowing the trees to grow continuously; the passive node is not as an active node that makes the tree growth, because they are completely surrounded by other nodes in the same tree. In addition, the active node can also contact the node of another tree. When the active node of the search tree detects that the neighboring node belongs to another tree, it can determine an augmented-path.

The algorithm repeats the following three phases.

(1)   "Growth" stage: the search trees S, T grow until it finds the sink.
(2)   "Augmentation" Stage: augmentation the path, the search trees become a forest.
(3)   "Adoption" Stage: adopt isolated nodes, the search trees become a forest again.

The implementation of the algorithm is as follows:

In the "Growth" stage, the tree constantly expanding, and the active node from the free node to get the child node, then the child node has become a new active node of the tree. When all the neighboring nodes of the active node are processed once, the active node becomes a passive node. If an active node encounters a neighboring node that belongs to another tree, it stops growing. In this way, a path from S to T is detected. As shown in Fig. 3.11, the active and passive nodes are labeled as A and P respectively, and the free node has no labels.

In the "Augmentation" Stage, the $S \rightarrow T$ path obtained during the growth process is extended. In the process of expansion, some edges become saturated, and the corresponding nodes in the search tree become isolated nodes, which leads to the division of the search tree into a forest. S and T are still the root nodes of the two trees, but the isolated nodes become the root nodes of the other trees.

In the "Adoption" Stage, trees S and T will be restored. Each isolated node will find a valid parent node. The parent node and the isolated node belong to tree S or T. The parent node is connected by an unsaturation edge. If the parent node is not satisfied, the isolated node is removed from the tree S or T, turning it into a free node,

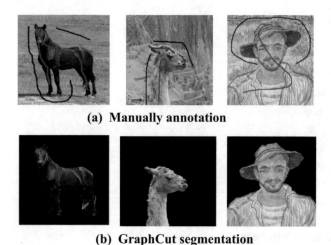

(a) **Manually annotation**

(b) **GraphCut segmentation**

Fig. 3.12 Segmentation results (**a** manually annotation, **b** GraphCut segmentation)

and all the child nodes before it become isolated nodes. When there is no isolated node, the adoption stage ends, and trees S and T will be restored.

When the adoption stage is completed, the algorithm returns to the growth stage and is executed until the search trees S and T are no longer growing. This will solve the maximum flow, the corresponding minimum cut also solved. Thus, the optimal segmentation of the image is completed.

As shown in Fig. 3.12, for the original image after the annotation (Fig. 3.12a). The light is the foreground target, and the dark is the background target. These seed points for the hard constraints and guide the division. Figure 3.12b is the result image for the GraphCut image segmentation.

The code is shown as PROGRAMME 3.8.

PROGRAMME 3.8: Improved GraphCut algorithm

```
function [SegIm, SegColorIm]=Graph_Cuts(im_org,im_cosal,ObjGMM,BkgGMM)
sigma=10;% Similarity variance
lambda=5;
c=1;
%mex maxflowmex.cpp maxflow-v3.0/graph.cpp maxflow-v3.0/maxflow.cpp % Mex
im=double(im_org);
```

```
m1=im(:,:,1);
m2=im(:,:,2);
m3=im(:,:,3);
[height,width] = size(im(:,:,1));
N = height*width;
disp('building graph');
% construct graph
% Calculate weighted graph
E = edges8connected(height,width);
dist1=(m1(E(:,1))-m1(E(:,2))).^2;
dist2=(m2(E(:,1))-m2(E(:,2))).^2;
dist3=(m3(E(:,1))-m3(E(:,2))).^2;
Smooth_Dist=(dist1+dist2+dist3);
% define n-links
V = c*exp((-abs(Smooth_Dist))./(2*sigma^2));
K=max(V)+1;
A = sparse(E(:,1),E(:,2),V,N,N,8*N);
% define t-links
T = Calc_Weights(im,im_cosal,K, lambda,ObjGMM,BkgGMM);
%Max flow Algorithm
disp('calculating maximum flow');
[flow,labels] = maxflow(A,T);
labels = reshape(labels,height,width);

SegColorIm=zeros(height,width);
for i = 1: height
    for j = 1: width
        if(labels(i,j)==0)
                SegColorIm(i,j)=255;
        else
                egColorIm(i,j)=0;
        end
    end
end
SegIm=labels;
```

3.7   Video Motion Region Extraction Method Based on Cumulative Difference

Moving object extraction is an important part of video object segmentation [6–8]. At the same time, moving object extraction is one of the key links in the recently developed dynamic visual attention research. Differential image method is a fast and simple method to realize moving target detection. Most of the related algorithms are based on differential image method. The difference image method includes continuous frame difference, frame difference, accumulated difference, background subtraction and so on. This section introduces a motion region extraction algorithm based on cumulative difference and mathematical morphology processing. In the time domain window, the image is degraded to obtain the gray scale image, the gray scale difference image is accumulated and the mathematical morphological processing to get the trajectory template of the moving target. The trajectory template is using "And operation" with differential image of a current frame, and the moving object pixels of the current frame are obtained. Finally, the moving region of the current frame is obtained by multi-level mathematical morphology. The algorithm is shown in Fig. 3.13.

Under the condition that the camera is motionless, the video sequence image is composed of the stationary background and the moving foreground target. However, due to the influence of imaging system noise, the nonzero gray values in the continuous frame difference image are not all caused by the target motion, but also a large part is due to the influence of noise. The noise can be modeled as Gauss noise, and the feature parameters (mean μ, variance σ^2) of the corresponding model can be obtained by multi frame analysis of video sequences. This method works better, but the calculation is complex. This section uses a simple and effective denoising method. The basic idea is to convert the original 256 gray-scale image into a low gray level (usually 8 levels) image, that is, a gray range of gray degree distribution is reduced to the same gray scale. At the same time, the gray scale of the current frame image takes into account the change of the grayscale value corresponding to the gray scale of the current frame and the gray scale of the pixels smaller than a certain threshold will not be changed. Because the frame difference image noise can be regarded as the average gauss noise of 0, its change is usually small, and the gray band number can be effectively selected to restrain the influence of noise.

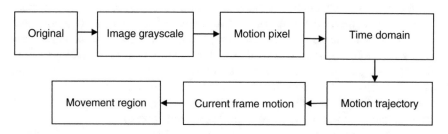

Fig. 3.13 Algorithm block diagram

Let $GL(x, y, t)$ be the pixel grayscale value of the t moment at (x, y). GL_{max} and GL_{min} are the maximum and minimum gray values of gray scale images. The symbol n is the number of grayscale. The symbol S is a continuous frame change threshold with varying grayscale. $GLB[x, y, t]$ and $GLB[x, y, t-1]$ are gray bands of t and $t-1$, and then have (3.23) and (3.24):

$$GL_{diff} = GL_{max} - GL_{min} + 1 \qquad (3.23)$$

$$GLB(x, y, t) = \begin{cases} GLB[x, y, t-1] & \text{if } \max(\frac{(GLB[x,y,t-1]-1)*GL_{diff}}{n} - S, GL_{min}) \\ \qquad \leq GL(x,y,t) < \min(\frac{GLB[x,y,t]*GL_{diff}}{n} + S, GL_{max}) \\ [\frac{GL[x,y,t]*n}{GL_{diff}}] + 1 & \text{otherwise} \end{cases}$$

$$(3.24)$$

The grayscale image of the original gray scale image is obtained, and the non-zero pixel in the frame image is directly taken as the moving pixel. Let $Mov(x, y, t)$ be 1 for (x, y) where the pixel is moving pixels and zero is the background pixel, then there are (3.25):

$$Mov(x, y, t) = \begin{cases} 0 \text{ if } GLB[x,y,t] = GLB[x,y,t-1] \\ 1 \text{ if } GLB[x,y,t] \neq GLB[x,y,t-1] \end{cases} \qquad (3.25)$$

In the two successive frames, the moving pixels are mostly isolated points, and if the internal texture of the moving target is not significant, the occlusion of the foreground frame and foreground object will cause the partial motion regions to be mistaken for the background regions. In addition, there are still some noise points after the gray band processing, so it is very limited to extract the moving region only according to the difference image between the front and rear frames. A feasible solution is to accumulate and accumulate multiple frames of difference images, and the trajectories of moving objects in the differential images can be concentrated. This is actually the time domain consistency and correlation of the moving target in the airspace, and the noise point is usually independent, even when the partial image is accumulated in the differential image, the relative motion trajectory region is also small, easy to filter out.

The traditional cumulative difference method usually first selects a reference frame, then compares the difference between each frame image and the reference frame, and accumulates the difference gray value. Finally, we can obtain the trajectory region by a certain threshold algorithm. In this section, we add the binary difference graph $Mov(x, y, t)$ of the adjacent two frames directly to the cumulative result. For cumulative results, the nonzero full mark $Mov_{sign}(x, y, t)$ of the image is 1, which are the pixels in the motion track region. $Mov_{men}(x, y, t)$ whose value is 0 are unchanged, which are regarded as the background pixels.

Let w be a cumulative frame (time domain window size), $pstart$ is the first frame gray difference image sequence. The window update equation is (3.26):

$$Mov_{mem}(x, y, t) = \begin{cases} \sum_{pstart}^{t} Mov(x, y, t) & \text{if } t \le w + pstart \\ Mov_{mem}(x, y, t-1) + Mov(x, y, t) - Mov(x, y, t-w) \text{ otherwise} \end{cases}$$

(3.26)

The region sign of motion track is (3.27):

$$Mov_{sign}(x, y, t) = \begin{cases} 0 \; Mov_{men}(x, y, t) = 0 \\ 1 \text{ otherwise} \end{cases}$$

(3.27)

After the binary marked image of the trajectory region is obtained, we can get the trajectory region by discarding those small regions. The implemental steps are shown as follows.

(1) Read video data. Read the adjacent two frames in the human eye of the most sensitive brightness data.
(2) Calculate the frame differences between the two frames by subtracting luminance matrix of the previous frame and the next frame.
(3) Calculates the mean and standard deviation of the noise in the frame difference iteratively.
(4) Get the change region by filtering noise according to the mean and standard deviation.
(5) Get the final template of the object by mathematical morphology.

The MATLAB code is shown as PROGRAMME 3.9.

PROGRAMME 3.9: The MATLAB implementation code is as follows

```
% Main program
clc;
close all;
clear all
info = imfinfo('3.gif');% first read, used to get attribute values
W = info.Width;
H = info.Height;
W = W(1);
H = H(1);
len = length(info);
figure('NumberTitle', 'off', 'ToolBar', 'none', 'Menu', 'none');
pos = get(gcf, 'position');
set(gcf, 'position', [pos(1) pos(2) W H]);
set(gca, 'position', [0 0 1 1]);
hold on;
for i = 1 : len
```

```
str=sprintf('photo%d.jpg',i);
[Ii, map] = imread('3.gif', 'frames', i);
imwrite(Ii,str,'jpg');
%F(:, i) = im2frame(flipud(Ii), map);
I=imread(str,'jpg');
I=mat2gray(I);    % any interval is mapped to [0,1];
OBsegment(I);
end
%movie(F, 20);
close;
```

Object segmentation subroutine:

```
function BWfinal=OBsegment(I)
%Step 1: Read
%figure, imshow(I), title('original image');
%Step 2: Detect
BWs = edge(I, 'sobel', (graythresh(I) * .1));
%figure, imshow(BWs), title('binary gradient mask');
%Step 3: Fill Gaps
se90 = strel('line', 3, 90);
se0 = strel('line', 3, 0);
%Step 4: Dilate the Image
BWsdil = imdilate(BWs, [se90 se0]);
%figure, imshow(BWsdil), title('dilated gradient mask');
% Step 5: Fill Interior Gaps
BWdfill = imfill(BWsdil, 'holes');
%figure, imshow(BWdfill),title('binary image with filled holes');
% Step 6: Remove Connected Objects on Border
BWnobord = imclearborder(BWdfill, 4);
%figure, imshow(BWnobord), title('cleared border image');
```

The algorithm is mainly applicable to the motionless video test sequence. The experimental results are shown in Fig. 3.14.

Apart from the frame difference method, the most important moving object extraction method is the background difference method. It is a moving object detecting method by subtracting background model in the image sequence, and its performance depends on the background modeling method. The accuracy of the modeling the background directly affects the detection effect. Due to the complexity of the scene, unpredictability, the various environmental disturbances and noises, such as the sudden change of light, the fluctuation of some objects in the actual background image, the jitter of the camera, the influence of the moving objects on the original scene, the modeling of the background is difficult. Common background modeling methods include median background modeling, mean method background modeling,

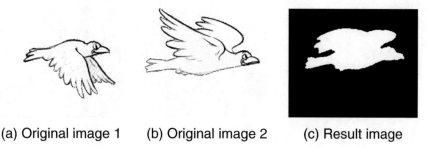

(a) Original image 1   (b) Original image 2   (c) Result image

Fig. 3.14 Screenshot and result (**a** original image 1, **b** original image 2, **c** result image)

Kalman filter model, single Gaussian distribution model, multi Gaussian distribution model, background modeling based on Codebook, and so on. The moving target extraction method will be described in Chap. 12.

References

1. Nguyen TNA, Cai J, Zheng J et al (2013) Interactive object segmentation from multi-view images. J Visual Commun Image Rep 24(4):477–485
2. Wang L, Lekadir K, Lee SL et al (2013) A general framework for context-specific image segmentation using reinforcement learning. IEEE Trans Med Imag 32(5):943
3. Park C, Huang JZ, Ji J et al (2013) Segmentation, inference and classification of partially overlapping nanoparticles. IEEE Trans Pattern Anal Mach Intel 35(3):669–681
4. Lin L, Du J (2017) Hyperspectral image segmentation method based on kernel method. In: International conference on intelligent information hiding and multimedia signal processing. Springer, Cham, pp 433–439
5. Thasneem AAH, Sathik MM, Mehaboobathunnisa R (2017) A fast segmentation and efficient slice reconstruction technique for head CT images. J Intel Syst
6. Suzuki CT, Gomes JF, Falcão AX et al (2013) Automatic segmentation and classification of human intestinal parasites from microscopy images. IEEE Trans Biomed Eng 60(3):803
7. Riaz F, Silva FB, Ribeiro MD et al (2013) Impact of visual features on the segmentation of gastroenterology images using normalized cuts. IEEE Trans Bio-med Eng 60(5):1191–1201
8. Beucher S, Lantuéj C (1979) Workshop on image processing, real-time edge and motion detection

Chapter 4
Feature Extraction and Representation

Abstract This chapter is focused on some classical feature representations for image and video analysis. In particular, we will introduce the histogram-based features, texture features, and some local point features.

4.1 Introduction

Image features [1, 2] is one of the most basic attributes used to distinguish different images. They may be natural features that can be identified by human vision, or some certain manmade parameters during the process of measuring and processing. The feature extraction is such a process of measuring the intrinsic, essential and important features or attributes of the research object and quantizing the result, or decomposing and symbolizing the object to form the feature vector or symbol string and relational map.

Common image features usually include: color features, texture features, shape features, spatial relations characteristics.

The color feature is a global feature that describes the surface properties of the object corresponding to the image or image region. Generally speaking, color feature is a characteristic based on pixels, where all the pixels belong to the image or image region have their contributions respectively. Since the color is not sensitive to the changes of direction, size in image or image region, color features cannot capture the local characteristics of the object good enough. Common features include: color histogram, color sets, color moments, color coherence vector, color correlogram, and so on.

The texture feature is also a global feature just like the color feature. However, the texture is only a characteristic of the surface of objects and it cannot fully reflect the essential attributes of objects. So, it is impossible to obtain high-level image content only by using texture features. Different from the color feature, the texture feature is not a characteristic based on pixels, and it needs to be calculated by statistical ways in image regions which include a lot of pixels. Common extraction methods of texture feature include: Gray level co-occurrence matrix, Tamura texture feature,

S. Gong et al., *Advanced Image and Video Processing Using MATLAB*,
Modeling and Optimization in Science and Technologies 12,
https://doi.org/10.1007/978-3-319-77223-3_4

113

simultaneous auto-regressive (SAR) texture model, wavelet transform [3] and so on. The gray level co-occurrence matrix is mainly to calculate four parameters of energy, inertia, entropy and correlation based on the gray level co-occurrence matrix. Tamura texture features put forward six kinds of attributes: roughness, contrast, orientation, line image degree, regularity and coarseness which are based on human visual perception psychology study of texture. The extraction of texture feature from the autocorrelation function of the image (i.e. the energy spectrum function of the image) extracts the characteristic parameters such as the thickness and the directionality of the texture through the calculation of the energy spectrum function. The SAR model takes parameters as texture feature based on the construction model of images. The typical method is the random field model method, such as Markov random field model method and Gibbs random field model method.

There are usually two types of representation methods of shape features, one is the contour feature, the other is the local feature. The contour feature mainly focuses on the outer boundary of the object, and the local feature is related to the entire local area. Typical shape feature description methods are as follows:

(1)   Boundary feature method

The classical Hough transform was concerned with the identification of lines in the image, but later the Hough transform has been extended to identifying positions of arbitrary shapes, most commonly circles or ellipses [4]. The Hough transform as it is universally used today was invented by Richard Duda and Peter Hart in 1972, who called it a "generalized Hough transform" [5] after the related 1962 patent of Paul Hough [6, 7]. The transform was popularized in the computer vision community by Dana H. Ballard through a 1981 journal article titled "Generalizing the Hough transform to detect arbitrary shapes". Boundary feature method obtains shape parameters of the image by describing the boundary feature. Among them, the detection method of parallel line with Hough transform and Edge direction histogram are classical methods. Hough transform is a method which connects the marginal pixels by using the global characteristics to a local closed boundary. The basic idea of which is the duality of the point and line. The edge direction histogram obtains the image border by calculus, then makes histograms of border's size and direction and the usual way is to make a grayscale gradient direction matrix.

(2)   Fourier shape descriptor method

The basic idea of Fourier shape descriptors is to use Fourier transform of the object boundary as the shape description, it converts two-dimensional problem into one-dimensional problem by utilizing the closure of regional boundaries. Three kinds of shape expressions are derived from the boundary points, they are as follows: curvature function, centroid distance and the complex coordinate function.

(3)   Geometric parameter method

It mainly includes the moment, area, perimeter, roundness, eccentricity, spindle direction and algebraic invariant moments of regions. It must be based on image processing and image segmentation when refers to the extraction of shape parameter, and the accuracy of parameter will be affected by the result of segmentation.

This section, we will introduce several features and their implementation briefly.

4.2 Histogram-Based Features

4.2.1 Grayscale Histogram

The grayscale histogram of a digital image is a discrete function of grayscale, and its definition can be expressed by Eq. (4.1).

$$H(i) = \frac{n_1}{N}, \; i = 0, \; 1, \; \ldots, \; L-1 \tag{4.1}$$

where i is the gray level, L is called the number of classes, n_i is the number of pixels with gray level i in the image, and N represents the total number of pixels in the image.

Common statistical features based on histogram are as follows:

(1) Mean value: The mean value reflects the average gray value of an image.

$$\mu = \sum_{i=0}^{L-1} i H(i) \tag{4.2}$$

(2) Variance: The variance reflects the discrete distribution of the grayscale of an image.

$$\sigma^2 = \sum_{i=0}^{L-1} (i - \mu)^2 H(i) \tag{4.3}$$

(3) Skewness: The skewness reflects the degree of asymmetry of the image's histogram distribution. The larger the degree of skewness is, the more asymmetric the histogram distribution is, otherwise, the more symmetrical it is.

$$\mu_s = \frac{1}{\sigma^3} \sum_{i=0}^{L-1} (i - \mu)^3 H(i) \tag{4.4}$$

(4) Kurtosis: The kurtosis reflects the approximate state of the grayscale distribution in an image when it is close to the mean value. And it is used to determine whether the grayscale distribution is concentrated near the average grayscale greatly. The smaller the kurtosis, the more concentrated it is, otherwise, the more dispersed.

$$\mu_k = \frac{1}{\sigma^4} \sum_{i=0}^{L-1} (i - \mu)^4 H(i) - 3 \qquad (4.5)$$

(5) Energy: The energy reflects the degree of uniformity of the grayscale distribution, and the energy is larger when the grayscale distribution is more uniform. On the contrary, the smaller it is.

$$\mu_N = \sum_{i=0}^{L-1} H(i)^2 \qquad (4.6)$$

(6) Entropy: The entropy also reflects the uniformity of histogram gray distribution.

$$\mu_E = - \sum_{i=0}^{L-1} H(i) \log_2[H(i)] \qquad (4.7)$$

The programme for extracting the mean feature of the histogram is shown as PROGRAMME 4.1.

PROGRAMME 4.1: **Grayscale Histogram Mean Feature Extraction**

```
tic
filename='1.jpg';
pi=imread(filename);
pix=double(pi);
s=double(zeros(254,1));
for i=1:512
    for j= 1:512
        for k=1:254
            switch pix(i,j)
                case k
                    s(k)=s(k)+double(1);
                otherwise
            end
        end
    end
end
sum=d ouble(0);

for k= 1:254

        sum=sum+s(k);

end

h=double(zeros(254,1));
```

for u = 1:254

       h(u)=s(u)/sum;

end

junzhi=double(0);

for i =1:254

junzhi= junzhi+ i*h(i);

end

4.2.2 Histograms of Oriented Gradients

HOG is a feature descriptor used to perform object detection in computer vision and image processing. The features are constructed by calculating and counting histograms of gradient directions in the local area of images. HOG feature is a statistical pedestrian characteristic in a local area, so it is not sensitive to illumination and positional offset, and it owns strong robustness. The classic HOG feature extraction and calculation process are as follows:

(1) Standardized Gamma space and color space: Adopt the Gamma correction method to normalize the input image in color space for adjusting image's contrast ratio and reducing the impact caused by the local shadow and illumination changes. Meanwhile, the noise interference can also be restrained. Here we set Gamma equal to 1/2.

$$I(x, \ y) = I(x, \ y)^{Gamma} \tag{4.8}$$

(2) Calculate the gradient of images: Calculate the image gradient of abscissa and ordinate direction $G_x(x, \ y)$ and $G_y(x, \ y)$, as shown in Eq. (4.9). Then calculate the gradient direction value $\theta(x, \ y)$ of each pixel position, as shown in Eqs. (4.10) and (4.11). The derivative operation not only capture the contours, silhouette and texture information, but also weaken the effect of illumination.

$$\begin{cases} G_x(x, \ y) = I(x + 1, \ y) - I(x - 1, \ y), \forall x, \ y \\ G_y(x, \ y) = I(x, \ y + 1) - I(x, \ y - 1), \forall x, \ y \end{cases} \tag{4.9}$$

$$M(x, \ y) = \sqrt{G_x(x, \ y)^2 + G_y(x, \ y)^2} \tag{4.10}$$

$$\theta(x, \ y) = \arctan(G_y(x, \ y)/G_x(x, \ y)) \tag{4.11}$$

(3) Make statistics of which bin every pixel belongs to: Pixels of the gradient direction in $[0, \ \pi]$ are divided into nine uniform small intervals, and we use $V_k(x, \ y)$ to represent the gradient intensity of pixel $(x, \ y)$ on the kth gradient direction.

$$V_k(x, y) = \begin{cases} M(x, y), & \theta(x, y) \in bin_k \\ 0, & \theta(x, y) \notin bin_k \end{cases}, \ 1 \le k \le 9 \qquad (4.12)$$

(4) Image block and cell unit: Dividing the picture into several blocks, and each of which is divided into several cell units, as shown in Fig. 4.1. Assuming that the size of the image is 64×128, the size of the block is 16×16, each block is divided into four cell units averagely, each cell is 8×8. Counting each direction's distribution of the gradient values of all the pixels in each cell and classifying them into nine bins between $[0, \pi]$, thus we can get the eigenvectors of cells. Four cell eigenvectors in each block are connected, then we can get the eigenvectors of each block. The cell's eigenvector is 9-dimension, and the block's eigenvector is 36-dimension. With 8 pixels for a step, there are 7 scan windows in the horizontal direction and 15 scan windows in the vertical direction. Then the eigenvector is 3780-dimension.

(5) Normalization: To further eliminate the influence of illumination, each cell is normalized:

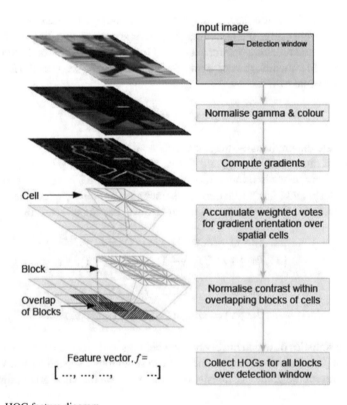

Fig. 4.1 HOG feature diagram

$$f(C_i, k) = \frac{\sum_{(x,y)\in C_i} V_k(x, y) + \varepsilon}{\sum_{(x,y)\in B} V_k(x, y) + \varepsilon}, \; i \in [1, 4] \cap k \in [1, 9] \qquad (4.13)$$

The programme of HOG feature extraction is shown as PROGRAMME 4.2.

PROGRAMME 4.2: **HOG Feature Extraction**

```
clear all; close all; clc;
img=double(imread('lena.jpg'));
imshow(img,[]);
[m n]=size(img);
img=sqrt(img);              % Gamma correction
% Edge computing is below
fy=[-1 0 1];                % Define vertical templates
fx=fy';                     % Define horizontal template
Iy=imfilter(img,fy,'replicate');       % The vertical edges
Ix=imfilter(img,fx,'replicate');       % The horizontal edge
Ied=sqrt(Ix.^2+Iy.^2);                 % Edge strength
Iphase=Iy./Ix;             % Edge slope, some of which are inf, -inf, nan, where nan need to deal with
%The following is the cell
step=16;                   % step*step pixels as a unit
orient=9;                  % The number of directions in the direction histogram
jiao=360/orient;           % Each direction contains the number of angles
Cell=cell(1,1);            % The cell can be dynamically increased for all the angle histogram, so first
    set up
ii=1;
jj=1;
for i=1:step:m             % If the processed m/step is not an interge,the best is i=1:step:m-step
    ii=1;
    for j=1:step:n                        % Comments as above
        tmpx=Ix(i:i+step-1,j:j+step-1);
        tmped=Ied(i:i+step-1,j:j+step-1);
        tmped=tmped/sum(sum(tmped));       % Local edge strength normalization
        tmpphase=Iphase(i:i+step-1,j:j+step-1);
        Hist=zeros(1,orient);                % The current step * step pixel block statistical
angle histogram is the cell
        for p=1:step
            for q=1:step
                if isnan(tmpphase(p,q))==1    % 0/0 obtains nan. If the pixel is nan,reset to 0
```

```
                    tmpphase(p,q)=0;
            end
            ang=atan(tmpphase(p,q));        % atan is between [-90 90] degrees
            ang=mod(ang*180/pi,360);        % It's all getting positive, -90 is going to be 270
            if tmpx(p,q)<0                   % Determine the true angle in the x direction
                if ang<90
                        ang=ang+180;        % Move to the third quadrant
                end
                if ang>270                   % If it's the fourth quadrant
                        ang=ang-180;        % Move to the second quadrant
                end
            end
            ang=ang+0.0000001;              % Prevent ang from 0
            Hist(ceil(ang/jiao))=Hist(ceil(ang/jiao))+tmped(p,q);
                end
            end
        %Hist=Hist/sum(Hist);                        % Direction histogram normalization, this step can not,
    because it is normalization after forming the block
            Cell{ii,jj}=Hist;               % Put it in Cell
            ii=ii+1;                        % Y coordinate loop variables for Cell
        end
        jj=jj+1;                            % X coordinate loop variables for Cell
end
% The following is to find feature, 2*2 cell to synthesize a block, without explicit block
[m n]=size(Cell);
feature=cell(1,(m-1)*(n-1));
for i=1:m-1
    for j=1:n-1
        f=[];
        f=[f Cell{i,j}(:)' Cell{i,j+1}(:)' Cell{i+1,j}(:)' Cell{i+1,j+1}(:)'];
    f=f./sum(f);% normalization
        feature{(i-1)*(n-1)+j}=f;
    end
end
% At the end of this, feature is the request
% The following is written to show
l=length(feature);
```

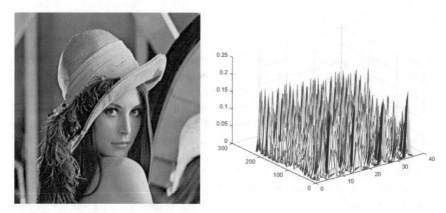

Fig. 4.2 Original drawing and HOG feature extraction result

```
f=[];
for i=1:l
     f=[f;feature{i}(:)'];
end
figure
mesh(f)
```

See Fig. 4.2.

4.3 Texture Features

Texture is a spatial distribution in a certain image area, where neighboring pixels' grayscale, tone, color etc. subject to some statistical arrangement rules. Not only it reflects the grayscale statistical information of the image, but also reflects the spatial distribution information and the structural information of the image. The texture of an image is an organized local feature that can be qualitatively represented by one or more of the following descriptions: coarseness, contrast, directionality, line-likeness, regularity, roughness, indention and so on. The basic characteristic of texture is shift invariance, that is, the visual perception of the texture is essentially unrelated to its position in images.

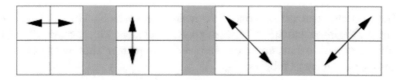

Fig. 4.3 The calculation of the four defined directions in the co-occurrence matrix

4.3.1   Haralick Texture Descriptors

Haralick et al. proposed Gray-Level Co-occurrence Matrices (GLCM) and their texture feature descriptors in 1973. Because of its good performance, this feature still owns a wide range of applications today. GLCM reflects the comprehensive information of the image grayscale which refer to the direction, the neighboring interval and the rangeability. It can be seen as the basis of analyzing the unit of images and the arrangement structure. As the characteristic of texture analysis, instead of applying GLCM directly, it often extracts texture feature on the basis of GLCM.

From the mathematical point of view, each element of the co-occurrence matrix is the calculation of the second joint conditional probability density $P(i, j|d, \theta)$ between the image's grayscales. Starting with the grayscale i, $P(i, j|d, \theta)$ indicates the probability of occurrence of the grayscale j, where d represents spatial distance, θ indicates the orientation, and it takes i as starting point. Generally speaking, d is different from θ. The direction can usually be defined into four types which named horizontal, vertical, left diagonal and right diagonal respectively, that is, $0°, 90°, 45°$ and $135°$, specifics as shown in Fig. 4.3.

According to the co-occurrence matrix $P(i, j|d, \theta) = \{P_0(i, j|d)\}_{N_G \times N_G}$ (Where N_G is grayscale), we can define a lot of texture features. Haralick et al. have defined 14 texture features, mainly in the following:

(a)   Energy: $E(P_0(d)) = \sum_{i=0}^{N_G-1} \sum_{j=0}^{N_G-1} [P_0(i, j|d)]^2$

(b)   Entropy: $H(P_0(d)) = -\sum_{i=0}^{N_G-1} \sum_{j=0}^{N_G-1} P_0(i, j|d) \log P_0(i, j|d)$

(c)   Correlation: $C(P_0(d)) = \left[\sum_{i=0}^{N_G-1} \sum_{j=0}^{N_G-1} (i - \mu_x)(j - \mu_y) P_0(i, j|d)\right]/\sigma_x \sigma_y$

(d)   Local uniformity: $L(P_0(d)) = \sum_{i=0}^{N_G-1} \sum_{j=0}^{N_G-1} \frac{1}{H(i-j)^2} P_0(i, j|d)$

(e)   Moment of inertia: $I(P_0(d)) = \sum_{i=0}^{N_G-1} \sum_{j=0}^{N_G-1} (i - j)^2 P_0(i, j|d)$

In the above formula,

$$\mu_x = \sum_{i=0}^{N_G-1} i \sum_{j=0}^{N_G-1} P_0(i, j|d)$$

$$\mu_y = \sum_{j=0}^{N_G-1} j \sum_{i=0}^{N_G-1} P_0(i, j|d)$$

$$\sigma_x^2 = \sum_{i=0}^{N_G-1} (i - \mu_x)^2 \sum_{j=0}^{N_G-1} P_0(i, j|d)$$

The co-occurrence matrix is one of the most common method in texture analysis. It indicates the interrelationship between grayscale models, which are not affected by monotonic grayscale transformation. The specific implementation steps of Haralick texture extraction are as follows:

Step 1: Read the image. If the original input is color image, convert the RGB image into the gray image for calculating gray-scale co-occurrence matrix in the next step.

Step 2: The complexity of gray-scale co-occurrence matrix is very huge. If the of the original image has a high grayscale value, we can compress grayscale value first to reduce gray levels.

Step 3: Select the distance and angle, then calculate the gray-scale co-occurrence matrix.

Step 4: Select the appropriate texture features, then calculate the texture parameters.

Step 5: The characteristics can be further extracted as needed, such as mean, variance et al. and select them as the final image characteristics.

The corresponding MATLAB programme is shown as PROGRAMME 4.3. The texture features selected here are energy, entropy, moment and correlation.

PROGRAMME 4.3: Haralick Texture Extraction

```
function TT = Texture(Image)

% Based on the texture feature extraction of co-occurrence matrix, d=1. Four matrices are θ=0° ,
   45° , 90° , 135°

% The image gray level is 256.

% function : TT=Texture(Image)

% Image      : Input image data

% T          : Returns the 8-dimensional texture feature row vector

Gray=Image;

[M,N,O] = size(Gray);

M = 128;

N = 128;

TT=[];

% 1. Convert each color component to grayscale, and do this if the image is a RGB image.

% Gray = double(0.3*Image(:,:,1)+0.59*Image(:,:,2)+0.11*Image(:,:,3));

% 2. In order to reduce the amount of computation, the original image is compressed by gray level to

   quantize the gray level into 16.

for i = 1:M

    for j = 1:N

        for n = 1:256/16

            if (n-1)*16<=Gray(i,j)&Gray(i,j)<=(n-1)*16+15

                Gray(i,j) = n-1;

            end

        end

    end

end

% 3. Calculate four co-occurrence matrix P, and take the distance as 1, and the angle as 0,45,90,135

P = zeros(16,16,4);
```

```
for m = 1:16
    for n = 1:16
        for i = 1:M
            for j = 1:N
                if j<N&Gray(i,j)==m-1&Gray(i,j+1)==n-1
                    P(m,n,1) = P(m,n,1)+1;
                    P(n,m,1) = P(m,n,1);
                end
                if i>1&j<N&Gray(i,j)==m-1&Gray(i-1,j+1)==n-1
                    P(m,n,2) = P(m,n,2)+1;
                    P(n,m,2) = P(m,n,2);
                end
                if i<M&Gray(i,j)==m-1&Gray(i+1,j)==n-1
                    P(m,n,3) = P(m,n,3)+1;
                    P(n,m,3) = P(m,n,3);
                end
                if i<M&j<N&Gray(i,j)==m-1&Gray(i+1,j+1)==n-1
                    P(m,n,4) = P(m,n,4)+1;
                    P(n,m,4) = P(m,n,4);
                end
            end
        end
        if m==n
            P(m,n,:) = P(m,n,:)*2;
        end
    end
end
% Normalization of co-occurrence matrix

for n = 1:4
    P(:,:,n) =P(:,:,n)/sum(sum(P(:,:,n)));
end
% 4. Calculate  4  texture parameters of energy, entropy, moment of inertia and correlation for the
    symbiotic matrix
H = zeros(1,4);
I = H;
Ux  = H;          Uy = H;
deltaX = H;   deltaY = H;
C = H;
forn = 1:4
```

```
    E(n) = sum(sum(P(:,:,n).^2));
    for i = 1:16
        for j = 1:16
            if P(i,j,n)~=0
                H(n) = -P(i,j,n)*log(P(i,j,n))+H(n);
            end
            I(n) = (i-j)^2*P(i,j,n)+I(n);
            Ux(n) = i*P(i,j,n)+Ux(n); % μ x in correlation
            Uy(n) = j*P(i,j,n)+Uy(n); % μ y in correlation
        end
    end
end
for n = 1:4
    for i = 1:16
        for j = 1:16
            deltaX(n) = (i-Ux(n))^2*P(i,j,n)+deltaX(n); % σ x in correlation
            deltaY(n) = (j-Uy(n))^2*P(i,j,n)+deltaY(n); % σ y in correlation
            C(n) = i*j*P(i,j,n)+C(n);
        end
    end
    C(n) = (C(n)-Ux(n)*Uy(n))/deltaX(n)/deltaY(n); % correlation
end
% The mean and standard deviation of energy, entropy, moment of inertia and correlation are taken as
    the final 8-dimensional texture features.
a1 = mean(E)
b1 = sqrt(cov(E))
a2 = mean(H)
b2 = sqrt(cov(H))
a3 = mean(I)
b3 = sqrt(cov(I))
a4 = mean(C)
b4 = sqrt(cov(C))
TT=[a1,b1,a2,b2,a3,b3,a4,b4];
sprintf('The energy in the direction of 0,45,90,135 is: %f, %f, %f, %f',E(1),E(2),E(3),E(4))
% The output data;
sprintf('The entropy in the direction of 0,45,90,135 is: %f, %f, %f, %f',H(1),H(2),H(3),H(4))
% The output data;
sprintf('The moment of ineria in the direction of 0,45,90,135 is: %f, %f, %f, %f',I(1),I(2),I(3),I(4))

% The output data;
sprintf('The correlation in the direction of 0,45,90,135 is: %f, %f, %f, %f',C(1),C(2),C(3),C(4))
% The output data;
```

Table 4.1 The selected four Haralick textures in 4 directions

	0°	45°	90°	135°
Energy	0.0735	0.0720	0.0919	0.0670
Entropy	3.2379	3.2488	3.0074	3.3636
Moment of inertia	0.9512	0.9564	0.6465	1.3582
Correlation	0.1910	0.1907	0.1978	0.1816

Fig. 4.4 The original image

Table 4.2 The further compressed Haralick texture

	Mean	Variance
Energy	0.0761	0.0109
Entropy	3.2144	0.1493
Moment of inertia	0.9781	0.2919
Correlation	0.1903	0.0066

Table 4.1 shows four characteristics (energy, entropy, moment of inertia and correlation) in four directions based on gray-scale co-occurrence matrix of Fig. 4.4. The grayscale of the image is compressed to 16. Based on this, the above texture features are further compressed, the mean value and variance are selected as characteristics. The results are shown in Table 4.2.

4.3.2 Wavelet Texture Descriptors

Wavelet transform is a linear operation that decomposes the signal into different scale components. In practice, we convolve signal by multi-scale filters to realize its implementation. Wavelet transform provides a tool to analyze image texture on different scales.

The wavelet transform is often compared with the Fourier transform, in which signals are represented as a sum of sinusoids. In fact, the Fourier transform can be viewed as a special case of the continuous wavelet transform with the choice of the mother wavelet. The main difference in general is that wavelets are localized in both time and frequency whereas the standard Fourier transform is only localized in frequency. The Short-time Fourier transform (STFT) is similar to the wavelet transform, in that it is also time and frequency localized, but there are issues with the frequency/time resolution trade-off.

The wavelet transform's multi-resolution properties enables large temporal supports for lower frequencies while maintaining short temporal widths for higher frequencies by the scaling properties of the wavelet transform. This property extends conventional time-frequency analysis into time-scale analysis [7].

If the function $\varphi_x \in L^2(R)$ and satisfies the condition $\int_{-\infty}^{\infty} \psi(x)dx = 0$, we call it a basic wavelet or mother wavelet. Telescope generated function cluster $\{\psi_{b,a}(x)\} = a^2\psi(\frac{x-b}{a})$, $(b,\ a \in R)$ by translating the mother wavelet constitute a group of wavelet base, where a is the scale parameter, and b is the position parameter. Then the continuous wavelet transforms of signal $f(x)$ on the scale $a \in R^+$, $a \neq 0$ and the position $b \in R^+$ is defined as:

$$(W_\varphi f)(b,\ a) = \int_\infty^\infty f(x)\overline{\varphi}_{b,a}dx = a^2 \int_\infty^\infty f(x)\varphi(\frac{b-x}{a})dx = <f(x),\ \varphi_{b,a}(x)>$$

(4.14)

where $\langle\rangle$ represents dot-product. The form of convolution is represented as $W_\phi f(b,\ a) = f * \varphi_{b,a}$. So, wavelet transform can be seen as a calculation of wave filtering between original signals and a set of multi-scale filters, then decomposes the signal into a series of bands for processing.

In practice, it is necessary to discretize the above continuous wavelet and its wavelet transform. The binary discretization of continuous wavelet transform's stretching and the discretization of convoluted translation are as follow:

$$a = 2^j,\ j \in Z,\ b = \frac{k}{2^j},\ j \in Z,\ k \in Z$$

(4.15)

In this way, it can obtain the wavelet transform which is discrete to scale-time, that is, multi-resolution analysis.

The wavelet function $\varphi(x)$ is generated by the linear combination of scaling and translation of the scaling function $\phi(x)$. And the scaling function $\phi(x)$ satisfies the two-scale difference equation, that is, the function with a certain scale can be derived from the linear combination of its next scale. They satisfy the following two-scale relationship equations:

$$\phi(x) = \sqrt{2} \sum_k h(k)\phi(2x - k) \tag{4.16}$$

$$\varphi(x) = \sqrt{2} \sum_k g(k)\phi(2x - k) \tag{4.17}$$

where h is a low-pass filter and g is a high-pass filter, h and g are the quadrature mirror filters. The relationship between them is as follow:

$$g(k) = (-1)^k h(1 - k)$$

For the two-dimensional wavelet transform, the wavelet basis function and the scaling function can be obtained by the vector product of the one-dimensional wavelet function $\varphi(x)$ and the scaling function $\phi(x)$.

$$\begin{cases} \phi(x, y) = \phi(x)\phi(y) \\ \varphi^1(x, y) = \phi(x)\varphi(y) \\ \varphi^2(x, y) = \varphi(x)\phi(y) \\ \varphi^3(x, y) = \varphi(x)\varphi(y) \end{cases} \tag{4.18}$$

$\phi(x, y)$ is a two-dimensional scaling function. And $\varphi^1(x, y)$, $\varphi^2(x, y)$, $\varphi^3(x, y)$ are two-dimensional wavelet functions.

Under the resolution of 2^j, the approximation $A_{2^j}^d f$ of the image signal $f(x, y)$ can be expressed as an inner product relationship:

$$A_{2^j}^d f = (< f(x, y), \phi_{2^j}(x - 2^{-j}n)\phi_{2^j}(y - 2^{-j}m) >)(m, n \in Z^2) \tag{4.19}$$

Under the different resolutions of 2^{j+1} and 2^j, the information of the image's approximations $A_{2^{j+1}}^d f$ and $A_{2^j}^d f$ is different. And the difference signals can be indicted by the detail signal D_{2^j}, which can be represented by three detail images $D_{2^j}^1$, $D_{2^j}^2$, $D_{2^j}^3$:

$$D_{2^j}^2 f = (< f(x, y), \psi_{2^j}^2(x - 2^{-j}n, y - 2^{-j}m) >)(m, n) \in z^2$$
$$D_{2^j}^3 f = (< f(x, y), \psi_{2^j}^3(x - 2^{-j}n, y - 2^{-j}m) >)(m, n) \in z^2$$
$$D_{2^j}^1 f = (< f(x, y), \psi_{2^j}^1(x - 2^{-j}n, y - 2^{-j}m) >)(m, n) \in z^2$$
$$D_{2^j}^2 f = (< f(x, y), \psi_{2^j}^2(x - 2^{-j}n, y - 2^{-j}m) >)(m, n) \in z^2$$
$$D_{2^j}^3 f = (< f(x, y), \psi_{2^j}^3(x - 2^{-j}n, y - 2^{-j}m) >)(m, n) \in z^2 \tag{4.20}$$

The detail subgraph is the high frequency component of the original image, which contains the main texture information. So, we take the energy of some detail subgraphs as the texture feature, and they reflect the energy distribution along the frequency axis about the scale and direction. The common method uses the texture

Fig. 4.5 The result of two-layer wavelet decomposition

energy macro feature $E(i, j)$ which defined in the window of $(2n + 1) \times (2n + 1)$ to extract features from multiple channels:

$$E(i, j) = \frac{1}{(2n + 1)^2} \sum_{k=i-n}^{i+n} \sum_{l=j-n}^{j+n} |w(k, l)| \tag{4.21}$$

where $E(i, j)$ is the eigenvalue of pixel (i, j), and $w(k, l)$ represents the first (k, l) wavelet coefficients of $(2n + 1) \times (2n + 1)$ window which centered on pixel (i, j).

(1) Color texture feature extraction in RGB space

The most direct way is to perform two-layer wavelet decomposition and extract wavelet energy on every channel of an image when refers to RGB images.

Figure 4.5 shows the result of the two-layer wavelet decomposition of the image.

In general, the RGB three channels of the color image are not independent, so, there is a correlation between the wavelet coefficients of different channels. Utilize the wavelet to perform two-layer decomposition of the color image and extract the wavelet covariance signal to get a 36-dimensional eigenvector.

(2) Color texture feature extraction in HIS space

The color value in RGB space cannot reflect human visual system characteristics. It is necessary to convert it into other color models for meeting the psychological sense of people. Here, we use the H: Hue, S: Saturation, I: Illumination (HSI) model as example. Perform the two-layer wavelet decomposition and extract its wavelet energy. E_{ni}^{xj} represents the wavelet energy of subgraph D_{ni} in the channel x_j:

$$E_{ni} = \int (D_{ni}(\vec{b}))^2 d\vec{b} \quad \vec{b} \in R^2 \tag{4.22}$$

where $1 \leq n \leq d, d$ is the number of decomposition layer, and $i = 1, 2, 3$ are the corresponding three details. $\{x_j\}$ $j = 1, 2, 3$ represent the H, S, I three channels of the image respectively.

Utilize the wavelet texture analysis method to perform wavelet decomposition of the HSI channels and extract its wavelet energy. Then we normalize the data and get the energy characteristics:

$$\left\{ E_{ni}^{xj} \right\}_{1 \leq n \leq d, i=1,2,3}^{j=1,2,3}$$

where $j = 1, 2, 3$ represents three high frequency details, and d is the number of decomposition layers. When d equals to 2, it will get an 18-dimensional feature about color and texture attributes which is more consistent with the human visual system.

As a multi-scale analysis method, wavelet transform can reserve the spatial frequency decomposition characteristics of signals better. And the characteristic energy signals extracted by the wavelet decomposition are fused with color and texture information, which can simulate the human visual system better.

The part of MATLAB code of wavelet texture feature extraction is shown in PROGRAMME 4.4:

PROGRAMME 4.4: **Wavelet Texture Feature Extraction**

```
i=imread('hello.jpg');
i=im2double(i);
i=rgb2hsv(i);
h=i(:,:,1);
s=i(:,:,2);
v=i(:,:,3);
[Ch,Sh]=wavedec2(h,2,'db1');
[Cs,Ss]=wavedec2(s,2,'db1');
[Cv,Sv]=wavedec2(v,2,'db1');
[CHD1h,CVD1h,CDD1h]=detcoef2('all',Ch,Sh,1);
[CHD1s,CVD1s,CDD1s]=detcoef2('all',Cs,Ss,1);
[CHD1v,CVD1v,CDD1v]=detcoef2('all',Cv,Sv,1);
[CHD2h,CVD2h,CDD2h]=detcoef2('all',Ch,Sh,2);
[CHD2s,CVD2s,CDD2s]=detcoef2('all',Cs,Ss,2);
[CHD2v,CVD2v,CDD2v]=detcoef2('all',Cv,Sv,2);
CHD1h=imresize(CHD1h,2,'nearest');
CHD1s=imresize(CHD1s,2,'nearest');
CHD1v=imresize(CHD1v,2,'nearest');
CVD1h=imresize(CVD1h,2,'nearest');
CVD1s=imresize(CVD1s,2,'nearest');
CVD1v=imresize(CVD1v,2,'nearest');
CDD1h=imresize(CDD1h,2,'nearest');
```

```
CDD1s=imresize(CDD1s,2,'nearest');
CDD1v=imresize(CDD1v,2,'nearest');
CHD2h=imresize(CHD2h,4,'nearest');
CHD2s=imresize(CHD2s,4,'nearest');
CHD2v=imresize(CHD2v,4,'nearest');
CVD2h=imresize(CVD2h,4,'nearest');
CVD2s=imresize(CVD2s,4,'nearest');
CVD2v=imresize(CVD2v,4,'nearest');
CDD2h=imresize(CDD2h,4,'nearest');
CDD2s=imresize(CDD2s,4,'nearest');
CDD2v=imresize(CDD2v,4,'nearest');
CHD1h=medfilt2(CHD1h,[5,5]);
CHD1s=medfilt2(CHD1s,[5,5]);
CHD1v=medfilt2(CHD1v,[5,5]);
CVD1h=medfilt2(CVD1h,[5,5]);
CVD1s=medfilt2(CVD1s,[5,5]);
CVD1v=medfilt2(CVD1v,[5,5]);
CDD1h=medfilt2(CDD1h,[5,5]);
CDD1s=medfilt2(CDD1s,[5,5]);
CDD1v=medfilt2(CDD1v,[5,5]);
CHD2h=medfilt2(CHD2h,[5,5]);
CHD2s=medfilt2(CHD2s,[5,5]);
CHD2v=medfilt2(CHD2v,[5,5]);
CVD2h=medfilt2(CVD2h,[5,5]);
CVD2s=medfilt2(CVD2s,[5,5]);
CVD2v=medfilt2(CVD2v,[5,5]);
CDD2h=medfilt2(CDD2h,[5,5]);
CDD2s=medfilt2(CDD2s,[5,5]);
CDD2v=medfilt2(CDD2v,[5,5]);
X=[CHD1h;CHD1s;CHD1v;CVD1h;CVD1s;CVD1v;CDD1h;CDD1s;CDD1v;CHD2h;...
CHD2s;CHD2v;CVD2h;CVD2s;CVD2v;CDD2h;CDD2s;CDD2v];
```

4.3.3 LBP Texture Descriptors

Local Binary Pattern (LBP) [8] is a kind of texture descriptors to describe the local texture feature of the image, which has obvious advantages such as rotation invariance and gray-scale invariance. By comparing the grayscale value of each pixel with the value of its neighborhoods, we can utilize the binary representation of the comparison to describe the texture of an image. It can perform efficient measurement and extraction of texture information on local neighboring area of the grayscale image.

4	2	4
8	3	1
7	1	5

1	0	1
1		0
1	0	1

1	2	4
128		8
64	32	16

Fig. 4.6 Schematic diagram of the basic LBP operator

LBP has been widely used in texture classification, image retrieval, facial image analysis and other fields.

The initial LBP algorithm compares the pixels in the 3×3 neighborhood with the center pixel. And if the value of neighboring pixel is larger than or equal to the grayscale value of center pixel, the grayscale value of neighboring pixel will be set to 1. Otherwise it will be set to zero. The 3×3 neighborhood will form the 8-bit binary number according some order after computation of LBP, and its range is between 0 and 255. Since the LBP owns good local characteristics, it still maintains visual characteristics of the original image even after transformation. Figure 4.6 shows the schematic diagram of the basic LBP. First of all, we select a point randomly from the original image and take its 3×3 neighborhood. Then set the value of center point as the threshold and compare the grayscale value of 8 pixels in its neighborhood with threshold. So, we can get the binary pattern of the region and the pattern is represented by binary code. Next, we convert binary code into decimal number, that is the LBP code of center point. The histogram of LBP code can be used to describe the texture structure of the region.

The monotonic change of the grayscale value will not cause the change of LBP code. At the same time, comparing the corresponding LBP code with the pixel value, it adds the correlation of the pixel and its surrounding pixels, which can fully characterize the image features and reduce the influence of illumination changes and angle changes on feature extraction.

The original image after LBP transformation should be transformed into the histogram, and the histogram can be calculated by using the Eq. (4.23).

$$H_i = \sum_{x,y} I\{f_1(x, y) = i\}, \quad i = 0, \ldots, 2^P \qquad (4.23)$$

where $I(X) = \begin{cases} 1, & X \text{ is true.} \\ 0, & X \text{ is false.} \end{cases}$

The experiment shows that it is not enough to extract the LBP histogram of an image in giant database. In the case of ear recognition, if the LBP histogram of the entire ear image is used as the final feature merely, the recognition rate is very low. To solve this problem, the original human ear image may be divided into some blocks, such as 5×5 or 3×3. Then we can calculate the LBP histogram of each block.

Specific implementation steps are as follows:

Step 1:   Divide the original image A into 4×4 blocks.
Step 2:   Divide the original image into 16 blocks of $n \times m$, reconstruct the above blocks respectively. Then, the image becomes $(n + 2) \times (m + 2)$ by adding the whole 0 layer to the outside of the image matrix.
Step 3:   Utilize the LBP algorithm to calculate the image after reconstruction and get the matrix LBP_A.
Step 4:   Count the histogram of 16 LBP_A, and take its vector as feature vector. So, we can get 16 feature vectors.
Step 5:   Merge the 16 feature vectors into SVM for recognition.

The code of main function is shown as PROGRAMME 4.5.

PROGRAMME 4.5: LBP Feature Extraction

```
function [ new_A] = func_lbp_gouzao( A )
% A new data matrix named new_A is obtained by inputing the original image
% FUNC_LBP_FOUZAO Summary of this function goes here
% Detailed explanation goes here
[n m] = size(A);              % Get the row and column dimensions of the image
new_A = zeros(n+2,m+2);       % Plus a circle of all zero ranks outside the original image
new_A(2:end-1,2:end-1) = A(:,:);
% new_A represents the data of A, and the outermost layer is a matrix of all zeros
end
 function [ out_A ] = func_lbp_fenkuai( p )
% The function is to block the original image and calculate the LBP value
% FUNC_LBP_FENKUAI Summary of this function goes here
% Detailed explanation goes here
 out_A = zeros(1,255*16);
    for i =0:3
        for j = 0:3
            A = zeros(28,23);
            A(:,:) = p(i*28+1:(i+1)*28,j*23+1:(j+1)*23);
            gouzao_A = func_lbp_gouzao(A);     % Get a constructed image matrix
            lbp_A = func_lbp_value(gouzao_A); % The function is to get the LBP value of the
    constructed function matrix
  H = reshape(lbp_A,1,644);% Turn LBP values into row vectors
            H = H ./ sqrt(H(1,:)*H(1,:)');
            a = i*(255*4)+j*255+1;
            b = i*(255*4)+(j+1)*255;
            out_A(1,a:b) = hist(H,255);       % Calculate the LBP histogram of the obtained LBP
        end
    end
```

```matlab
  out_A;
end
function [ A ] = func_lbp_value(A )            % The function is to get the LBP value
% FUNC_LBP_VALUE Summary of this function goes here
% Detailed explanation goes here
for i=2:size(A,1)-1   % The outermost layer is all zero, which is convenient to calculate the LBP value
    for j=2:size(A,2)-1
        A(i,j) =jisuan(getNeighbor(i,j,A));   % Get the neighborhood of the determined pixel to
    calculate the LBP value
        end
end
A = A(2:end-1,2:end-1);
end
function LBP_value = jisuan(neighbor)          % The function is the exact calculation of LBP
LBP_value=0;
for ii =1:size(neighbor,2)-1
    if   neighbor(1,ii)<neighbor(1,9)      % Compare the neighbor values with the determined pixels
        neighbor(1,ii)=0;            % If the pixel is less than the determined pixel, it is assigned to 0
    else
        neighbor(1,ii)=1;            % If pixel is greater than the determined pixel, it is assigned to 1
    end
end
neighbor;
for jj =1:size(neighbor,2)-1
    LBP_value = neighbor(1,jj).*2^(jj-1)+LBP_value;
end
LBP_value;
end
function neighbor = getNeighbor(i,j,A)
% The function uses the position information of up and low and left and right to carry out the nearest
    neighbor of a pixel
neighbor = zeros(1,9);
% Where 1 to 8 is the nearest neighbor information, and the ninth is to determine the pixel information
neighbor(1,:) = [A(i-1,j-1) A(i-1,j) A(i-1,j+1) A(i,j+1) A(i+1,j+1) A(i+1,j) A(i+1,j-1) A(i,j-1) A(i,j)];
end
```

4.4 Corner Feature Extraction

4.4.1 Moravec Algorithm

The corner detection operator proposed by Moravec in 1981 is a method of corner detection based on grayscale variance. The operator calculates the grayscale variance of a pixel in the image along the horizontal, vertical, diagonal and anti-diagonal directions, where the minimum value is chosen as Corner Response Function (CRF). Then estimate the corner by local non-maximum suppression. Specific implementation steps are as follows:

Step 1: In the $w \times w$ window centered on the image pixel (x, y), utilize the following equations to calculate the grayscale variance in the four directions of each pixel, that is, the average grayscale change.

$$V_h = \sum_{i=-k}^{k-1} (I_{x+i,y} - I_{x+i+1,y})^2 \tag{4.24}$$

$$V_v = \sum_{i=-k}^{k-1} (I_{x,y+i} - I_{x,y+i+1})^2 \tag{4.25}$$

$$V_d = \sum_{i=-k}^{k-1} (I_{x+i,y+i} - I_{x+i+1,y+i+1})^2 \tag{4.26}$$

$$V_a = \sum_{i=-k}^{k-1} (I_{x+i,y-i} - I_{x+i+1,y-i-1})^2 \tag{4.27}$$

where $k = [w/2]$, $[w/2]$ represents rounding. $I_{x+i,y}$ represents the grayscale value at $(x + i, y)$ and so on. The smallest one of the four values above will be selected as CRF for the pixel (x, y):

$$CRF_{x,y} = \min\{V_h, V_v, V_d, V_a\} \tag{4.28}$$

Step 2: According to the threshold set by the actual image, use the window to traverse grayscale image and select the point whose CRF is greater than the threshold as the candidate corner. The principle of choosing the threshold is that the candidate corner should contain enough real corners, and the number of false corners should be as few as possible.

Step 3: Select the corner by local non-maximum suppression. Within the window of a certain size, take out the corner whose CRF is not a maximum value from the candidate corners in the second step and retain the maximum value to be the corner.

The most notable characteristic of Moravec operator is that the algorithm is simple, and the computation speed is fast. The related code is shown as PROGRAMME 4.6:

PROGRAMME 4.6: **Moravec Corner Detection**

```
clear all;close all;clc
img=double(imread('lena.jpg'));
[h w]=size(img);
imshow(img,[])
imgn=zeros(h,w);
n=4;
for y=1+n:h-n
    for x=1+n:w-n
        sq=img(y-n:y+n,x-n:x+n);
        V=zeros(1,4);
        for i=2:2*n+1          %The sum of squares of the grayscale variance in four directions:
vertical, horizontal, diagonal, and anti-diagonal
            V(1)=V(1)+(sq(i,n+1)-sq(i-1,n+1))^2;
            V(2)=V(2)+(sq(n+1,i)-sq(n+1,i-1))^2;
            V(3)=V(3)+(sq(i,i)-sq(i-1,i-1))^2;
            V(4)=V(4)+(sq(i,(2*n+1)-(i-1))-sq(i-1,(2*n+1)-(i-2)))^2;
        end
        pix=min(V);              % Select the minimum value in four directions
        imgn(y,x)=pix;
    end
end
T=mean(imgn(:));              % Set the threshold to zero if it is less than the mean
ind=find(imgn<T);
imgn(ind)=0;
for y=1+n:h-n                % Local maximum and nonzero values are selected as feature points
for x=1+n:w-n
    sq=imgn(y-n:y+n,x-n:x+n);
    if max(sq(:))==imgn(y,x) && imgn(y,x)~=0
        img(y,x)=255;
        end
    end
end
figure;
imshow(img,[]);
```

The corner detection result extracted by Moravec operator is shown in Fig. 4.7.

Fig. 4.7 The results of Moravec operator extracting corner

4.4.2 Harris Corner Detection Operator

Harris corner detection operator was presented by Chris Harris and Mike Stephens in 1988. It is a feature extraction algorithm for point based on still images. The operator is inspired by the auto-correlation function in the signal processing, given the matrix associated with the auto-correlation function. The eigenvalue of the matrix is the first order curvature of the auto-correlation function. For any points in the image, if its horizontal and vertical curvature values are higher than its local neighborhood, the point will be considered to be a feature point.

Actually, Harris corner detection operator is the improvement and optimization to Moravec operator. Based on the Moravec corner detection operator, the Harris operator is obtained:

(1) Moravec operator studies the average change of the brightness value of the image's local window after a little deviation in different directions. So, we need to consider 3 kinds of situations:

    (a) If the brightness value of the image in the window blocks are constant, then the deviation of all the different directions only leads to a small change.

    (b) If the window spans an edge, the deviation on the edge will result in a small change, but the vertical deviation will result in a big change.

    (c) If the window block contains a corner or isolated point, then the deviation of all different directions will cause a big change. Therefore, the point is the corner where the minimum change value caused by the deviation in any direction is greater than a certain threshold.

(2) The problem of Moravec operator and the solution of Harris et al.:

(a) Only considering 8 directions of every 45°, the small deviation of all directions can be reflected by extending the area change E:

$$E_{x,y} = \sum_{u,v} w_{u,v}[I_{x+u,y+v} - I_{u,v}]^2$$

$$= \sum_{u,v} w_{u,v}[xX + yY + O(x^2, y^2)]^2 \qquad (4.29)$$

where the first derivative is approximate to:

$$X = I \otimes (-1, 0, 1) \approx \partial I / \partial x$$
$$Y = I \otimes (-1, 0, 1)^T \approx \partial I / \partial y$$

$\otimes$ is the Kronecker product, an operation on two matrices of arbitrary size resulting in a block, So small deviation can be written as:
So small deviation can be written as:

$$E(x, y) = Ax^2 + 2Cxy + By^2 \qquad (4.30)$$

where

$$A = X^2 \otimes w$$
$$B = Y^2 \otimes w$$
$$C = (XY) \otimes w$$

(b) Moravec operator has no noise reduction, so it is sensitive to noise. We can use Gaussian smooth to denoise:

$$w_{u,v} = \exp -(u^2 + v^2)/2\sigma^2 \qquad (4.31)$$

(c) Only consider the minimum value of E, the Moravec operator is sensitive to the edge response. So, the corner rule is redefined to solve the problem and E can be written as:

$$E(x, y) = (x, y)M(x, y)^T \qquad (4.32)$$

where $M = \begin{bmatrix} A & C \\ C & B \end{bmatrix}$ is a 2 × 2 symmetric matrix.

Note: E is closely related to the local auto-correlation function, and M describes the shape of E which at the original point. Let α and β be the 2 eigenvalues of M, α and β are proportional to the main curvature of the local auto-correlation function, which can be used to describe the rotation invariance of M. We consider 3 kinds of situation for α and β:

(a) Assuming that both α and β are small, so that the local auto-correlation function is flat. Then the brightness of the window area in the image is approximately invariant.
(b) Assuming that one of the values is large and the other small, so that the local auto-correlation function presents a ridge shape. Then it shows one edge.
(c) Assuming that both α and β are large, so that the local auto-correlation function presents a shape of the mutant peak. Then any direction's deviation will be added by E and it must be the corner.

(3) To avoid asking for eigenvalues, calculate:

$$Tr(\mathbf{M}) = \alpha + \beta = \mathbf{A} + \mathbf{B}$$
$$Det(\mathbf{M}) = \alpha\beta = \mathbf{AB} - \mathbf{C}^2 \tag{4.33}$$

Then the Harris corner detection operator is defined as:

$$\mathbf{R} = Det - kTr^2 \tag{4.34}$$

If R > 0, it is the corner. If R < 0, it is the edge. The invariant area R is very small.

Harris corner detection operator is not sensitive to noise. But it gets a good performance on detection of L-shape corner. Due to the three Gaussian filtering, the speed of detection is slow.

Harris corner detection subroutine is shown as PROGRAMME 4.7.

PROGRAMME 4.7: **Harris Corner Detection Function**

```
function [H,J] = tharris(I,sig,approach)
if nargin==1
        sig = 1;
end
if nargin<3
        approach='old';
end
% Find the gradient of each direction
[g,gx] = gaussDeriv(sig);
Ix = conv2(gx,g,I);
Iy = conv2(g,gx,I);
xx = conv2(g,g,Ix.*Ix);
xy = conv2(g,g,Iy.*Ix);
yy = conv2(g,g,Iy.*Iy);
D = xx.*yy-xy.^2;
T = xx+yy;
```

```
if 0
    imagesc(D);
    pause;
    imagesc(T)
    pause
    sd = std(D(:))/std(T(:))
    H = D-0.02*sd*T;
end
alpha = 0.06;
if all(approach=='old');
 %    T,
      H = D-alpha*T.^2;
      %    H = D./(T);
else
      H = D./(T);
end
if   0
    figure(2);
    imagesc(H)
    pause
end
H = NMS(H);
b = (size(H,1)-size(I,1))/2;
H = H((1+2*b):(end-2*b),(1+2*b):(end-2*b));
if 0
    figure(1);
    subplot(2,1,1), imagesc(Ix);
    subplot(2,1,2), imagesc(Iy);
end
if   0
    figure(2);
    imagesc(log(max(H,1e-6)));
    pause
end
if 0
topN = 50;
U = H(:); U = U(find(U>0));
s=sort(U);
N = length(U)-topN;
```

```
        tau = s(N);
        [Y,X] = find(H>tau);
        H = [Y+b X+b];
return;
end
if nargout==2
        J = I((1+b):(end-b),(1+b):(end-b));
else
        J = I((1+b):(end-b),(1+b):(end-b));
        U=H;
        U = U(:); U = U(find(U>0));
        U=sort(U);
        tau = median((U));
         N = length(U);
        [m, n]=size(U);% m*n,
    % m,
        while m>=500
        U = U(find(U>tau));
        tau=median(U);
        [m, n]=size(U);
        end
        [Y,X] = find(H>tau);
        H = [Y+b X+b];
end
```

4.4.3 SUSAN Corner Detection Algorithm

SUSAN algorithm is an algorithm proposed by Smith et al. in 1997 to calculate the corner features in an image. SUSAN algorithm uses a circular template (as shown in Fig. 4.8). Define the center pixel to be detected as the core points, then the neighborhood of core point is divided into two regions: one named Univalue Segment Assimilating Nucleus (USAN) where the luminance value is similar to the core point and another area where the luminance value is not similar to the core point.

The typical USAN area is shown in Fig. 4.9. When the template moves on the image, as shown in Fig. 4.9a, if the circle template is completely covered by background or target area, its USAN area is the largest. When the core point located at the edge, the USAN area is reduced by half which is shown in Fig. 4.9c. When the core point located at the corner, the USAN area is the smallest, as shown in Fig. 4.9d. Based on this principle, Smith proposed the SUSAN corner detection algorithm.

The specific steps of SUSAN corner detection algorithm are as follows:

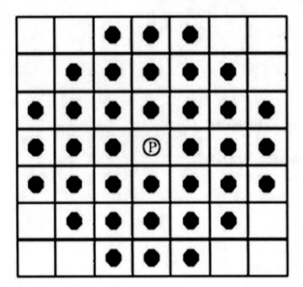

Fig. 4.8 Circular template

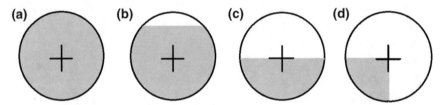

Fig. 4.9 The typical area: **a** the circular template is in the same area, **b** the core is in the area, **c** the core is on the edge of the area, **d** the core at the regional corner

Step 1:   A circular template with the size of 37 pixels slides on the image. Compare the gray level of each pixel in the template to template kernel, and determine whether it belongs to the USAN area. The discriminant function is as follows:

$$c(r, r_0) = \begin{cases} 1 & |I(r) - I(r_0)| \le t \\ 0 & |I(r) - I(r_0)| > t \end{cases} \tag{4.35}$$

Step 2:   Count the number $n(r_0)$ of pixels which have the similar brightness values with the kernel point in the circular template.

$$n(r_0) = \sum_{r \in D(r_0)} c(r, r_0) \tag{4.36}$$

where $D(r_0)$ is the circular template centered on r_0.

Step 3: We use the following corner response function. If the USAN value of a pixel is less than a certain threshold, the point is considered as the initial corner. Where g can be set to half of the maximum area of USAN.

$$R(r_0) = \begin{cases} g - n(r_0) & n(r_0) < g \\ 0 & n(r_0) \geq g \end{cases} \tag{4.37}$$

Step 4: Perform non-maximum suppression on the initial corner to get the final corner.

The implementation is shown as PROGRAMME 4.8.

PROGRAMME 4.8: **SUSAN Corner Detection**

```
clear all;
close all;
clc;
img=imread('i.jpg');
img=rgb2gray(img);
imshow(img);
[m n]=size(img);
img=double(img);
t=45;      % Set the threshold where the pixel grayscale of the template center is different from the
           surrounding gray level
usan=[]; % The difference between the current pixel and the surrounding pixel, which is less than t
% There are 37 pixels in the template
for i=4:m-3      % There is no extension of the image on the periphery, and the final image will shrink
    for j=4:n-3
        tmp=img(i-3:i+3,j-3:j+3);      % Construct the template of 7*7 first, that is, 49 pixels
        c=0;
        for p=1:7
            for q=1:7
                if (p-4)^2+(q-4)^2<=12    % In the selection, the final template is similar to a circle
                    % usan(k)=usan(k)+exp(-(((img(i,j)-tmp(p,q))/t)^6));
                    if abs(img(i,j)-tmp(p,q))<t      % Judge whether the grayscale is similar, and set
t by yourself
                        c=c+1;
                    end
                end
            end
```

```
                      end
                  end
                  usan=[usan c];
          end
  end
  g=2*max(usan)/3; % Confirm the number of corner extraction by yourself. When the value is relatively
      high, we will extract the edge.
  for i=1:length(usan)
      if usan(i)<g
              usan(i)=g-usan(i);
      else
              usan(i)=0;
      end
  end
  imgn=reshape(usan,[n-6,m-6])';
  figure;
  imshow(imgn)
  % Non-maximum suppression
  [m n]=size(imgn);
  re=zeros(m,n);
  for i=2:m-1
      for j=2:n-1
              if imgn(i,j)>max([max(imgn(i-1,j-1:j+1)) imgn(i,j-1) imgn(i,j+1) max(imgn(i+1,j-1:j+1))]);
                  re(i,j)=1;
              else
                  re(i,j)=0;
              end
      end
  end
  figure;
  imshow(re==1);
```

4.5   Local Invariant Feature Point Extraction

Local feature description is a basic research problem in computer vision. It plays an important role on finding the corresponding points in the image and describing the object features. It is a basis of many methods, so it is also a hot spot in the field of vision research with a wide range of applications.

The fundamental problem of local feature description is invariance (robustness) [8–13] and distinguishability. It is usually for dealing with various image transformations robustly when use the local feature descriptors. Therefore, the invariance is the first problem to be considered in constructing or designing feature descriptors. However, the distinguishability of feature descriptors is often inconsistent with its invariance. In other words, it is weaker for a feature descriptor with a large number of invariance to distinguish the local content of the image. However, if a feature descriptor is easy to distinguish different local contents of the image, its robustness is often not enough. On the other hand, if the local gray histogram is used to describe the feature, this description has strong invariance and it is robust to rotation changes of the local image content. But its ability of discrimination is weak. For example, it is impossible to distinguish two local image blocks with the same gray histogram but different contents.

Scale Invariant Feature Transform (SIFT) is the most widely used one among the local feature descriptors. It was first proposed in 1999 and refined by 2004. Not only SIFT owns invariance of changes such as scale, rotation, and a certain angle of view and illumination change, but also has a strong discriminability, so it has been applied widely in object recognition, 3D reconstruction and image retrieval since it was put forward.

Speeded Up Robust Features (SURF) is an improved version of SIFT. It uses Haar wavelet to approximate the gradient operation in SIFT and the integral graph technique for fast computation. SURF is 3–7 times faster than SIFT, and in most cases, it has the same performance as SIFT. So, it has been used in many applications, especially for occasions with high demand of time.

4.5.1 Local Invariant Point Feature of SURF

When it refers to pedestrian detection and tracking in intelligent monitoring system, the size of the pedestrian often has different scales with the camera angle and distance. If we do not timely correct the size of feature point, it will not make pedestrians match. To solve this problem, Bay et al. proposed scale invariant SURF feature detection. The feature of SURF is calculated mainly by the following steps:

(1) Construct Hessian matrix

SURF algorithm uses Hessian matrix to extract feature points, so Hessian matrix is the core of SURF algorithm. Assuming that the image is I, it is necessary to perform the Gauss filtering before constructing the Hessian matrix, and the calculation is shown in the formula (4.38).

$$L(x, \sigma) = G(\sigma) \cdot I(x, \sigma) \tag{4.38}$$

where the dot product is an algebraic operation that takes two equal-length vectors and returns a single number.

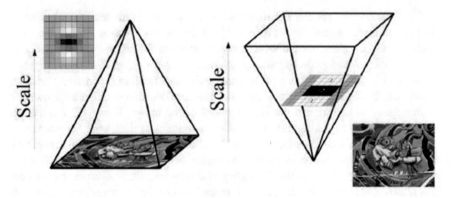

Fig. 4.10 The comparison of SIFT and SURF scale spanning spaces

Where $L(x, \sigma)$ is a convolution of images at different scales of σ, which can be realized by convolution of Gauss kernel $G(\sigma)$ and image function I at the point x. The Hessian matrix discriminant of each pixel is defined as:

$$H(x, \sigma) = \begin{bmatrix} L_{xx}(x, \sigma) \, L_{xy}(x, \sigma) \\ L_{xy}(x, \sigma) \, L_{yy}(x, \sigma) \end{bmatrix} \tag{4.39}$$

(2)   Generate scale space

The scale space of images is the representation of images at different scales. Pyramid images are divided into many layers, and each layer has several images at different scales. During the Gaussian Blur, the size of SIFT filter is always stay the same. We only change the size of each layer of images and the size of filters to get different size of the picture, it can reduce the sampling time greatly as shown in following Fig. 4.10.

(3)   Select feature points

Each pixel point processed by the Hessian matrix is compared with the 26 neighboring points of the 3-dimensional neighborhood. If it is the maximum or minimum of the 26 points, it is retained as the feature point, as shown in Fig. 4.11.

(4)   Select the main direction of feature points

For ensure the rotation invariance, SURF counts Haar wavelet features in neighborhood of the feature points. Within the radius of $6s$ centered by feature point, we calculate the Haar wavelet response sum of all points in the horizontal and vertical directions within 60° sector. And we select the main direction of the feature point where the response sum is the maximum, as shown in Fig. 4.12.

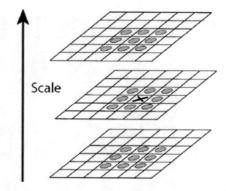

Fig. 4.11 The selection of SURF feature point

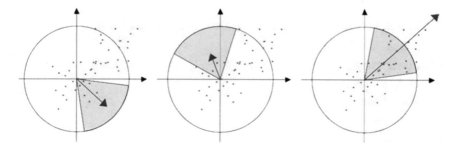

Fig. 4.12 Determination of the main direction of the feature point

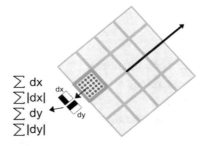

Fig. 4.13 SURF feature point description

(5) Construct SURF feature point description operator

SURF takes a square box in the main direction around the feature points. As shown in Fig. 4.13, we divide the frame into 4 * 4 sub regions and count the Haar wavelet features of the horizontal and vertical directions of all pixels in sub regions.

The program of SURF feature detection and matching is shown as PROGRAMME 4.9 (Fig. 4.14).

Fig. 4.14 SURF matching
results

PROGRAMME 4.9: SURF Feature Detection and Matching

clear;close all

%Read the two images.

I1= imread('images/girl.jpg');

I1=imresize(I1,0.5);

I1=rgb2gray(I1);

I2= imread('images/head.jpg');

I2=imresize(I2,0.5);

I2=rgb2gray(I2);

%Find the SURF features

points1 = detectSURFFeatures(I1);

points2 = detectSURFFeatures(I2);

%Extract the features and calculate the description vector

[f1, vpts1] = extractFeatures(I1, points1);

[f2, vpts2] = extractFeatures(I2, points2);

%Retrieve the locations of matched points. The SURF feature vectors are already normalized.

%Match

indexPairs = matchFeatures(f1, f2, 'Prenormalized', true) ;

matched_pts1 = vpts1(indexPairs(:, 1));

matched_pts2 = vpts2(indexPairs(:, 2));

%Display the matching points. The data still includes several outliers,

%but you can see the effects of rotation and scaling on the display of matched features.

% The matching results are displayed, and you can see that there are some outliers

figure('name','result'); showMatchedFeatures(I1,I2,matched_pts1,matched_pts2);

legend('matched points 1','matched points 2');

4.5.2 SIFT Scale-Invariant Feature Algorithm

Scale-Invariant Feature Transform (SIFT) is a local descriptor proposed by David G. Lowe in 1999. It has the invariance of scale, rotation and translation and it is robust to illumination change, affine transformation and 3-D projection transformation. The algorithm is used for object recognition and image matching. The main idea of SIFT algorithm is to find the extreme points in scale space, and then filter the extreme points to find the stable feature points. Finally, the local characteristics of the image are extracted around each stable feature point and form a local descriptor used in later matching. The concrete implementation process is as follows:

1. Construct scale space

    (1) Multi-resolution image pyramid

    The early multi-scale image is represented as the form of an image pyramid. The image pyramid is a set of results obtained by the same image at different resolutions. The generation process usually includes two steps:

a. Smooth the original image
b. Downsample the processed image (Usually 1/2 of the horizontal and vertical direction)

    After downsampling, it can get a series of constantly shrinking images. Obviously, each layer of the image is half the length and height of the upper layer in a traditional pyramid. Although the generation of multi-resolution image pyramid is simple, it is difficult to maintain the local features of the image for its essence. In other words, it is hard to maintain the scale invariance of the feature.

(2) Gaussian scale space

    We can also simulate the imaging process of the object on the retina through the fuzzy degree. The closer the distance is, the larger the size is, and the image is blurrier. That is the Gauss scale space, using different parameters to blur image (resolution keep the same) is another form of scale space.

    As we know, the convolution of image and Gauss function can blur the image. Different 'Gauss kernels' can be used to get different blurred images. The Gauss scale space of an image can be obtained by different Gauss convolution:

$$L(x, y, \sigma) = G(x, y, \sigma) * I(x, y) \tag{4.40}$$

where $G(x, y, \sigma)$ is Gaussian kernel function.

$$G(x, y, \sigma) = \frac{1}{2\pi\sigma^2} e^{\frac{x^2+y^2}{2\sigma^2}} \tag{4.41}$$

σ is called the scale space factor, which is the standard deviation of the Gaussian normal distribution. It reflects the degree of the blurred image. The larger the value is, the more blurred the image is, and the corresponding scale is larger. $L(x, y, \sigma)$ represents the Gaussian scale space of the image.

2. Approximate calculation of LoG

The purpose of constructing scale space is to detect the feature points at different scales. The better operator to detect feature points is $\Delta^2 G$ (Gauss-Laplace, LoG).

$$\Delta^2 = \frac{\partial^2}{\partial x^2} + \frac{\partial^2}{\partial y^2} \tag{4.42}$$

Although LoG can detect the feature points better, its computation is too large. Generally, Difference of Gaussian (DoG) can be used to calculate LoG approximately.

Let k be the scaling factor of two adjacent Gauss scale spaces, then DoG is defined:

$$\begin{aligned} D(x, y, \sigma) &= [G(x, y, k\sigma) - G(x, y, \sigma)] * I(x, y) \\ &= L(x, y, k\sigma) - L(x, y, \sigma) \end{aligned} \tag{4.43}$$

where $L(x, y, \sigma)$ is the gauss scale space of image.

3. Extremum detection in DoG Space

In order to find the extreme point of the scale space, each pixel should be compared with all the adjacent points of its image domain (same scale space) and scale field (adjacent scale space). When it is greater than (or less than) all adjacent points, the point is the extreme point. The first and last layers of each set of images cannot obtain the extremes by comparison. In order to satisfy the continuity of scale transformation, three images continue to be generated using Gaussian Blur on the top layer of each set of images. Each group of the gauss pyramid has $S + 3$ layers of the image, and each group of the DoG pyramid has $S + 2$ layers.

Let $S = 3$, that is, each group has 3 layers, then $k = 2^{\frac{1}{S}} = 2^{\frac{1}{3}}$. Each group in pyramid has 3 layers of images, and each group in DoG pyramid has 2 layers of images. The first group in DoG pyramid has two layers of scales: σ and $k\sigma$. The second group has two layers of scales: 2σ and $2k\sigma$. Only two items are unable to get the extreme values by comparison (if the left and right side both have the value, we can get the extreme values). Since we cannot compare the extremum, we need to continue to perform the Gauss Blur for images of each group, so that the scale can be formed as σ, $k\sigma$, $k^2\sigma$, $k^3\sigma$ and $k^4\sigma$. Then we choose three items in the middle $k\sigma$, $k^2\sigma$ and $k^3\sigma$. The three items of next corresponding group obtained by down-sampling in the previous group is $2k\sigma$, $2k^2\sigma$ and $2k^3\sigma$. The first item is $2k\sigma = 2 \times 2^{\frac{1}{3}}\sigma = 2^{\frac{4}{3}}\sigma$, which coincides with the scale of the last item $k^3\sigma = 2^{\frac{3}{3}}\sigma$ in the last group (Fig. 4.15).

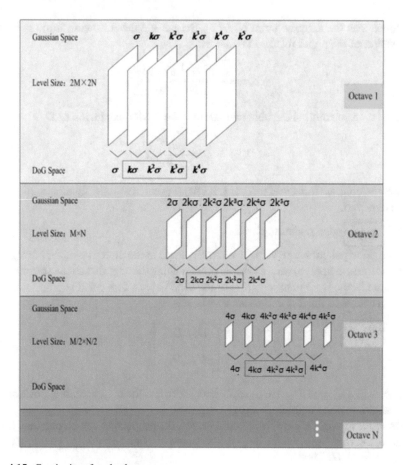

Fig. 4.15 Continuity of scale change

4. Remove the bad feature points

The local extreme point of the DoG obtained by comparative detection is found in the discrete space search. Since the discrete space is the result of the continuous space sampling, the extreme point found in the discrete space may be not the true point. So, we try to remove the point that does not satisfy the condition. The extreme point can be found through the scale space DoG function curve fitting. The essence of the step is to remove the asymmetric point of DoG local curvature.

There are two kinds of point that do not meet the requirements:

(1) Low contrast feature points

Candidate feature point x, whose offset is defined as Δx, and contrast is $|D(x)|$. Applying Taylor expansion to $D(x)$:

$$D(x) = D + \frac{\partial D^T}{\partial x}\Delta x + \frac{1}{2}\Delta x^T \frac{\partial^2 D}{\partial x^2}\Delta x \qquad (4.44)$$

Since x is the extreme point of $D(x)$, we can get the following formula by the derivation of the upper form and make it 0.

$$\Delta x = -\frac{\partial^2 D^{-1}}{\partial x^2} \frac{\partial D(x)}{\partial x} \qquad (4.45)$$

And then substituted the obtained Δx into the Taylor expansion of D (x):

$$D(\hat{x}) = D + \frac{1}{2} \frac{\partial D^T}{\partial x} \hat{x} \qquad (4.46)$$

Set the contrast threshold to T. If $|D(\hat{x})| \geq T$, then retain the feature point, otherwise removed.

(2)   Unstable edge response points

The principal curvature value is relatively large in the direction of the edge gradient, and the principal curvature value is smaller along the edge direction. The principal curvature of the DoG function $D(x)$ of the candidate feature point is proportional to the eigenvalue of the 2×2 Hessian matrix H.

$$H = \begin{bmatrix} D_{xx} & D_{yx} \\ D_{xy} & D_{yy} \end{bmatrix} \qquad (4.47)$$

where D_{xx}, D_{xy}, and D_{yy} are the difference between the corresponding positions of the candidate point neighborhood.

In order to avoid asking for specific values, the proportion can be obtained by H feature. Let $\alpha = \lambda_{max}$ be the largest eigenvalue of H and $\alpha = \lambda_{min}$ be the smallest eigenvalue of H. Then

$$Tr(H) = D_{xx} + D_{yy} = \alpha + \beta \qquad (4.48)$$

$$Det(H) = D_{xx} + D_{yy} - D_{xy}^2 = \alpha \cdot \beta \qquad (4.49)$$

where $Tr(H)$ is the trace of matrix H, and $Det(H)$ is determinant of matrix H.

Let $\gamma = \frac{\alpha}{\beta}$ represent the ratio of the maximum eigenvalue to the minimum eigenvalue, then we can find:

$$\frac{Tr(H)^2}{Det(H)} = \frac{(\alpha + \beta)^2}{\alpha\beta} = \frac{(\gamma\beta + \beta)^2}{\gamma\beta^2} = \frac{(\gamma + 1)^2}{\gamma} \qquad (4.50)$$

The result of the upper form is related to the proportion of the two eigenvalues, and it has nothing to do with the concrete size. When the two eigenvalues are equal, the value is minimum, and it increases with γ. Therefore, in order to detect whether the principal curvature is smaller than a threshold T_y, we only need to detect:

$$\frac{Tr(H)^2}{Det(H)} > \frac{(T_\gamma + 1)^2}{T_\gamma}$$

If the upper form is established, the feature point will be removed, otherwise reserved.

5. Determine the principal direction of feature points

Through the above steps, we have found the feature points at different scales. In order to achieve the invariance of image rotation, it is necessary to assign the direction of feature points. The direction parameter is determined by the gradient distribution of the neighbor pixels of feature points. Then, the stable direction of local structure of key point is obtained by using gradient histogram of image.

By finding the feature points, the scale σ of the feature points can be obtained, and then the scale images where feature points exit can be obtained.

$$L(x, y) = G(x, y, \sigma) * I(x, y) \qquad (4.51)$$

Compute the regional angle and amplitude of the which take feature points as center and take $3 \times 1.5\sigma$ as radius. Each point's modulus $m(x, y)$ and the direction $\theta(x, y)$ of the gradient $L(x, y)$ can be obtained by the following formula:

$$m(x, y) = \sqrt{[L(x + 1, y) - L(x - 1, y)]^2 + [L(x, y + 1) - L(x, y - 1)]^2} \quad (4.52)$$

$$\theta(x, y) = \arctan \frac{L(x, y + 1) - L(x, y - 1)}{L(x + 1, y) - L(x - 1, y)} \qquad (4.53)$$

After the computation of gradient direction, the gradient direction and amplitude corresponding to the pixels in the neighborhood of the feature points will be counted by using histogram. The horizontal axis of the histogram in the gradient direction is the angle of gradient direction (the range of the gradient direction is 0°–360°, and the histogram is 10 columns per 36°, or 8 columns per 45°). The vertical axis is the gradient direction corresponding to the accumulation of the gradient amplitude, and the peak value of the histogram is the main direction of the feature point. In order to obtain more accurate direction, we can fit the discrete gradient histogram by the interpolation. In particular, the direction of the key points can be obtained by parabola interpolation with the three column values closest to the main peak value. In the gradient histogram, if a column value corresponding to the 80% energy of the peak value, this direction can be considered as the auxiliary direction of the feature point. Therefore, a feature point may detect multiple directions (In other words, a feature point may generate multiple points with identical coordinates, and the same scale, but different directions).

After getting the principal direction of the feature points, three information (x, y, σ, θ) can be obtained, i.e. position, scale and direction. Thus, a SIFT feature region can be determined. A SIFT feature region is represented by three values, the center represents the position of the feature point, the radius represents the scale of the key point, and the arrow represents the main direction. The key points with multiple directions can be duplicated into multiple copies. Then the direction values are assigned to the copied feature points respectively. Eventually, a feature point will generate multiple points with identical coordinates, and the same scale, but different directions.

6. Generate feature descriptors

The location, scale and direction of SIFT feature points have been found by the above steps, and a group of vectors is needed to describe the key points. This descriptor contains not only the feature points, but also the pixels that contribute to the points around the feature points. The descriptor should have high independence to ensure the matching rate.

The generation of feature descriptors has three steps:

(1) Correcting the principal direction of rotation to ensure rotation-invariance;
(2) Generate descriptors, and finally form a 128-dimensional feature vector;
(3) Carry out normalization, the length of the feature vector is normalized to further remove the influence of illumination.

In order to ensure the rotation-invariance of the feature vector, we should take feature point as the center, and rotate the axis by θ in neighboring area. In other words, the coordinate axis rotates to the principal direction of the feature point. The new coordinate of neighbor pixels after rotation is:

$$\begin{bmatrix} x' \\ y' \end{bmatrix} = \begin{bmatrix} \cos\theta & -\sin\theta \\ \sin\theta & \cos\theta \end{bmatrix} \begin{bmatrix} x \\ y \end{bmatrix} \tag{4.54}$$

After rotation, take the window of 8×8 with the principal direction center. As Fig. 4.16 shown on the left, the current position of the key point is in the center. Each cell represents a pixel in the scale space where the neighborhood of the key point is. Compute the gradient amplitude and gradient direction of each pixel. The direction of the arrow represents the gradient direction of the pixel, and the length represents the gradient amplitude. Then the weighted computation is carried out by using Gauss window. Finally, the gradient histograms of 8 directions are plotted on each 4×4 blocks, and the accumulated values of each gradient direction are calculated to form a seed point, as shown in Fig. 4.16 on the right. Each feature point is composed of 4 seed points, and each seed point carry 8 directions of vector messages. The neighbor directional information is combined to enhance the anti-noise capability of the algorithm. At the same time, it also provides more rational fault tolerance for feature matching with positional error.

Different from the main direction, the gradient histogram of each seed region is divided into 8 directions from 0 to 360. Each interval is 45°, that is, each seed point has 8 directions of gradient intensity information.

By dividing the pixels around the feature points, the gradient histogram in the block is calculated to generate the vector with uniqueness. This vector is an abstraction of the image information in the region and it is unique.

The programme of SIFT feature description is shown as PROGRAMME 4.10.

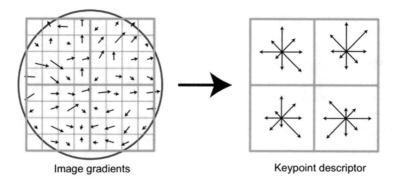

Image gradients   Keypoint descriptor

Fig. 4.16 Formation of seed points

PROGRAMME 4.10: SIFT Feature Description

```
function [image, descriptors, locs] = sift(imageFile)
% Load image
image = imread(imageFile);
% if isrgb(image)
%       image = rgb2gray(image);
% end
[rows, cols] = size(image);
% Convert into PGM imagefile, readable by "keypoints" executable
f = fopen('tmp.pgm', 'w');
if f == -1
      error('Could not create file tmp.pgm.');
end
fprintf(f, 'P5\n%d\n%d\n255\n',   cols, rows);
fwrite(f, image', 'uint8');
fclose(f);
% Call keypoints executable
if isunix
      command = '!./sift ';
else
      command = '!siftWin32 ';
end
command = [command ' <tmp.pgm >tmp.key'];
eval(command);
```

```
% Open tmp.key and check its header
g = fopen('tmp.key', 'r');
if g == -1
    error('Could not open file tmp.key.');
end
[header, count] = fscanf(g, '%d %d', [1 2]);
if count ~= 2
    error('Invalid keypoint file beginning.');
end
num = header(1);
len = header(2);
if len ~= 128
    error('Keypoint descriptor length invalid (should be 128).');
end
% Creates the two output matrices (use known size for efficiency)
locs = double(zeros(num, 4));
descriptors = double(zeros(num, 128));
% Parse tmp.key
for i = 1:num
    [vector, count] = fscanf(g, '%f %f %f %f', [1 4]); %row col scale ori
    if count ~= 4
        error('Invalid keypoint file format');
    end
    locs(i, :) = vector(1, :);
    [descrip, count] = fscanf(g, '%d', [1 len]);
    if (count ~= 128)
        error('Invalid keypoint file value.');
    end
    % Normalize each input vector to unit length
    descrip = descrip / sqrt(sum(descrip.^2));
    descriptors(i, :) = descrip(1, :);
end
fclose(g);
function showkeys(image, locs)
disp('Drawing SIFT keypoints ...');
% Draw image with keypoints
figure('Position', [50 50 size(image,2) size(image,1)]);
colormap('gray');
imagesc(image);
hold on;
```

```
imsize = size(image);
for i = 1: size(locs,1)
        % Draw an arrow, each line transformed according to keypoint parameters.
        TransformLine(imsize, locs(i,:), 0.0, 0.0, 1.0, 0.0);
        TransformLine(imsize, locs(i,:), 0.85, 0.1, 1.0, 0.0);
        TransformLine(imsize, locs(i,:), 0.85, -0.1, 1.0, 0.0);
end
hold off;
% ------ Subroutine: TransformLine -------
% Draw the given line in the image, but first translate, rotate, and
% scale according to the keypoint parameters.
%
% Parameters:
%     Arrays:
%       imsize = [rows columns] of image
%       keypoint = [subpixel_row subpixel_column scale orientation]
%
%     Scalars:
%       x1, y1; begining of vector
%       x2, y2; ending of vector
function TransformLine(imsize, keypoint, x1, y1, x2, y2)
% The scaling of the unit length arrow is set to approximately the radius
%     of the region used to compute the keypoint descriptor.
len = 6 * keypoint(3);
% Rotate the keypoints by 'ori' = keypoint(4)
s = sin(keypoint(4));
c = cos(keypoint(4));
% Apply transform
r1 = keypoint(1) - len * (c * y1 + s * x1);
c1 = keypoint(2) + len * (- s * y1 + c * x1);
r2 = keypoint(1) - len * (c * y2 + s * x2);
c2 = keypoint(2) + len * (- s * y2 + c * x2);
line([c1 c2], [r1 r2], 'Color', 'c');
```

The SIFT function is responsible for reading the picture and returning the SIFT feature point. The showkeys is responsible for displaying the feature points. The original image and the execution results are as follows (Fig. 4.17).

Fig. 4.17 The original image and SIFT feature description

 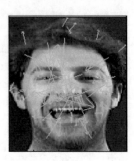

References

1. Marques O (2011) Feature extraction and representation. Practical image and video processing using MATLAB. Wiley-IEEE Press, pp 447–474
2. Guyon I, Nikravesh M, Gunn S et al (2006) Feature extraction. Springer Berlin Heidelberg
3. Virmani J (2016) Breast tissue density classification using wavelet-based texture descriptors. In: Proceedings of the second international conference on computer and communication technologies. Springer India, pp 539–546
4. Duda RO, Hart PE (1972) Use of the hough transformation to detect lines and curves in pictures. Commun ACM **15**, 11–15 (1972)
5. Hough PVC (1959) Machine analysis of bubble chamber pictures. In: Proceedings of international conference on high energy accelerators and instrumentation (1959)
6. Hough PVC (1962) Method and means for recognizing complex patterns. US Patent 3,069,654, 18 Dec 1962
7. Mallat S (1998) A wavelet tour of signal processing, pp 250–252
8. Nanni L, Lumini A, Brahnam S (2012) Survey on LBP based texture descriptors for image classification. Pergamon Press, Inc.
9. Stephens M, Harris C (1989) 3D wire-frame integration from image sequences. Image Vis Comput 7(1):24–30
10. Smith S, Lange TD (1997) TRF1, a mammalian telomeric protein. Trends Genet Tig 13(1):21
11. Bay H, Tuytelaars T, Gool LV (2006) SURF: speeded up robust features[J]. Comput Vis Image Underst 110(3):404–417
12. Lowe DG, Lowe DG (2004) Distinctive image features from scale-invariant keypoints. Int J Comput Vis 60(2):91–110
13. Dalal N, Triggs B (2005) Histograms of oriented gradients for human detection. In: 2005 IEEE computer society conference on CVPR. IEEE, pp 886–893

Part II
Advances in Image Processing

Chapter 5
Image Correction

5.1 Introduction

In the process of image generation, transmission and recording, the quality of images will decrease due to various reasons, which will lead to the degradation of the image. Image correction refers to the restoration of distorted images. The causes of image distortion including: aberration, distortion and limited bandwidth of the imaging system; geometry distortion caused by photographic attitude and nonlinear sweep scanning of the imaging device; motion blur, radiation distortion and the noise-corruption. The basic idea of image correction is to establish the corresponding mathematical model based on the distortion reasons, extracting the needed information from the contaminated or distorted image signals, restore the image to its original appearance along the reverse process that distorts the image. The actual correction process is to design a filter to estimate pixel value of the original image from the distorted image which maximum close to the original images according to the prescribed error criterion.

5.2 Noise Reduction Using Spatial-Domain Techniques

Noise is one of the most important and common causes of image degradation. The noise of digital image mainly comes from the formation and transmission process of images. For example, using Charge-coupled Device (CCD) camera to obtain an image, the illumination degree and sensor temperature are the main factors of noise forming, and the image will be polluted by noise due to the interference to the transmission channel in the process of transmission. For instance, images transmitted over a wireless network could be polluted by interference from light or other atmospheric factors. Due to the influence of noise, the grayscale of image pixel will change, and the grayscale of the noise can be regarded as a random variable represented by the

© Springer International Publishing AG, part of Springer Nature 2019
S. Gong et al., *Advanced Image and Video Processing Using MATLAB*,
Modeling and Optimization in Science and Technologies 12,
https://doi.org/10.1007/978-3-319-77223-3_5

probability density function (PDF). Therefore, by analyzing the gray-scale statistical properties of the noise components, we can filter out the noise more effectively.

5.2.1   Selected Noise Probability Density Functions

The following are the most common noise found in image correction, including Gaussian noise, Rayleigh noise, Erlang(gamma) noise, exponential noise, uniform noise, impulse noise, etc [1].

(1)   Gaussian noise

It is the noise whose distribution satisfies the Gaussian distribution. The PDF of a Gaussian random variable z, it can be expressed as:

$$p(z) = \frac{1}{\sqrt{2\pi}\sigma}e^{\frac{-(z-\mu)^2}{2\sigma^2}} \tag{5.1}$$

where z represents the grayscale value, μ is the mean (average) value or the mathematical expectation of z, and σ is standard deviation, the standard deviation squared σ^2 is called the variance of z. Figure 5.1 shows the curve of the Gaussian PDF.

The grayscale distribution of Gaussian noise is concentrated in the vicinity of the mean, which decreases with the increase of the distance from the mean. It is about 68.3% of z's value will be in the range $[(\mu - \sigma), (\mu + \sigma)]$, and approximately 95.4% will be in the range $[(\mu - 2\sigma), (\mu + 2\sigma)]$.

(2)   Gamma noise

The probability density of Gamma distribution noise can be given by the following formula:

$$p(z) = \left[\begin{array}{ll} \frac{a^b z^{b-1}}{(b-1)!}e^{-az} & z \geq 0 \\ 0 & z < 0 \end{array} \right. \tag{5.2}$$

Fig. 5.1 The probability density function of Gaussian noise

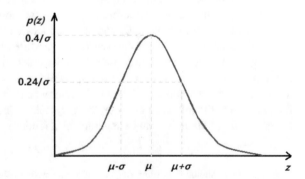

Fig. 5.2 The Gamma
distribution PDF

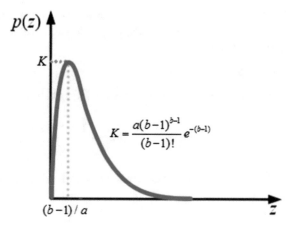

where $a > 0$, b is a positive integer, and '!' indicates factorial. The mean and variance of this density are given by the following formula:

$$\mu = \frac{b}{a} \tag{5.3}$$

$$\sigma^2 = \frac{b}{a^2} \tag{5.4}$$

Figure 5.2 shows a plot of the Gamma distribution PDF.

(3) Uniform noise

The PDF of uniform noise is given by:

$$p(z) = \begin{cases} \frac{1}{b-a} & a \leq z \leq b \\ 0 & otherwise \end{cases} \tag{5.5}$$

The mean and variance of this uniform noise can be calculated by the formulas (5.6) and (5.7):

$$\mu = \frac{a+b}{2} \tag{5.6}$$

$$\sigma^2 = \frac{(b-a)^2}{12} \tag{5.7}$$

The uniform distribution of noise is random, and each pixel in a noise-corrupted image may be affected and the grayscale value will be changed. Figure 5.3 shows a plot of the uniform density.

(4) Exponential noise

The probability density of exponential distribution noise can be given by Eq. (5.8):

$$p(z) = \begin{cases} ae^{-az} & z \geq 0 \\ 0 & z < 0 \end{cases} \tag{5.8}$$

The mean and variance of exponential distribution are:

$$\mu = \frac{1}{a} \tag{5.9}$$

$$\sigma^2 = \frac{1}{a^2} \tag{5.10}$$

Figure 5.4 shows the curve of the exponential distribution PDF.

(5) Impulse noise

The PDF of impulse distribution noise can be given by the following formula:

$$p(z) = \begin{cases} P_a & \text{if } z = a \\ P_b & \text{if } z = b \\ 0 & \text{otherwise} \end{cases} \tag{5.11}$$

Equation (5.11) indicates that the impulse noise can be positive or negative if neither P_a nor P_b is zero. If $b > a$, the gray value b will appear as a bright spot in

Fig. 5.3 The PDF of uniform noise

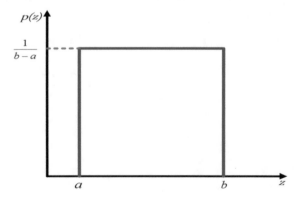

Fig. 5.4 The PDF of
exponential noise

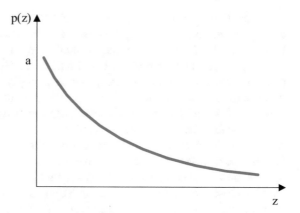

Fig. 5.5 The probability
density function of pulse
noise

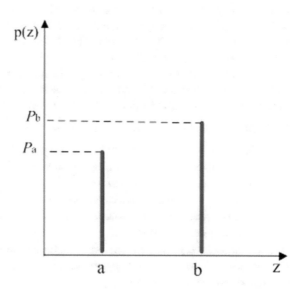

the image, whereas the value of a will be shown as a dark spot. Especially, if they
are approximately equal, the impulse noise value is similar to the salt-and-pepper
particles randomly distributed over the image. The impulse noise is also called the
salt-and-pepper noise due to this reason [2, 3]. The pepper noise corresponds to the
value of a, while the salt noise corresponds to the state that the value of noise is equal
to b. If we show the image, the negative impulses appear as black (pepper point),
while a positive pulse is displayed as white (salt) in the image. If either P_a or P_b is
zero, the impulse noise is called unipolar impulse. Figure 5.5 shows the curve of the
impulse density.

Noise PDF parameters can generally be obtained from sensor specifications. However, for some special imaging devices, the parameters often need to be estimated by users. In this case, only the images obtained by imaging devices can be useful. For this reason, it is often possible to estimate the PDF parameter by selecting small patches of the region with constant grayscale intensity from the image. By showing the histogram of the small area, it shows that the shape of the histogram is very close to the corresponding PDF described above. Therefore, we can use the data in this small area to calculate the parameters, such as the mean and variance of the grayscale. For example, if the shape of histogram is close to the Gaussian distribution, indicating that the image is disturbed by Gaussian noise, the variance and mean are the parameters of the Gaussian function. If the histogram is any other shape, you can choose the PDF closest to the formulas (5.2)–(5.8), and use the mean and variance to solve the parameters a and b. But impulse noise is handled differently because it needs to estimate the probability of the occurrence of black and white pixels. Therefore, in order to compute the histogram, the image must have a relatively constant medium grayscale region, where the spike of the black and white pixels corresponding to the estimated values of p_a as well as p_b.

Since the above functions are simple to implement with MATLAB, In this section,we only give the PDF based on Gaussian and impulse noise. The code for adding noise to the image is shown in PROGRAMME 5.1:

PROGRAMME 5.1: Add noise to the image

```
      function b = imnoise(varargin)    %IMNOISE Add noise to image.
%     J = IMNOISE(I,TYPE,...) Add noise of a given TYPE to the intensity image
%     I. TYPE is a string that can have one of these values:
%          'gaussian'           Gaussian white noise with constant   mean and variance
%          'localvar'    Zero-mean Gaussian white noise     with an intensity-dependent variance
%          'salt & pepper'    "On and Off" pixels
%     Depending on TYPE, you can specify additional parameters to IMNOISE. All
%     numerical parameters are normalized; they correspond to operations with
%     images with intensities ranging from 0 to 1.
%     J = IMNOISE(I,'gaussian',M,V) adds Gaussian white noise of mean M and
%     variance V to the image I. When unspecified, M and V default to 0 and
%     0.01 respectively.
%     J = imnoise(I,'localvar',V) adds zero-mean, Gaussian white noise of
%     local variance, V, to the image I.   V is an array of the same size as I.
%     J = imnoise(I,'localvar',IMAGE_INTENSITY,VAR) adds zero-mean, Gaussian
%     noise to an image, I, where the local variance of the noise is a
%     function of the image intensity values in I.   IMAGE_INTENSITY and VAR
%     are vectors of the same size, and PLOT(IMAGE_INTENSITY,VAR) plots the
%     functional relationship between noise variance and image intensity.
%     IMAGE_INTENSITY must contain normalized intensity values ranging from 0 to 1.
%     J = IMNOISE(I,'poisson') generates Poisson noise from the data instead
```

```
%      of adding artificial noise to the data.   If I is double precision,
%      then input pixel values are interpreted as means of Poisson
%      distributions scaled up by 1e12.   For example, if an input pixel has
%      the value 5.5e-12, then the corresponding output pixel will be
%      generated from a Poisson distribution with mean of 5.5 and then scaled
%      back down by 1e12.   If I is single precision, the scale factor used is
%      1e6.   If I is uint8 or uint16, then input pixel values are used
%      directly without scaling.   For example, if a pixel in a uint8 input
%      has the value 10, then the corresponding output pixel will be
%      generated from a Poisson distribution with mean 10.
%      J = IMNOISE(I,'salt & pepper',D) adds "salt and pepper" noise to the
%      image I, where D is the noise density.   This affects approximately
%      D*numel(I) pixels. The default for D is 0.05.
%      J = IMNOISE(I,'speckle',V) adds multiplicative noise to the image I,
%      using the equation J = I + n*I, where n is uniformly distributed random
%      noise with mean 0 and variance V. The default for V is 0.04.
%      The mean and variance parameters for 'gaussian', 'localvar', and
%      'speckle' noise types are always specified as if for a double image
%      in the range [0, 1].   If the input image is of class uint8 or uint16,
%      the imnoise function converts the image to double, adds noise
%      according to the specified type and parameters, and then converts the
%      noisy image back to the same class as the input.
%      Class Support
%      For most noise types, I can be uint8, uint16, double, int16, or
%      single.   For Poisson noise, int16 is not allowed. The output
%      image J has the same class as I.   If I has more than two dimensions
%      it is treated as a multidimensional intensity image and not as an RGB image.
%      Example
%            I = imread('eight.tif');
%            J = imnoise(I,'salt & pepper', 0.02);
%              figure, imshow(I), figure, imshow(J)
%      See also RAND, RANDN.
And then invoke the function PROGRAMM5-1:
I = imread('lena.jpg');
I1 = imnoise(I,'gaussian');
I2 = imnoise(I,'salt & pepper');
subplot(1,3,1),imshow(I);% Show the original image
title('orginal image');
subplot(1,3,2),imshow(I1); % add Gaussian noise
title('gaussian image');
subplot(1,3,3),imshow(I2); % Salt and pepper noise
title('salt & pepper');
```

The pictures below show the effect of adding Gaussian noise and salt-and-pepper noise on the original image (Fig. 5.6).

Fig. 5.6 The image corrupted by Gaussian noise and salt-and-pepper noise

5.2.2   Filtering

(1)   Mean filter

Mean filtering is a technique that is smoothing directly over a spatial domain. The technique is based on the assumption that the image consists of many small, gray, constant patches, with very high spatial correlations between adjacent pixels while the noise being relatively independent. Based on the above hypothesis, the average value of all neighbor pixels of a pixel can be assigned to the corresponding pixels in the smoothed image, so as to achieve the goal of smoothing.

There are two forms of mean filtering, the unweighted neighborhood averaging method and the weighted neighborhood averaging method.

Given an input image $f(x, y)$, where, as usual, M and N are the raw and column dimensions of the image, the smoothed image obtained by the neighborhood averaging method is $g(x, y)$, then:

$$g(x, y) = \frac{1}{MN} \sum_{i,j \in s} f(i, j) \tag{5.12}$$

For $x = 0, 1, 2, 3, \ldots, M - 1; y = 0, 1, 2, 3, \ldots, N - 1;$ S is a collection of pixel coordinates in the (x, y) neighborhood, which does not include (x, y); MN represents the total number of pixels in set S.

The unweighted neighborhood averaging method can be described in the form of a mask, and then calculated by convolution, that is, moving the mask point by point in the image to find the sum of the products of the filter and the corresponding pixel encompassed by the filter.

The specific implementation process is, when the filter and image values are convoluted, the coefficient $w(0, 0)$ in the filter should be located at the position where the image corresponds to the pixel (x, y). For a mask with a size of $m \times n$, we assume that $m = 2a + 1$ and $n = 2b + 1$, where a and b are non-negative integers. The length and width of the mask are usually odd, like $3 \times 3, 5 \times 5, 7 \times 7$ and so on.

Figure 5.7 shows the mechanics of linear spatial filtering with a 3×3 filter mask, at the point (x, y) in the image, the response obtained with this mask is:

$$R = w(-1, -1)f(x - 1, y - 1) + w(-1, 0)f(x - 1, y) + \cdots$$
$$+ w(0, 0)f(x, y) + \cdots + w(1, 0)f(x + 1, y) + w(1, 1)f(x + 1, y + 1) \quad (5.13)$$

In the unweighted neighborhood averaging method, each coefficient in the mask is 1. Figure 5.8 shows the unweighted neighborhood average 3×3 mask, while Fig. 5.9 shows the enhancement effect of the neighborhood averaging method.

Another neighborhood mean method is called weighted average, where all mask coefficients could have different weights. Figure 5.10a shows a weighted averaging filter mask, Fig. 5.10b is an example.

Given an image $f(x, y)$ of size $m \times n$, the process of weighted averaging filtering through a filter of size $m \times n$ (m and n are odd) can be given in the following formula:

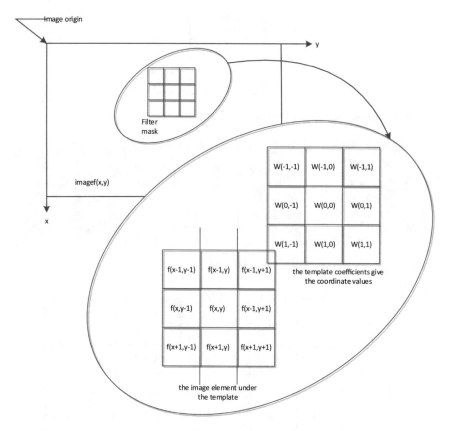

Fig. 5.7 The mechanics of linear spatial filtering with a 3×3 filter mask

$$g(x, y) = \frac{\sum_{s=-a}^{a} \sum_{t=-b}^{b} w(s, t) f(x+s, y+t)}{\sum_{s=-a}^{a} \sum_{t=-b}^{b} w(s, t)} \tag{5.14}$$

In the equation, $a = \frac{(m-1)}{2}$ and $b = \frac{(n-1)}{2}$, the denominator is the sum of all the coefficients of the mask, which is a constant. In order to obtain a complete filtered image, it is necessary to applying Eq. (5.14) for $x = 0, 1, 2, \ldots, M - 1$ and $y = 0, 1, 2, \ldots, N - 1$.

For the mask shown in Fig. 5.10b, the weight of the pixel at the center of the mask is higher than the weight of any other pixel, so the pixel given in the average calculation is more important, while the other pixels farther away from the center of the mask is less important. Since the diagonal terms are farther from the center

$$\frac{1}{9} \times \begin{array}{|c|c|c|} \hline 1 & 1 & 1 \\ \hline 1 & 1 & 1 \\ \hline 1 & 1 & 1 \\ \hline \end{array}$$

Fig. 5.8 The unweighted neighborhood average 3 × 3 mask

Fig. 5.9 The enhancement effect of the neighborhood averaging method

(a)

$$\begin{array}{|c|c|c|} \hline W1 & W2 & W3 \\ \hline W4 & W5 & W6 \\ \hline W7 & W8 & W9 \\ \hline \end{array}$$

(b)

$$\frac{1}{16} \times \begin{array}{|c|c|c|} \hline 1 & 2 & 1 \\ \hline 2 & 4 & 2 \\ \hline 1 & 2 & 1 \\ \hline \end{array}$$

Fig. 5.10 3 × 3 weighted averaging filter mask: **a** the general form **b** a specific example

adjacent to the orthogonal direction, so they are less important than the four pixels that are directly adjacent to the center. The center point is enhanced to the highest value, and as the distance from the center point increases the coefficient value is reduced, which is to reduce the blurring in smoothing. Certainly, we could have taken other weights to achieve the same purpose. However, the sum of all the coefficients in the mask of Fig. 5.10b is 16, which is convenient for the computer to implement, since it is an integer power of 2. MATLAB implementation of average filtering is shown in PROGRAMME 5.2:

PROGRAMME 5.2: Average filtering

```
I=imread('star.JPEG');
J =rgb2gray(I);
I1=imnoise(J,'salt & pepper'); %add noise
subplot(121),imshow(I1);title(' a graph containing salt and pepper noise ');
J1=filter2(fspecial('average',3),I1)/255; % average filtering
subplot(122),imshow(J1);title(' the graph after the average filtering ');
```

(2) Order Statistic Filters

Although the neighborhood averaging method can smooth the image, it will blur some details in the image while eliminating the noise. The order statistic filter is a non-linear filter. In order to perform order statistic filtering in an image, we will select a window W with an odd number of pixels at first, each pixel in the window is sorted according to the grayscale value from small to large, then replace the original grayscale value with the grayscale value of the kth position. For the given values $\{a_1, a_2, \ldots, a_n\}$, these values are sorted in order of size, the element in the kth position is used as the image filter output, which is the two-dimensional statistical filter of the serial number k. MATLAB implementation of order statistic filtering is shown in PROGRAMME 5.3:

PROGRAMME 5.3: Order statistic filters

```
I=imread('peppers.jpg');
J=imnoise(I,'salt & pepper');
H=rgb2gray(J);
B=ordfilt2(H,5,true(3)); % order statistic filters
subplot(121),imshow(H);title('the graph before filtering');
subplot(122),imshow(B);title('the graph after filtering');
```

Figure 5.11 shows the original image with salt and pepper noise, and the processed image using average filtering method. The image is smooth after filtering and the noise are fully removed.

(3) Adaptive Filters

The adaptive filter is a kind of filter relative to the fixed one, and fixed filter belongs to the classic filter, whose frequency is fixed, while the frequency of the adaptive

Fig. 5.11 Noise reduction with order statistic filters

filter can change automatically according to the input signal, so it has a wider range of applications. Without any prior knowledge of the signal and noise, the adaptive filter will automatically adjust the filter parameters using the obtained parameters from a previous time in order to adapt to statistical characteristics of unknown or random variations in signals and noises, so as to realize the optimal filtering. The adaptive filter is essentially a wiener filter which can adjust its own transmission characteristics to achieve optimization. The wiener filter [4] wiener2 estimates the local mean and variance around each pixel.

$$\mu = \frac{1}{NM} \sum_{n_1,n_2 \in \eta} a(n_1, n_2) \tag{5.15}$$

and

$$\sigma^2 = \frac{1}{NM} \sum_{n_1,n_2 \in \eta} a^2(n_1, n_2) - \mu^2, \tag{5.16}$$

where η is the N-by-M local neighborhood of each pixel in the image A. wiener2 then creates a pixelwise Wiener filter using these estimates,

$$b(n_1, n_2) = \mu + \frac{\sigma^2 - v^2}{\sigma^2}(a(n_1, n_2) - \mu), \tag{5.17}$$

where $v2$ is the noise variance. If the noise variance is not given, wiener2 uses the average of all the local estimated variances.

This filtering is implemented in MATLAB image processing toolbox using function wiener2, when the local change of the image is small, the function can be processed in a relatively large way, whereas the smaller smoothing is performed. Compared with others, the adaptive filter can preserve the boundaries and the high frequency components of the image, but it consumes much more time. In MATLAB, the invoke of wiener2 function can be shown in PROGRAMME 5.4:

PROGRAMME 5.4: The Wiener2 function

J=wiener2(I,[m n],noise);

J=wiener2(I,[m n]);

[J,noise]=wiener2(I,[m n]);

The first function outputs the filtered image based on the original noise image, the size of the specified filter window is $m \times n$, the default value is 3×3; the second function returns the estimated value of the noise power while performing image filtering. MATLAB implementation of wiener2 adaptive filtering is shown in PROGRAMME 5.5.

PROGRAMME 5.5: Wiener2 adaptive filter

```
I = imread('rice.jpg');
K1 = imnoise(I,'gaussian',0,0.02);% add Gaussian noise
subplot(221);imshow(K1);
title(' white Gaussian noise ');
M1 = wiener2(K1,[5 5]);%wiener2 adaptive filter
subplot(222);imshow(M1);
title(' filtered Gauss white noise ');
K2 = imnoise(I,'salt & pepper',0.02);
subplot(223);imshow(K2);
title(' salt-and-pepper noise ');
M2 = wiener2(K2,[9 9]);%wiener2 adaptive filter
subplot(224);imshow(M2);
title(' filtered salt-and-pepper noise ');
```

The results of the filtering are shown in Fig. 5.12. It shows that Wiener2 filter has a relatively better effect on white Gaussian noise filtering. When the noise of the salt-and-pepper is filtered, the edge information of the image is blurred with the increase of filter window.

5.3 Image Deblurring

The blurred images restoration is the process of restructuring the original image from the blurred images [5]. And the image deblurring is the basis of images processing and pattern recognition, which is often applied to judging or appraisal of defocus images in practice works. Therefore, it is also a hot topic of research in recent years. In order to deblur the image, it is usually necessary to know the reason of image degradation, and reconstruct or recover the original image by using some prior knowledge of the image degradation. If we can accurately calculate the Point Spread Function (PSF) of blurred images, on this basis, adopt a variety of anti-degradation processing methods, such as inverse filter, wiener filtering and so on to restore the image, which is a typical image restoration method. For various blurred images, the degradation reasons may

white Gaussian noise   filtered Gauss white noise

salt-and-pepper noise   filtered salt-and-pepper noise

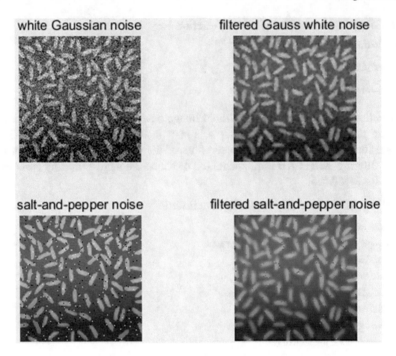

Fig. 5.12 Image contrast before and after adaptive filtering

be different, but as the image restoration problem, they are essentially the same: the formation of a blurred image can be described by a convolution process. Thus, the problem of image restoration is actually a deconvolution problem. In order to calculate conveniently, frequency domain filtering is often used to solve the problem of deconvolution.

The following types of blurred images are usually discussed in image restoration: the first category is defocus blurred, during the process of shooting, the recorded subject will be blurred and form the so-called defocus blurred image due to the deviation of an imaging plane from the focus of an optical lens or other reasons. The second category is motion blurred, which is caused by the relative motion between target objects and imaging devices during the process of image acquisition. For image restoration, it can be processed either by continuous mathematics or by discrete mathematics.

5.3.1   The Restoration of Defocus Blurred Image

Among all kinds of blur, the defocus blur exists widely in satellite remote sensing imaging, space exploration imaging, medical diagnosis and so on. Defocus blur can also be caused by poor focusing, hand shaking or poor quality of imaging system in daily life.

Usually in the linear translation space invariant motion blur system, the blurred image $g(x, y)$ can repressed as the two-dimensional convolution of the original image $f(x, y)$ and PSF $h(x, y)$:

$$g(x, y) = f(x, y) * h(x, y) + n(x, y) \qquad (5.18)$$

where $n(x, y)$ is the additive noise, take the Fourier transform of Eq. (5.18), the corresponding frequency domain expression is:

$$G(u, v) = F(u, v)H(u, v) + N(u, v) \qquad (5.19)$$

For defocus blur, the PSF can be expressed as follows:

$$h_r(x, y) = \begin{cases} \frac{1}{\pi r^2}, & x^2 + y^2 \leq r^2 \\ 0 \end{cases} \qquad (5.20)$$

r is the radius of the defocus blur, which is the only parameter needs to be obtained, after r is determined, the degradation function can be obtained and the image can be corrected. The calculation process of r is given below:

After performing the Fourier transform on the degenerated model, it follows that

$$H(u, v) = \frac{2J_1\left[r\sqrt{(\frac{2\pi}{M}u)^2 + (\frac{2\pi}{N}v)^2}\right]}{\sqrt{(\frac{2\pi}{M}u)^2 + (\frac{2\pi}{N}v)^2}} \qquad (5.21)$$

$J_1[\cdot]$ represents the first order Bessel function of the first kind, while M, N is the parameter of the two-dimensional Fourier transform, according to the properties of the function, the first dark ring of $H(u, v)$ in the frequency domain, in another word, the trajectory of the first zero point is:

$$2\pi r\sqrt{(\frac{u}{M})^2 + (\frac{v}{N})^2} = 3.85 \qquad (5.22)$$

When the noise is relatively small, we can see from the above Eq. (5.19) that if we find the corresponding u and v of the first zero position (dark ring) of the Fourier transform in the defocused image, the required r can be obtained by this equation.

The restoration algorithm may be summarized as follows:

Step1: Fourier transform is applied to the defocus blurred image, extracting the tangent plane through the center of the concentric circle;

Step2: Fourier transform is applied to the tangent plane curve, the period of the curve can be extracted from the Fourier spectrum due to the periodic nature of the tangent curve, and the length of the period is equivalent to the distance from the center of the spectrum to the first dark ring in the frequency domain;

Step3:   Calculate the blur radius and generate PSF, Wiener filtering and other methods can be applied to recover the degraded image, ensuring that the processed
image as near as possible to the original image.

MATLAB implementation of defocus blurred image restoration is shown
in PROGRAMME 5.6.

PROGRAMME 5.6: Defocus blurred image restoration

```
r = 10; % blur radius r=10
psf = fspecial('disk',r); % get the Point Spread Function,PSF
f = imread('lena.jpg'); % input the original image
f = rgb2gray(f);
f = im2double(f);
figure; subplot(2,3,1); imshow(f); title('Original Image');
g = imfilter(f,psf,'circular','conv'); % realize the defocus blur
subplot(2,3,2); imshow(g); title('Out-of-focus Blur Image');
G = fftshift(log(abs(fft2(g))));
subplot(2,3,3); imshow(G,[],'InitialMagnification','fit'); title('Fourier  Spectrum');%draw  the  Fourier
   Spectrum
[R,C] = size(G);
section = G(R/2,:);
subplot(2,3,4); plot(section); title('Cross Section of Logsrithm Spectrum');
FSection = abs(fft(section));
subplot(2,3,5); plot(FSection); title('Spectrum of Cross Section');
```

Figure 5.13 shows the results of the image deblurring of the blurred image. It can be
seen from the Fourier spectrum section diagram that if remove the DC component, the
maximum peak is located at the position of 20, so we could calculate the blur radius
r is equal to 10. Generate the PSF, using Lucy-Richardson filtering, the restoration
effect is shown in Fig. 5.14 when the number of iterations is 50.

5.3.2   Restoration of Motion Blurred Image

In the research field about motion deblur, the motion blur caused by the uniform
linear motion has both universal and special properties, since non-uniform linear
motion can be approximated as a uniform linear motion under certain conditions
or can be decomposed into a combination of multiple uniform linear motion. For a
uniform linear motion blur image, the PSF can be described as:

$$h(x, y) = \begin{cases} \frac{1}{L} \sqrt{x^2 + y^2} \le L \ and \frac{y}{x} = \tan \theta \\ \\ 0 \ otherwise \end{cases} \tag{5.23}$$

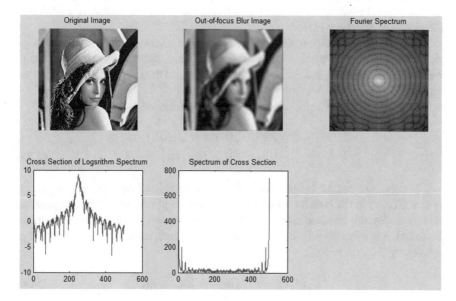

Fig. 5.13 The results of the image deblurring

Fig. 5.14 The result of Lucy-Richardson filtering, the number of iterations is 50

Here L is the blur scale, while θ corresponds to the angle between the direction of motion and the positive x-axis. It can be seen that the motion blur PSF depends on two parameters—the blur scale L and the direction of motion θ. Therefore, the estimation of motion blur PSF is equivalent to estimate these two parameters. Thus, the Fourier transform of Eq. (5.23) is:

$$H(u, v) = \frac{\sin(\pi wL)}{\pi wL} e^{-j\pi wL} \tag{5.24}$$

where $w = u \cos\theta + v \sin\theta$.

Therefore, the spectrum $G(u, v)$ of a uniform linear motion blurred image [6, 7] has a series of parallel dark lines, these dark stripes are perpendicular to the horizontal direction and the position corresponds to the zero point of the function $\sin(\pi wL)/\pi wL$.

The cepstrum of the image $g(x, y)$ is defined as follows:

$$C_g(p, q) = F^{-1}\left[\log|F\{g(x, y)\}|\right]$$
$$= F^{-1}\left[\log|G(u, v)|\right] \qquad (5.25)$$

The cepstrum can be understood as the transformation from $g(x, y)$ to $C_g(p, q)$. Where $G(u, v)$ is the Fourier transform of the image $g(x, y)$ and $F^{-1}[\bullet]$ represents the inverse Fourier transform. In practical engineering applications, in order to make the function is meaningful when $G(u, v)$ has zero value, the cepstrum of the image is given by the expression

$$C_g(p, q) = F^{-1}\{\log[1 + |G(u, v)|]\} \qquad (5.26)$$

When there is no noise, the cepstrum of the image degradation used was

$$C_g(p, q) = C_h(p, q) + C_f(p, q) \qquad (5.27)$$

It shows that the blurred image is the convolution of the original image and the blur kernel in the time domain, and then it is expressed as the sum of the cepstrum of the original image as well as the PSF after transform to the cepstrum domain. Thus, we could separate the blur information from the original image easily. For motion blurred images, there is a bright band along the direction of motion blur in the cepstrum, and the angle between the bright band and the horizontal direction is the angle of motion blur. When it comes to motion blur direction, the cepstrum three-dimensional map consists of two parts; one part is the positive peak component, which reflects the characteristics of the non-degraded image; the other is the negative peak component, which indicates the characteristics of the blur system. These two parts are different from the area occupied by the graph, the distance between the two negative peak points is two times that of the motion blur scale.

The restoration algorithm may be summarized as follows:

(1) Calculate the cepstrum, of motion blur image
(2) Find the two positions of maximal negative peaks, and because the negative peak is located on the straight line of the bright band, the blur angle and length are calculated according to the negative peak position
(3) Generate PSF, using Wiener filtering and other methods to restore the image

MATLAB implementation of motion blurred image restoration is shown in PROGRAMME 5.7.

PROGRAMME 5.7: Motion blurred image restoration

```
len = 40;% motion displacement
theta = 30;% motion angle
psf = fspecial('motion',len,theta); %motion psf
f = imread('lena.jpg');% input the original image
f = rgb2gray(f);
f = im2double(f);
figure; subplot(2,3,1); imshow(f); title('Original Image');
g = imfilter(f,psf,'circular','conv'); % realize the defocus blur
subplot(2,3,2); imshow(g); title('Motion Blur Image');
G = abs(fft2(g));
subplot(2,3,3); imshow(fftshift(log(G)),[],'InitialMagnification','fit'); title('Fourier Spectrum');
Cep = fftshift(ifft2(log10(abs(1+G)))));
subplot(2,3,4); imshow(Cep,[],'InitialMagnification','fit'); title('Cepectrum');
minmum = min(Cep(:));
[R,C] = find(Cep==minmum);
[rows,cols] = size(Cep);
row1s = rows;
col1s = cols;
m = 0.5;
if(length(R)==1&&length(C)==1)
    R(2) = 0;
    C(2) = 0;
    row1s = 0;
    col1s = 0;
    m = 0;
end
retrive_len = (1      -m)*sqrt((R(1)-rows/2)^2+(C(1)-cols/2)^2 )+m*sqrt((R(2)      -row1s/2)^2+(C(2)-
    col1s/2)^2 );
retrive_theta = (1      -m)*acot(abs(C(1)-cols/2-1)/abs(R(1)-rows/2-1))*180/pi+m*acot(abs(C(1)-cols/2-
    1)/abs(R(1)-rows/2-1))*180/pi;
ipsf = fspecial('motion',retrive_len,retrive_theta);
f_restored_LR = deconvlucy(g,ipsf,50);
subplot(2,3,5); imshow(f_restored_LR); title('Lucy-Richardson Method');
f_restored_WNR=deconvwnr(g,ipsf); % perform Wiener filtering restoration
subplot(2,3,6); imshow(f_restored_WNR); title('Wiener Filter Method');
```

Figure 5.15 shows the process of restoration of motion blurred images.
Here is the result: the blur scale *len* $= 39.9625$ and the blur angle *theta* $= 29.7449$.

Original Image

Motion Blur Image

Fourier Spectrum

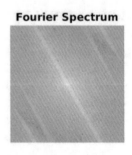

Cepectrum

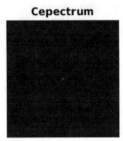

Lucy-Richardson Method

Wiener Filter Method

Fig. 5.15 The restoration of motion blurred images

5.4   Fisheye Distortion Correction Using Spherical Coordinates Model

The viewing angle of the fisheye lens is about 180°–270°, and it works in a staring manner without using the machinery rotating and scanning, which is of small volume, low cost, and low light energy loss. At present, many computer vision areas such as mobile robot automatic navigation, video conferencing, monitoring and virtual reality applications require the use of wide-angle or fisheye cameras with large field of view, so the fisheye lens has become more and more popular. However, the images captured by fisheye camera will have very serious distortion. Therefore, it is necessary to correct the images acquired by fisheye lens in most applications.

The fisheye lens correction algorithm is based on the distortion model of fisheye lens, considering various distortion types of fish eye lens, such as the common radial deformation, decentering distortion, thin prism deformation and so on to formulate an accurate calibration model, and then get the internal and external parameters of fisheye lens through the experimental and objective function, so as to achieve the purpose of accurate restoration of fisheye image deformation. Fisheye image restoration algorithms generally include two categories:

Firstly, the analysis of fisheye lens imaging was conducted from the angle of two projection models—spherical projection model and paraboloid imaging model.

(1) The spherical projection model regards the fisheye lens imaging surface as a spherical surface. This method requires knowing the optical center of the fisheye image and the radius of the transformed sphere in advance. Therefore, it is only applicable to fisheye images with circular areas.

(2) The paraboloid imaging model regards the imaging surface of the fisheye lens as a paraboloid. More precise effects can be obtained when the depth of the scene is restored, which is generally used to restore the depth information from fisheye photos due to the calculation is too complex.

The second type of analysis is carried out from the perspective whether the fisheye distortion correction is in 2D or 3D space [8–10].

(1) 2D fisheye image distortion correction determines the transformation of the coordinates between the deformed image and the corresponding points on the image to be corrected directly, and then carries out the gray interpolation of pixels. This method includes spherical coordinate positioning, polynomial coordinate transformation and its improvement, projective invariance and the correction of fisheye distortion with polar radius mapping.

(2) 3D fisheye image distortion correction includes two methods: projection transformation and fisheye lens calibration. Projection transformation algorithm is to map each 2D image plane point (x, y) on the fisheye image to the 3D plane (x, y, z), then projected to the point (x', y') on the 2D plane. Finally, the correction is implemented according to the relationship between the pixels of the image and the 3D vector of the corresponding ray.

This section mainly introduces the 2D spherical coordinate positioning method, this method is a typical fast two-dimensional fisheye image correction algorithm. Firstly, the center point and the standard circle transform of the fisheye image are calculated and then the spherical coordinate is positioned. The distorted scenes in the fisheye image can be represented by the longitude in Fig. 5.15, where the different pixels on each longitude have the same column coordinate values in the distortion corrected image, as the figure shows, although point h and point k are in a different vertical and horizontal coordinates, they have the same x coordinate after the correction. The greater the longitude of the warp, the greater the degree of distortion is. For any y_i coordinate in the vertical direction of the image, the angle difference between the left and right sides of the sphere is the same, and the corresponding line segment d_x divides the longitude uniformly in the x-axis direction, which makes the distance between the x-axis direction on the longitude of different y_i is equal. Thus, we could obtain the x coordinate of point h by point k according to the scaling relations among images.

Fig. 5.16 Obtain the radius
and center point of the
circular area

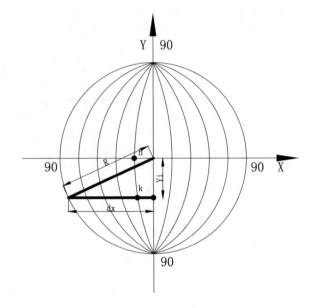

$$\frac{x_k}{d_x} = \frac{x_h}{R} \Rightarrow x_k = \frac{x_h}{R} \times \sqrt{R^2 - Y_i^2} \qquad (5.28)$$

where R is the radius of fisheye distortion image, x_h is the distance between point
h and point O, the center of the image, in x-axis direction, while x_k is the distance
between point k and point O. For those fisheye images whose horizontal view is not
180°, it can also be corrected by the above method after the correction of the standard
circle. The steps may be summarized by the following procedure:

(1)   In order to obtain the radius and center point of the circular area, we need to
      determine the edge and divide the edge out firstly. Then calculate the brightness
      of all image pixels, and set the threshold value, looking for the upper and lower
      boundary through the loop, identifying the scope of the circle and the center
      coordinates and radius can be obtained eventually (Fig. 5.16).
(2)   Find the corresponding point of the distorted plane center in the correction plane,
      then calculate the coordinates of any point on the circle corresponding to the
      correction plane according to Eq. (5.28) (Fig. 5.17).

   The flow chart of fisheye image correction algorithm based on spherical coordinate
positioning is as follows:
   MATLAB implementation of fisheye distortion correction based on spherical
coordinate positioning is shown in PROGRAMME 5.8.

Fig. 5.17 The flow chart of
fisheye image correction
algorithm based on spherical
coordinate positioning

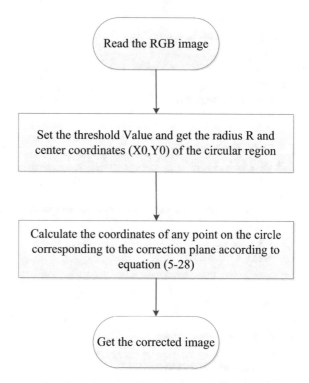

Read the RGB image

Set the threshold Value and get the radius R and
center coordinates (X0,Y0) of the circular region

Calculate the coordinates of any point on the circle
corresponding to the correction plane according to
equation (5-28)

Get the corrected image

PROGRAMME 5.8: The fisheye distortion correction based on spherical coordinate positioning

```
rgb = imread('1.jpg');
%rgb = imread('10.jpg');
[ height, width, v ] = size(rgb);
[I,X0,Y0,R] = Yuan(rgb); % return the coordinates and radius of the circular area
%% find the corresponding point of the distorted plane center in the correction plane
x0=floor(X0);
y0=floor(Y0);
r = round( R);
u0 = x0;
v0 = y0;
Image = zeros ( height,width,v );
Image = uint8 ( Image );
%%take any point p (x, y) on the imaging plane image, and x and y are all integers
for y = 1 : height
    Delta = floor( sqrt( r ^ 2 - ( y - y0 ) ^ 2 ) );
    Delta=real(Delta);
  for x = x0 - Delta : x0 + Delta
    det=floor(sqrt(r^2-(y-y0)^2));
```

```
dx=x-x0;
if dx==0% there is no distortion on the y-axis
    v = y;
    u = x;
else if dx<0
    u=round(x0-((x0-x)/det)*r);
    v=y;
    else
    u=round(x0+((x-x0)/det)*r);
    v=y;
    end
end
        if v > height || v < 1 || u > width || u < 1 % if the parameters is out of the picture, then leave it
continue
        end
    Image( v, u, :)=rgb( y, x,:);
end
end
%figure, imshow ( Image) ;
subplot(121)
imshow(rgb), title(' the original image ')
subplot(122)
imshow(Image),title(' corrected image ');
function [I,X0,Y0,R]=Yuan(rgb)    % Define the function to obtain the center coordinates and radius of
    the circular area
r=rgb(:,:,1);
g=rgb(:,:,2);
b=rgb(:,:,3);
I=0.59*r+0.11*g+0.3*b; % the pixel brightness calculation formula
I=uint8(I);
[Height,Width]=size(I);
Thre=46;   % set the threshold value
for Row1=1:(Height/2);   % look for the upper boundary of circle region through the loop
    CurRow_Bright=I(Row1,:);
    Max=max(CurRow_Bright); % find the maximum brightness value
    Min=min(CurRow_Bright); % find the minimum brightness value
   Lim=Max-Min;   % the limiting luminance difference of the scan line
      if (Lim>Thre),
          Ybot=Row1;
          break;
```

```
        end
    end
    for Row2=Height:-1:(Height/2);   % look for the lower boundary of circle region through the loop
        CurRow_Bright=I(Row2,:);
        Max=max(CurRow_Bright); % find the maximum brightness value
        Min=min(CurRow_Bright); % find the minimum brightness value
        Lim=Max-Min;   % the limiting luminance difference of the scan line
        if (Lim>Thre)
            Ytop=Row2;
            break;
        end
    end
    for Col1=1:(Width/2);   % look for the left boundary of circle region through the loop
        CurCol_Bright = I(:,Col1);
        Max=max(CurCol_Bright); % find the maximum brightness value
        Min=min(CurCol_Bright); % find the minimum brightness value
        Lim=Max-Min;   % the limiting luminance difference of the scan line
        if (Lim>Thre),
            Xleft=Col1;
            break;
        end
    end
    for Col2=Width:-1:Width/2;   % look for the right boundary of circle region through the loop
        CurCol_Bright = I(:,Col2);
        Max=max(CurCol_Bright); % find the maximum brightness value
Min=min(CurCol_Bright); % find the minimum brightness value
        Lim=Max-Min;   % the limiting luminance difference of the scan line
        if (Lim>Thre),
            Xrig=Col2;
            break;
        end
    end
    X0=(Xleft+ Xrig)/2;
    Y0=(Ytop+Ybot)/2;
    Rx=floor((Xrig-Xleft)/2);
    Ry=floor((Ytop-Ybot)/2);
    R=max(Rx,Ry);
end
```

Figures 5.18 and 5.19 simulate the distortion correction of two original fisheye lens images. The spherical coordinate positioning algorithm is a correction method relatively rough, and the final processing result is not satisfactory.

orginal image

corrected image

Fig. 5.18   The corrected image (a): based on Spherical coordinate positioning

orginal image

corrected image

Fig. 5.19   The corrected image (b): based on Spherical coordinate positioning

5.5   Skew Correction of Text Images

The premise of skew correction of text images is to detect the skew of the document image correctly, which is the inclination angle of the document image. At present, the following four methods are used to detect the inclination angle: projection profile analysis, connected component analysis, Fourier transform and Hough transform. The projection profile analysis method calculates the inclination angle by computing the cost function of the projection histogram from different angles of the document image; the connected component analysis divides the document image into different connected components, and determines the inclination angle by analyzing the characteristics of different connected components; the Fourier transform method determine the inclination angle by calculating the maximum direction of spatial density in Fourier transform of the document image, while the Hough transform algorithm selects the peak in Hough space to determine the inclination angle.

In this section, the Hough transform method is used to determine the inclination angle, so as to achieve the purpose of skew correction.

5.5.1 Feature Analysis of Text Images

Before making the skew correction, it is necessary to make clear the characteristics of the text image:

(1) The background of the image is the paper grayscale pattern, and the foreground is the text of image information.
(2) There are two situations of image skew: foreground skew or foreground and background skew simultaneously, which need to be considered comprehensively;
(3) The layout is dominated by horizontal lines, and there may be some defects and adhesions between characters.
(4) There are many symbols coexist in the text, as well as different fonts whose sizes are different.
(5) The character spacing between the same text lines may be somewhat jumping, but the upper and lower boundaries of each character are consistent.

5.5.2 The Basic Idea of Hough Transform

Assume that we draw a straight line on a black-and-white image and we need to know the specific location of the line. Obviously, the equations of the straight line can be expressed by $y = kx + b$, where the parameters k and b represent the slope and intercept, respectively. The parameters of all the lines passing through a point (x_0, y_0) satisfy the equation $y_0 = kx_0 + b$, that is, the point (x_0, y_0) determines a cluster of straight lines. The equation $y_0 = kx_0 + b$ is a straight line on the parameter plane $k - b$ (or the straight line corresponding to the equation $b = -kx_0 + y_0$). Thus, a foreground pixel on the image plane $x - y$ corresponds to a straight line on the parameter plane. Here is an example to illustrate the principle of solving the above problem: suppose that the line on the image is $y = x$, take three points: $A(0, 0)$, $B(1, 1)$, $C(2, 2)$, it can be found that the parameters of the straight line passing through the point A should satisfy the equation $b = 0$, and the line passing through point B has to satisfy the equation $1 = k + b$, while the line passing through point C has to satisfy the equation $2 = 2k + b$, these three equations correspond to three straight lines on the plane of the parameter and they intersect at one point where $k = 1$ and $b = 0$, and the straight line on the parameter plane corresponding to the other points on the line $y = x$ (such as $(3, 3)$, $(4, 4)$ and so on) will also pass through the point $(k = 1, b = 0)$. This property provides a way to solve the problem by mapping the points on the image plane to the line on the parameter plane, and finally solving the problem by statistic characteristics. If there are two straight lines on the image plane, then there will be two peaks in the parameter plane, and the rest may be deduced by analogy.

The basic idea of Hough transform is to convert the line detection problem in image space into the local maximum search problem in parameter space (ρ, θ) by using the duality of point and line. The basic strategy of Hough transform is to use the coordinates of the image space target pixel to calculate the possible trajectory of the reference point (ρ_j, θ_j) in the parameter space. Then, the reference points are counted in an accumulated matrix $A_{n \times m}$, and the straight lines in the image space are determined by the determination of the counter. The element in the accumulated matrix $A_{n \times m}$ corresponds to the reference point (ρ_j, θ_j) in the parameter space, denoted by $A(\rho_j, \theta_j)$.

If the element $A(\rho_j, \theta_j)$ in the accumulated matrix $A_{n \times m}$ satisfies the preset threshold condition, the value (ρ_j, θ_j) of the element will define a straight line in the image space. If there is a straight line in the image, there must exist an element in the accumulation matrix corresponding to this line, which is the maximum local value—$A_{n \times m}(\rho_j, \theta_j)$. In the text image, the text line has a strong direction. Thus, in the cumulative matrix, there is a column that has a local maximum. In text images, each text line has a strong directionality. Thus, in the accumulated matrix $A_{n \times m}$, there is a θ_j column that enables $\sum_{j=1}^{n} A(\rho_j, \theta_j)$ to get the local maximum value.

5.5.3   The Implementation Steps of Text Images Skew Correction

The implementation steps can be divided into two parts: the first part is the image preprocessing, including: image text dilation and thinning; the second part is the Hough transform and obtaining the inclination angle (Fig. 5.20).

MATLAB implementation of text images skew correction is shown in PROGRAMME 5.9.

PROGRAMME 5.9: Text images skew correction

```
I=imread('E:\1001.jpg'); % input the original image
A=im2bw(I);
se=strel('ball',12,0); % Defining the structural elements
BW=imdilate(A,se); % image text dilation
BW2=bwmorph(BW,'thin',Inf); % image text thinning
[H,T,R]=hough(BW2); %Hough transform
P=houghpeaks(H,5); %extract the peak point
lines=houghlines(BW2,T,R,P); % extract the straight line
for k=1:length(lines)
    z=[lines(k).point1;lines(k).point2]; % extract the coordinate
end
m=(z(2,2)- z(1,2))/(z(2,1)-z(1,1)); % calculate the slope of the line
M=atan(m); % calculate the angle
M=M*180/3.14;
C=imrotate(A,M); % Image correction
subplot(1,2,1);
imshow(A);
title(' original image');
subplot(1,2,2);
imshow(C);
title(' the corrected image ');
```

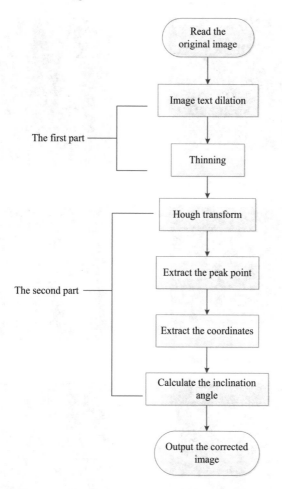

Fig. 5.20 The implementation flow chart of text images skew correction

The experiment is simulated in the MATLAB platform, with a simulation of the rubbings as an example to implement the text image skew correction. For the inclination angle of the image, there has the following four conditions (Figs. 5.21, 5.22, 5.23 and 5.24).

From the above four situations, it can be seen that the correction angle of this method is between $-90° < \theta < 90°$, the correction effect and the characteristics of the image have a certain relationship, which is also one of the shortcomings of Hough transform skew correction.

Fig. 5.21   The inclination angle $0° < \theta < 90°$

Fig. 5.22   The inclination angle $-90° < \theta < 0°$

Fig. 5.23   The inclination angle $90° < \theta < 180°$

Fig. 5.24 The inclination angle $180° < \theta < 270°$

5.6 Image Dehazing Correction

Haze [11–13] can significantly degrade the imaging quality of outdoor visible light sensor due to a series of reactions, such as scattering, refraction, and absorption between particles or water droplets and light from the atmosphere. Image dehazing is an important issue in many scene understanding applications such as surveillance systems, intelligent vehicles, satellite imaging, or target identification and feature extraction.

Image dehazing remains a challenge due to the unknown scene depth information. Early works treated the problem of weather-degraded image restoration as yet another instance of image contrast enhancement. Conventional contrast enhancement filters such as histogram stretching and equalization, linear mapping, or gamma correction are limited to perform the dehazing task, introducing halos artifacts and distorting the color.

Recently, various methods have been proposed to enhance the visibility of the hazy image. Those methods can be classified into two categories: multiple images processing and single image processing. In many cases, it is impossible to acquire multiple images. Thus, single image dehazing methods have been attracted increasing attention in recent years. However, it is a great challenge for single image dehazing due to its ill-posed nature. Many single image dehazing approaches were proposed, yet they required additional information about the input scene.

5.6.1 Single Image Dehazing

Significant progress has been made on single image dehazing in recent years. Tan removed the haze in image by maximizing local contrast of the restored image and his results were visually compelling. However, it is tended to be over-saturated and not be physically valid. Fattal estimated the scene albedo and then inferred the medium transmission, under the assumption that the transmission and surface shading should be locally uncorrelated. However, this approach cannot well handle heavy haze images. He proposed the Dark Channel Prior (DCP) model to estimate optical transmission based on the observation that a haze-free pixel generally contains one or more RGB color channels being black or nearly black.

After the work of DCP by He, many approaches under the DCP framework were developed rapidly because of their simple implementation and satisfactory performance. However, those methods have several disadvantages.

Firstly, those methods may suffer from color bias. Due to the Rayleigh's law, scattering is more intense in the blue band, which makes the dehazed image appear to have a blue hue. Li proposed a prior named change of detail for single image dehazing, which was based on the local detail information rather than the color information. Zhu proposed a simple but powerful color attenuation prior for single image dehazing.

Secondly, those methods tend to overestimate the thickness of haze, causing the dehazed images to be too dark, especially in sky regions. Tang systematically investigated different haze relevant features in a learning framework to identify the best feature combination for single image dehazing.

Thirdly, patch-based approach such as the DCP model alleviates the white object problem of pixel-wise approach16, but induces halo effects. Ancuti and Wang proposed multi-scale fusion methods that could compensate the results for both pixel-wise and patch-based approaches.

Besides, those methods have block effect since the assumption of the DCP model, that the transmissions are constant in a patch, is not always true. And the transmissions between adjacent blocks are in discontinuity. The block effect may lead to erroneous result, especially in the region of sudden change depth. He proposed soft matting to refine the transmissions. However, it is quite time-consuming. Therefore, He proposed the real-time guided filtering to refine the transmissions.

5.6.2   Dark Channel Prior

In the field of computer vision and computer graphics, Narasimhan's lighting model widely used to describe the formation of a hazy image is

$$I(x) = J(x)t(x) + A(1 - t(x)) \tag{5.29}$$

where $I(x)$ is the hazy image, $J(x)$ is the scene radiance, A is the global atmospheric light, and $t(x)$ ($0 \leq t(x) \leq 1$) is the scene transmission.

He proposed the dark channel prior for single image dehazing, in which the prior comes from the observation that most non-sky patches in outdoor haze-free images have at least one color channel with some low intensity pixels. For an arbitrary image $J(x)$, its dark channel is given by

$$J^{dark}(x) = \min_{c \in (r,g,b)} (\min_{y \in \Omega(x)} (min(J^c(y))) \tag{5.30}$$

where J^c is a color channel of $J(x)$, and $\Omega(x)$ is a local window patch centered at pixel x. Dark channel is the outcome of two minimum operators: $\min_{c \in (r,g,b)}$ is performed on each pixel in the RGB color space, and $\min_{y \in \Omega(x)}$ is a minimum filter. If $J(x)$ is an outdoor haze-free image, then the intensity of $J(x)$'s dark channel is very low and tends to zero: $J^{dark}(x) \to 0$.

It assumes that the atmospheric light A would be a given constant value. First the top 0.1% brightest pixels in the dark channel are picked, and then the pixels with highest intensity in the input image I are selected as the atmospheric light.

According to Eq. (5.1), the hazed image can be normalized by A

$$\frac{I^c(x)}{A^c} = t(x)\frac{J^c(x)}{A^c} + 1 - t(x) \tag{5.31}$$

It assumes that the transmission in a local patch $\Omega(x)$ would be constant $\tilde{t}(x)$. The dark channel is calculated as follows

$$\min_{y \in \Omega(x)} (\min_c \frac{I^c(y)}{A^c}) = \tilde{t}(x) \min_{y \in \Omega(x)} (\min_c \frac{J^c(y)}{A^c}) + 1 - \tilde{t}(x) \tag{5.32}$$

The transmission can be estimated by

$$\hat{t}(x) = 1 - \min_{y \in \Omega(x)} (\min_c \frac{I^c(y)}{A^c}) \tag{5.33}$$

According to the DCP model, the transmission can be estimated by

$$\hat{t}(x) = 1 - \omega \min_{y \in \Omega(x)} (\min_{c \in (R,G,B)} \frac{I^c(y)}{A^c}) \tag{5.34}$$

where $\Omega(x)$ is a local window patch centered at pixel x, ω is a constant parameter ($0 \le \omega \le 1$) to keep a small amount of haze and A^c is the RGB color channel of the atmospheric light.

With the transmission map, we can recover the scene radiance according to Eq. (5.29). But the direct attenuation term $J(x)t(x)$ can be very close to zero when the transmission $t(x)$ is close to zero. The directly recovered scene radiance J is prone to noise. Therefore, we restrict the transmission $t(x)$ to a lower bound t_0, which means that a small certain amount of haze are preserved in very dense haze regions.

The final scene radiance $J(x)$ is recovered by:

$$J(x) = \frac{I(x) - A}{\max(t(x), t_0)} + A \tag{5.35}$$

A typical value of t_0 is 0.1. Since the scene radiance is usually not as bright as the atmospheric light, the image after haze removal looks dim.

5.6.3   *Implementation Steps of DCP*

Implementation of DCP needs following steps:

(1)   compute the dark channel with Eq. (5.30);
(2)   computer the atmosphere light A;
(3)   estimate the transmission with Eq. (5.34);
(4)   recover the scene radiance with Eq. (5.35);

PROGRAMME 5.10: Image Dehazing with DCP

```
%%%%%% the main program for DCP dehazing
warning('off','all');
image = double(imread('forest.jpg'))/255;
image = imresize(image, 0.1);
result = dehaze(image, 0.95, 15);
figure, imshow(image)
figure, imshow(result)
warning('on','all');
%%%%%% call different functions to dehaze
function [ radiance ] = dehaze( image, omega, win_size, lambda )
%DEHZE Summary of this function goes here
if ~exist('omega', 'var')
    omega = 0.95;
end
if ~exist('win_size', 'var')
    win_size = 15;
end
if ~exist('lambda', 'var')
    lambda = 0.0001;
end
[m, n, ~] = size(image);
dark_channel = get_dark_channel(image, win_size);
atmosphere = get_atmosphere(image, dark_channel);
trans_est = get_transmission_estimate(image, atmosphere, omega, win_size);
radiance = get_radiance(image, trans_est, atmosphere);
end


%%%%%% computer the atmosphere light A
function atmosphere = get_atmosphere(image, dark_channel)
    [m, n, ~] = size(image);
    n_pixels = m * n;
    n_search_pixels = floor(n_pixels * 0.01);
    dark_vec = reshape(dark_channel, n_pixels, 1);
    image_vec = reshape(image, n_pixels, 3);
    [~, indices] = sort(dark_vec, 'descend');
    accumulator = zeros(1, 3);
```

```
        for k = 1 : n_search_pixels
            accumulator = accumulator + image_vec(indices(k),:);
        end
        atmosphere = accumulator / n_search_pixels;
end
%%%%%% compute the dark channel
function dark_channel = get_dark_channel(image, win_size)
    [m, n, ~] = size(image);
    pad_size = floor(win_size/2);
    padded_image = padarray(image, [pad_size pad_size], Inf);
    dark_channel = zeros(m, n);
    for j = 1 : m
        for i = 1 : n
            patch = padded_image(j : j + (win_size-1), i : i + (win_size-1), :);
            dark_channel(j,i) = min(patch(:));
        end
    end
end
%%%%%% estimate   the transmission
function trans_est = get_transmission_estimate(image, atmosphere, omega, win_size)
    [m, n, ~] = size(image);
    rep_atmosphere = repmat(reshape(atmosphere, [1, 1, 3]), m, n);
    trans_est = 1 - omega * get_dark_channel( image ./ rep_atmosphere, win_size);
end
%%%%%% recover image
function radiance = get_radiance(image, transmission, atmosphere)
    [m, n, ~] = size(image);
    rep_atmosphere = repmat(reshape(atmosphere, [1, 1, 3]), m, n);
    max_transmission = repmat(max(transmission, 0.1), [1, 1, 3]);
    radiance = ((image - rep_atmosphere) ./ max_transmission) + rep_atmosphere;
end
```

Figure 5.25c is the estimated transmission map from an input haze image (Fig. 5.25a) using the patch size 15×15. It is roughly good but contains some block effects since the transmission is not always constant in a patch. The recovered scene radiance in Fig. 5.25b is not smooth. We need refine this map for better image quality.

5.6.4 Refine Transmission Map Using Soft Matting

We notice that the haze imaging Eq. (5.29) has a similar form with the image matting equation. A transmission map is exactly an alpha map. Therefore, we apply a soft matting algorithm to refine the transmission.

Denote the refined transmission map by $t(x)$. Rewriting $t(x)$ and $\tilde{t}(x)$ in their vector form as t and $\tilde{t}$, minimize

$$E(t) = t^T L t + \lambda (t - \tilde{t})^T (t - \tilde{t}) \qquad (5.36)$$

Fig. 5.25 Dehazing result
with DCP

where L is the Matting Laplacian matrix, and λ is a regularization parameter. The first term is the smooth term and the second term is the data term.

The (i, j) element of the matrix L is defined as:

$$\sum_{k|(i,j)\in w_k} \left(\delta_{ij} - \frac{1}{|w_k|}\left(1 + (I_i - u_k)^T\left(\Sigma_k + \frac{\varepsilon}{|w_k|}U_3\right)^{-1}(I_j - u_k)\right)\right) \tag{5.37}$$

where I_i and I_j are the colors of the input image I at pixels i and j, δ_{ij} is the Kronecker delta, u_k and Σ_k are the mean and covariance matrix of the colors in window w_k, U_3 is a 3×3 identity matrix, ε is a regularizing parameter, and $|w_k|$ is the number of pixels in the window w_k.

The optimal t can be obtained by solving the following sparse linear system:

$$(L + \lambda U)t = \lambda \tilde{t} \tag{5.38}$$

where U is an identity matrix of the same size as L.

Implementation of DCP + soft Matting needs following steps:

(1) Compute the dark channel with Eq. (5.30);
(2) Compute the atmosphere light A;
(3) Estimate the transmission with Eq. (5.34);
(4) Refine the transmission with the solution of Eq. (5.38);
(5) recover the scene radiance with Eq. (5.35);

PROGRAMME 5.11: Image Dehazing with DCP + Soft Matting

```
%%%%%% the main program for DCP dehazing
    warning('off','all');
    image = double(imread('forest.jpg'))/255;
    image = imresize(image, 0  .1);
    result = dehaze(image, 0.95, 15);
    figure, imshow(image)
    figure, imshow(result)
    warning('on','all');
%%%%%% call different functions to dehaze
function [ radiance ] = dehaze( image, omega, win_size, lambda )
    %DEHZE Summary of this function goes here
    if ~exist('omega', 'var')
        omega = 0.95;
    end
    if ~exist('win_size', 'var')
        win_size = 15;
    end
    if ~exist('lambda', 'var')
        lambda = 0.0001;
    end
    [m, n, ~] = size(image);
    dark_channel = get_dark_channel(image, win_size);
```

```
atmosphere = get_atmosphere(ima      ge, dark_channel);
trans_est = get_transmission_estimate(image, atmosphere, omega, win_size);
L = get_laplacian(image);
A = L + lambda * speye(size(L));
b = lambda * trans_est(:);
x = A \ b;
transmission = reshape(x, m, n);
radiance = get_radiance(image, transmission, atmosphere);
end

function [ L ] = get_laplacian( image )
%GET_LAPLACIAN
[m, n, c] = size(image);
img_size = m*n;
win_rad = 1;
epsilon = 0.0000001;
max_num_neigh = (win_rad*2+1)^2;
ind_mat = reshape( 1:img_size, m, n);
indices = 1 : (m*n);
num_ind = length(indices);
max_num_vertex = max_num_neigh * num_ind;
row_inds = zeros( max_num_vertex, 1 );
col_inds = zeros( max_num_vertex, 1 );
vals = zeros( max_num_vertex, 1 );
len = 0;
for k = 1 : length(indices);
    ind = indices(k);
    [i, j] = ind2sub( [m n], ind );
    m_min = max( 1, i - win_rad );
    m_max = min( m, i + win_rad );
    n_min = max( 1, j - win_rad );
    n_max = min( n, j + win_rad );
    win_inds = ind_mat( m_min : m_max, n_min : n_max );
    win_inds = win_inds(:);
    num_neigh = size( win_inds, 1 );
    win_image = image( m_min : m_max, n_min : n_max, : );
    win_image = reshape( win_image, num_neigh, c );
    win_mean = mean( win_image, 1 );
    win_var = inv( (win_image' * win_image / num_neigh) - (win_mean' * win_mean) +
(epsilon / num_neigh * eye(c) ) );
    win_image = win_image - repmat( win_mean, num_neigh, 1 );
    win_vals = ( 1 + win_image * win_var * win_image' ) / num_neigh;
    sub_len = num_neigh*num_neigh;
    win_inds = repmat(win_inds, 1, num_neigh);
    row_inds(1+len: len+sub_len) = win_inds(:);
    win_inds = win_inds';
    col_inds(1+len: len+sub_len) = win_inds(:);
```

```
    vals(1+len: len+sub_len) = win_vals(:);
    len = len + sub_len;
end

A = sparse(ro w_inds(1:len),col_inds(1:len),vals(1:len),img_size,img_size);
D = spdiags(sum(A,2),0,n*m,n*m);
L = D - A;
End
```

Figure 5.26a is the soft matting result using Fig. 5.25c as the date term. As we can see, the refined transmission map manages to capture the sharp edge discontinuities and outline the profile of the objects. The recovered scene radiance with DCP + Soft Matting in Fig. 5.26b is better than the result with DCP in Fig. 5.25b.

(a)

(b)

Fig. 5.26 Dehazing result with DCP + soft matting

5.7   Image Deraining Correction

Under rainy conditions, the impact of rain streaks on images and video is often undesirable. In addition to a subjective degradation, the effects of rain can also severely affect the performance of outdoor vision systems, such as surveillance systems. Effective methods for removing rain streaks are needed for a wide range of practical applications.

5.7.1   Related Work

To date, many methods have been proposed for removing rain from images. These methods fall into two categories: video-based methods and single-image based methods.

For video-based methods, rain can be more easily identified and removed using inter-frame information. Many of these methods work well, but are significantly aided by the temporal content of video.

Single-image based methods are significantly more challenging since much less information is available for detecting and removing rain. Kim J. H. proposed a method based on kernel regression and a non-local mean filtering to detect and remove rain streaks. Chen Y. L. proposed a generalized model in which additive rain is assumed to be low rank. In general, however, success has been less noticeable than in video-based algorithms and there is still much room.

5.7.2   Single Image De-rain with Deep Detail Network

Fu X. Y. proposed a deep network architecture for removing rain streaks from individual images based on the deep convolutional neural network (CNN). Inspired by the deep residual network (ResNet) that simplifies the learning process by changing the mapping form, Fu proposed a deep detail network to directly reduce the mapping range from input to output, which makes the learning process easier.

We denote the input rainy image and corresponding clean image as X and Y, respectively. When compared to the clean image Y, the residual of the rainy image $Y - X$ has a significant range reduction in pixel values. This implies that the residual can be introduced into the network to help learn the mapping. Thus we use the residual as the output of the parameter layers, as shown in Fig. 5.27. This skip connection can also directly propagate lossless information through the entire network, which is useful for estimating the final de-rained image. Because rain tends to appear in

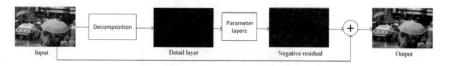

Fig. 5.27 The proposed framework for single-image rain removal

images as white streaks, most values of $Y - X$ tend to be negative. Thus we refer to this as "negative residual mapping" (neg-mapping for short). We train a deep CNN architecture h(X) on multiple images to minimize the objective function

$$L = \sum_i ||h(X_i) + X_i - Y_i||_F^2 \tag{5.39}$$

We first model the rainy image as

$$X = X_{detail} + X_{base} \tag{5.40}$$

where the subscript 'detail' denotes the detail layer, and 'base' denotes the base layer. The base layer can be obtained using low-pass filtering of X after which the detail layer $X_{detail} = X - X_{base}$. After subtracting the base layer from the image, the interference of background is removed and only rain streaks and object structures remain in the detail layer. The detail layer is sparser than the image since most regions in the detail layer are close to zero.

The input of the de-rain system is a rainy image X and the output is an approximation to the clean image Y. Based on the previous discussion, we define the objective function to be

$$L = \sum_{i=1}^{N} ||f(X_{i,detail}, W, b) + X_i - Y_i||_F^2 \tag{5.41}$$

where N is the number of training images, $f(\bullet)$ is ResNet. W and b are network parameters that need to be learned. For X_{detail}, we first use guided filtering as a low-pass filter to split X into base and detail layers.

Network architecture for the rain removal problem is shown in Fig. 5.28. Removing image indexing, our basic network structure can be expressed as,

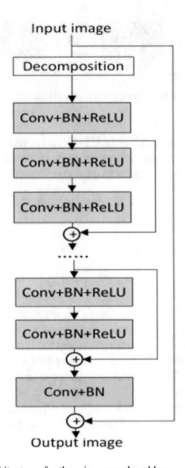

Fig. 5.28 The network architectures for the rain removal problem

$$X^0_{detail} = X - X_{base}$$
$$X^1_{detail} = \sigma(BN(W^1 * X^0_{detail} + b^1))$$

$$\vdots$$

$$X^{2l}_{detail} = \sigma(BN(W^{2l} * X^{2l-1}_{detail} + b^{2l})) \tag{5.42}$$
$$X^{2l+1}_{detail} = \sigma(BN(W^{2l+1} * X^{2l}_{detail} + b^{2l+1})) + X^{2l-1}_{detail}$$

$$\vdots$$

$$Y_{approx} = BN(W^L * X^{L-1}_{detail} + b^L) + X$$

where $l = 1, \ldots, \frac{L-1}{2}$ with L the total number of layers, $*$ indicates the convolution operation, W contains weights and b biases, $BN(\bullet)$ indicates batch normalization to alleviate internal covariate shift, $\sigma(\bullet)$ is a Rectified Linear Unit (ReLU) for non-linearity. In this network, all pooling operations are removed to preserve spatial information.

For the first layer, we use filters of size $c \times s_1 \times s_1 \times a_1$ to generate a_1 feature maps; s represents filter size and c represents the number of image channels, e.g., c = 1 for gray-scale and c = 3 for color image. For layers 2 through $L - 1$, filters are size $a_1 \times s_2 \times s_2 \times a_2$. For the last layer, use filters of size $a_2 \times s_3 \times s_3 \times c$ to estimate the negative residual. The de-rained image is obtained by directly adding the estimated residual to the rainy image X.

5.7.3 Implementation of Image Deraining with Deep Network

Set the detail network depth to L = 26, and use SGD (Stochastic Gradient Descent) with weight decay of 10^{-10}, momentum of 0.9 and a mini-batch size of 20. Start with a learning rate of 0.1, dividing it by 10 at 100K and 200K iterations, and terminate training at 210K iterations. Set the filter sizes $s_1 = s_2 = s_3 = 3$ and filter numbers $a_1 = a_2 = 16$. during experiments, Fu X. Y. found that 3×3 filter size generates results that are representative of deep network structure, while still being computationally efficient. Since the process is applied on color images, set c = 3, the radius of the guided filter for low-pass filtering is 15.

PROGRAMME 5.12: Single image de-rain with deep detail network

```
clc;
clear;
addpath '. \fast- guided - filter- code - v1'
run '.\matconvnet \matlab \vl_setupnn'
load('network.mat'); % load trained model
use_gpu = 1; % GPU: 1, CPU: 0
%%% parameters of guidedfilter
r = 16;
eps = 1 ;
s = 4;
%%%
input = im2double(imread('.\image\real_world\3.jpg'));
```

```
base_layer = zeros(size(input)); % base layer
base_layer(:, :, 1) = fastguidedfilter(input(:, :, 1), input(:, :, 1), r, eps, s);
base_layer(:, :, 2) = fastguidedfilter(input(:, :, 2), in put(:, :, 2), r, eps, s);
base_layer(:, :, 3) = fastguidedfilter(input(:, :, 3), input(:, :, 3), r, eps, s);

detail_layer = input  - base_layer; % detail layer

output   = processing( input, detail_layer, model, use_gpu ); % perform de      - raining

%%%% If your Nvidia GPU encounters an "out of memory", try below code and make a small
"max_patch_size"
% max_patch_size = 120;
% output = processing_patch( input,detail_layer, model, use_gpu, max_patch_size );
%%%%
figure,imshow([input,output]);
title('Left: rainy image       Right: de  - rained result');
%%%%%%%%%%%%%%%%%%
function q = fastguidedfilter(I, p, r, eps, s)
%     GUIDEDFILTER     O(1) time implementation of guided filter.
%
%     - guidance image: I (should be a gray    - scale/single channel image)
%     - filtering input image: p (should be a gray    - scale/single channel image)
%     - local window radius: r
%     - regularization parameter: eps
%     - subsampling ratio: s (try s = r/4 to s=r)

    I_sub = imresize(I, 1/s, 'nearest'); % NN is often enough
    p_sub = imresize(p, 1/s  , 'nearest');
    r_sub = r / s; % make sure this is an integer

    [hei, wid] = size(I_sub);
    N = boxfilter(ones(hei, wid), r_sub); % the size of each local patch; N=(2r+1)^2 except
    for boundary pixels.

    mean_I = boxfilter(I_sub, r_sub) ./ N;
    mean_p = boxfilter (p_sub, r_sub) ./ N;
    mean_Ip = boxfilter(I_sub.*p_sub, r_sub) ./ N;
    cov_Ip = mean_Ip  - mean_I .* mean_p; % this is the covariance of (I, p) in each local
    patch.

    mean_II = boxfilter(I_sub.*I_sub, r_sub) ./ N;
var_I = mean_II - mean_I .* mean_I;


a = cov_Ip ./ (var_I + eps);
b = mean_p  - a .* mean_I;
```

```
        var_I = mean_II - mean_I .* mean_I;

        a = cov_ Ip ./ (var_I + eps);
        b = mean_p  - a .* mean_I;

        mean_a = boxfilter(a, r_sub) ./ N;
        mean_b = boxfilter(b, r_sub) ./ N;

        mean_a = imresize(mean_a, [size(I, 1), size(I, 2)], 'bilinear'); % bilinear is recommended
        mean_b = imresize(mean_b, [size(I, 1), size( I, 2)], 'bilinear');

        q = mean_a .* I + mean_b;
end

%%%%  function boxfilter
function imDst = boxfilter(imSrc, r)

%    BOXFILTER   O(1) time box filtering using cumulative sum
%
%    - Definition imDst(x, y)=sum(sum(imSrc(x-r:x+r,y-r:y+r)));
%    - Running time independent of r;
%    - Equivalent to the function: colfilt(imSrc, [2*r+1, 2*r+1],
%  'sliding', @sum);
%    - But much faster.

    [hei, wid] = size(imSrc);
    imDst = zeros(size(imSrc));

    %cumulative sum over Y axis
    imCum = cumsum(imSrc, 1);
    %difference over Y axis
    imDst(1:r+1, :) = imCum(1+r:2*r+1, :);
    imDst(r+2:hei-r, :) = imCum(2*r+2:hei, :) - imCum(1:hei-2*r-
    1, :);
    imDst(hei-r+1:hei, :) = repmat(imCum(hei, :), [r, 1]) -
    imCum(hei-2*r:hei-r-1, :);

    %cumulative sum over X axis
    imCum = cumsum(imDst, 2);
    %difference over Y axis
    imDst(:, 1:r+1) = imCum(:, 1+r:2*r+1);
    imDst(:, r+2:wid-r) = imCum(:, 2*r+2:wid) - imCum(:, 1:wid-2*r-
    1);
    imDst(:, wid-r+1:wid) = repmat(imCum(:, wid), [1, r]) - imCum(:,
    wid-2*r:wid-r-1);
end
```

Figure 5.29 shows an example of a real-world test image and the deraining result with PROGRAMME 5.11.

(a)

(b)

Fig. 5.29 Deraining result with deep network

References

1. Ott HW (1976) Noise reduction techniques in electronic systems. Wiley
2. Vaseghi SV (2000) Advanced digital signal processing and noise reduction. In: Advanced digital signal processing and noise reduction. Wiley, pp 187–192
3. Zhang P, Li F (2014) A new adaptive weighted mean filter for removing salt-and-pepper noise. IEEE Signal Process Lett 21(10):1280–1283

4. Gonzalez RC, Woods RE (2002) Digital image processing. Prentice-Hall, Upper Saddle River, NJ
5. Yuan L, Sun J, Quan L et al (2007) Image deblurring with blurred/noisy image pairs. In: ACM SIGGRAPH. ACM, p 1
6. Yitzhaky Y, Mor I, Lantzman A et al (1998) Direct method for restoration of motion-blurred images. J Opt Soc Am A 30200100(100):1512–1519
7. Wang X, Zhao R (2002) Restoration of motion-blurred images. Proc SPIE Int Soc Opt Eng 4875(1):413–421
8. Rui M, Barreto JP, Falcao G (2012) A new solution for camera calibration and real-time image distortion correction in medical endoscopy-initial technical evaluation. IEEE Trans Biomed Eng 59(3):634–644
9. Ip HHS, Chen Y (2005) Planar rectification by solving the intersection of two circles under 2D homography. Pattern Recognit 38(7):1117–1120
10. Haneishi H, Yagihashi Y, Miyake Y (1995) A new method for distortion correction of electronic endoscope images. IEEE Trans Med Imaging 14(3):548–555
11. He K, Sun J, Tang X (2009) Single image haze removal using dark channel prior. In: IEEE conference on computer vision and pattern recognition, 2009, CVPR 2009. IEEE, pp 1956–1963
12. He K, Sun J, Tang X (2011) Single image haze removal using dark channel prior. IEEE Trans Pattern Anal Mach Intell 33(12):2341–2353
13. Fu X, Huang J, Zeng D, Huang Y et al (2017) Removing rain from single images via a deep detail network. In: IEEE conference on computer vision and pattern recognition, CVPR, 2017. IEEE, pp 3855–3863

Chapter 6
Image Inpainting

In this chapter, the image inpainting algorithms are discussed. We first introduce the principle of image inpainting, and then two types of inpainting algorithms are detailed, including the vibrational PDE-based and exemplar-based inpainting.

6.1 Introduction

Image inpainting [1, 2] refers to the process of filling in the missing areas or modifying the damaged ones in an image, which aims to restore the image in a form that is not detectable by an ordinary observer. In fact, image restoration can only use part of the residual information in the image to approximate the original intact image, and the image obtained by this estimate is just an approximate image of the vision psychology of human eye, which does not really restore the original appearance of the image. Therefore, image inpainting itself is a subjective process, which may generate different results depends on different images or different restoration algorithms and different restorers. Applications of this technique include the restoration of old photographs; cultural relics protection; virtual reality, extra objects removal (removal of superimposed text like dates, subtitles, or publicity, etc.), data compression, network data transmission, etc., which has aroused a lot of attention from scholars in the world.

The conventional schemes that are proposed for image inpainting can be divided into two categories: structure-oriented method and texture-based image inpainting technology.

© Springer International Publishing AG, part of Springer Nature 2019
S. Gong et al., *Advanced Image and Video Processing Using MATLAB*,
Modeling and Optimization in Science and Technologies 12,
https://doi.org/10.1007/978-3-319-77223-3_6

6.1.1   Structure Oriented Image Inpainting Technology

For structure image inpainting, a primary category of the technique is to build up a Partial Differential Equation (PDE). The main idea is to make full use of the edge information around the damaged area, and sperate the information into the area to be restored with the propagating mechanism, so as to obtain a better result of repair. In fact, the PDE-based image inpainting utilizes the thermal diffusion equation in physics to propagate the information around the area to be repaired into the patch area. It transforms the image inpainting process into a series of partial differential equations or energy functional models which can be processed by numerical iterations and intelligent optimization.

The typical image inpainting algorithms based on structure include: Bertalmio-Sapiro-Caselles-Ballester (BSCB) model, Curvature Driven Diffusions (CDD) model [3, 4], Total Variation (TV) model, Euler's elastica model [5], Mumford-Shah model [6], Mumford-Shah-Euler model [7] and so on. Like everything else, the structural inpainting methods have both advantages and disadvantages, this kind of algorithms is only suitable for piecewise smooth images, or for filling images with a small area of damage.

The BSCB model aims to establish an image restoration model with isophote line as extension direction, which keeps the angle between the isophote line and the edge, and fills in the areas to be inpainted by propagating information smoothly from the surrounding areas in the isophotes direction at the same time. This algorithm can get good repair effect when it is damaged or broken in a narrow region. However, due to the characteristics of the algorithm itself, this kind of methods operate slowly and the inpainting image is sometimes blurring. The TV model uses an Euler-Lagrange equation to inpaint the image by minimizing the TV energy functional, coupled with anisotropic diffusion to preserve the direction of isophotes. It works remarkably well for local inpainting such as digital zoom-in and text removal [8–10]. However, the TV model uses the shortest straight line to connect the broken bar structure, it does not connect fracture edges well, so it is easy to destroy the connectivity of the vision during the inpainting process. The CDD model extended the TV algorithm to consider geometric information by defining the strength of isophotes, which enhances the visual connectivity to a certain extent. So, it can inpaint larger damaged regions. Since the CDD model still adopts linear approximation to the damaged area, thus, the damaged boundary will still have the phenomenon of fuzzy or even not smooth. Both the Mumford-Shah model and the Mumford-Shah-Euler model will build the data model and priori model of the image, so that the problem of image painting can be converted into a functional extremum problem, which can be solved with the variational method so as to restore the damaged image.

Since the PDE method itself does not take into account the order of the inpainting sequence, and lack of consideration about high-frequency part of the image. So, it will introduce ambiguity in the propagation process, especially in repairing large damaged area. Besides, as the PDE-based repair method only considers the structure layer of the image, the valuable information in the texture is often blurred by the PDE model, which cannot get good results in the restoration of the texture area.

6.1.2 Texture-Based Image Inpainting Technology

Texture-based image inpainting [11, 12] can grasp the structure and texture details of the image as a whole, and the restoration quality is relatively ideal. Besides, the speed is also superior to the algorithm based on the variational PDE obviously [13], which is mainly used to fill in large patches of missing information in the image. There are two approaches to this type of technology: The first one is based on image decomposition, which decompose the image into structural part and texture parts, then using the BSCB model to restore the structural part and nonparametric sample texture synthesis technique to fill the texture part. At last, the results of these two parts are stacked up, which is the final restored image; The other one is the exemplar-based technique which generates new texture by sampling and copying colour values from the source. Firstly, selecting a pixel point from the boundary of the patch to be repaired as the initial seed, taking this point as the center, select the appropriate texture block according to the texture features of the image, and look for the closest texture matching block around the area to be mended to replace the texture block. Among them, the most representative and creative one of the exemplar-based inpainting algorithms is the Criminisi model. On the basis of the structure and texture information, the algorithm determines the sequence of inpainting according to the value of the priority function, and the value is determined by the confidence items and data items (structure function) of the image patch. Finally, find the optimal matching block in the known part of the image according to certain criteria, update the pixel block to be repaired with the information in the optimal matching block until the entire damaged area is repaired.

6.2 The Principle of Image Inpainting

Image inpainting is a technology based on human visual psychology. According to the edge information of the damaged area, it extends in a certain direction and filling the obscured parts to simulate the effect of artificial inpainting. As most objects are opaque, people often rely on experience to guess which object is obscured. Since the world is considered to be made up of an orderly, complete way, rather than scattered individuals, the modeling process of image inpainting usually relies on the Helmholtz best guessing principle: for the given sensor data, what we feel is the best assumption based on the state of the real world.

Refer to the previous definition, image inpainting is the use of damaged image U_0 to restore original image U. According to the Helmholtz best guessing principle, the inpainting is to find the maximum posterior probability of Bayesian [14–16], that is, to make the largest U of prob($U|U_0$).

According to the Bayesian formula:

$$P(U|U_0) = \frac{P(U)}{P(U_0)} \cdot P(U_0|U) \tag{6.1}$$

If the image U_0 is given, then P (U_0) is a fixed constant which set to C. Where

$$P(U|U_0) = C \cdot P(U) \cdot P(U_0|U) \tag{6.2}$$

From the above, we can see that the estimation of the image U depends on two conditions, namely the relation between the observed images U_0 and U as well as the prior probability of the image based on the best guess. These two conditions correspond to two physical models in the image inpainting respectively:

$P(U|U_0)$: The data model, that is, how does the observed image U_0 get from the original image U.

$P(U)$: A priori model, that is, what does the real image look like.

On the other hand, in most image inpainting problems, the geometric information of the image will get lost, such as the boundary. In order to restore these kinds of information, the model should take advantage of the geometric information of the image, but most conventional probabilistic models fail to do so. However, since some energy models in image processing are driven by geometric information, we could establish the relationship between the probability formula and the energy according to the Gibbs rule.

The specific Gibbs rule is as follows:

$$probability(U) = \frac{1}{z}\exp[-\beta E(U)] \tag{6.3}$$

where E(U) is the energy of U, β denotes the reciprocal of the absolute temperature, z is the distribution function, so the Bayesian formula is expressed in the form of energy or variational:

$$E(U|U_0) = E(U) + E(U_0|U) + const \tag{6.4}$$

When you take the minimum of energy, the constant term can be discarded. E(U) and E($U|U_0$) are equivalent to the prior model and the data model in the probability formula respectively.

6.3   Variational PDE-Based Image Inpainting

From the viewpoint of mathematics, image inpainting is to fill the image in the area
to be mended according to the area to be restored, which belongs to the field of image
restoration. The following degradation model is often adopted:

$$I^0 = I + N \tag{6.5}$$

where I^0 is the observed image, I is the original image ($I = \{I(x)\}$), while N is the
additive white noise. For the most image inpainting problem, the data model has the
following form:

$$I^0|_{\Omega \setminus D} = [I + N]_{\Omega \setminus D} \tag{6.6}$$

where Ω denotes the entire image region [17], D represents the area where the lost
information needs to be patched, $\Omega \setminus D$ denotes the area where no information is lost,
I^0 is the available image portion on $\Omega \setminus D$, while I is the target image that needs to be
restored.

Assuming that N obeys the Gaussian distribution, then the energy function E of
the data model can be defined by the minimum mean square error:

$$E[I^o|I] = \frac{\lambda}{2} \int_{\Omega \setminus D} (I - I^o)^2 dx \tag{6.7}$$

Since there is no data available to D, the image (priori) model is more important
to image inpainting algorithm than other traditional restoration problems (such as
denoising, deblurring). The image model can be obtained from the image data through
filtering, parametric or nonparametric estimation and entropy methods. These statistical methods are important to repair images with rich texture. However, for the
majority of the restoration problems, the important geometric information (such as
the boundary) of the image is often lost in the area to be reconstructed. In order to
restore this geometric information, the image model needs to get these geometric
features in advance, while most traditional probability models lack such characteristics. Fortunately, in many kinds of literature, the 'energy' form inspired by geometric
information does exist, such as the Rudin-Osher-Fatermi model, and the Mumford-
Shah model, the so-called variational method.

In the variational approach, the image inpainting problem is transformed into a
constrained optimization problem:

$$\min E[I]$$

$$s.t. \quad E[I^0|I] \le \sigma^2 \tag{6.8}$$

where E[I] is the energy form of the image prior model, σ^2 indicates the variance of the Gaussian white noise, which can be estimated with an appropriate statistical estimator. Using Lagrange multiplier method, the constraint problem can be transformed into the following unconstrained problem:

$$\min E[I] + \lambda E[I^0|I] \qquad (6.9)$$

Generally, λ is used to equalize the matching term $E[I^0|I]$ and the regularization term $E[I]$. For the regularization term $E[I]$, that is, the prior model of the image is often implemented by the 'energy' functional. Such as Sobolev norm $E[I] = \int_\Omega |\nabla I|^2 dx$, Rudin et al. total variational models $E[I] = \int_\Omega |\nabla I| dx$, Mumford-Shah model $E[I|\Gamma] = \frac{\gamma}{2} \int_{\Omega \backslash \Gamma} |\nabla I|^2 dx + \beta H^1(\Gamma)$, where H^1 represents the 1-dimensional Hausdroff measure and Γ is the edge set of the image.

This section mainly introduces two kinds of the most important variational techniques based on geometric image models and their modified models. In the following content, ∇, div, ∇^2 represent the gradient, divergence and Laplace operator, respectively.

6.3.1 Image Inpainting Algorithm Based on Total Variational Model

Rudin et al. consider the image as a piecewise smooth function, and model the image on bounded variational space. Since the proposed total variation model can extend the image boundary, this model is very suitable for image restoration [18–20]. Tony Chan et al. extended the model to image inpainting, and established a total variational image inpainting model as follows:

$$\min |J[I]| = \int_\Omega |\nabla I| dx + \frac{\lambda}{2} \int_{\Omega \backslash D} |I - I^0|^2 dx \qquad (6.10)$$

where λ plays the role of the Lagrange multiplier. According to the variational [21–24] principle, the corresponding Euler-Lagrange equation for the energy functional J can be obtained as

$$-div[\nabla I / |\nabla I|] + \lambda_D(x)(I - I^0) = 0 \qquad (6.11)$$

where $\lambda_D(x) = \lambda \cdot 1_{\Omega \backslash D}(x) = \begin{cases} \lambda & x \in \Omega \backslash D \\ 0 & x \in D \end{cases}$. Thus it can be seen that the minimum value of solving functional (formula 6.10) is equivalent to solving partial differential equation (formula 6.11). In addition, a time variable t can be introduced, and then the infinitesimal steepest descent equation is given by

Fig. 6.1 Visual connectivity
principle

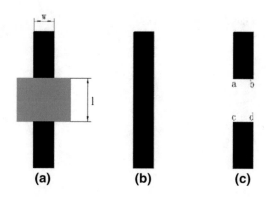

(a)   (b)   (c)

$$\frac{\partial I}{\partial t} = div[\nabla I/|\nabla I|] - \lambda_D(x)(I - I^0) \tag{6.12}$$

That is to say, with the change of the time variable t, when $\frac{\partial I}{\partial t} \to 0$, the minimum value of the required I is obtained.

From the point of view of the numerical calculation, since $|\nabla I|$ will be very small, or even close to zero in the smooth area, so, in the above two differential equations, to avoid the denominator to be zero, generally replace $div[\nabla I/|\nabla I|]$ by $div[\nabla I/|\nabla I|_\varepsilon]$, where $|\nabla I|_\varepsilon = \sqrt{\varepsilon^2 + |\nabla I|^2}$, ε is a small positive parameter. Thus the problem of optimization becomes

$$\min J_\varepsilon[I] = \int_\Omega |\nabla I|_\varepsilon dx + \frac{\lambda}{2} \int_{\Omega \backslash D} |I - I^0|^2 dx \tag{6.13}$$

As in most processing tasks that containing threshold (like denoising and edge detection), parameter ε can usually be considered as a threshold. In the smooth region where $|\nabla I| \leq \varepsilon$, the model tries to imitate the harmonic inpainting, while in the border area where $|\nabla I| \geq \varepsilon$, TV model can be used for processing.

The main advantages of the TV model are its maintenance of the edge and convenient numerical PDE implementation, but it also destroys the connectivity principle of the human disocclusion process. As shown in Fig. 6.1, where ω represents the width of an object, ω_1 represents the width of the damaged area, no matter what the ratio of ω to ω_1 is, the whole bar is shown in Fig. 6.1b seems to be the best guess to most of us, psychologically. However, for the TV model, when $\omega_1 < \omega$, the inpainting result is shown in Fig. 6.1b, while $\omega_1 > \omega$, the inpainting result is shown in Fig. 6.1c, which destroys the connectivity principle. As in the TV model, diffusion strength depends only on the contrast or strength of the isophote line, and it is reflected by the conduction coefficient $v = 1/|\nabla I|$, therefore, the intensity of diffusion is not dependent on the geometric information of the isophote line. For a plane curve, the scalar curvature k can reflect its geometric information. When $\omega_1 > \omega$, from the result of TV model inpainting, k is equal to $\pm\infty$ at 4 corners a, b, c and d.

Fig. 6.2 A target pixel O
and its neighbors

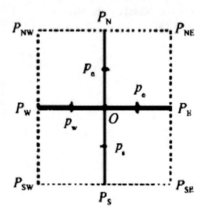

The key to the numerical implementation of the TV inpainting model lies in the approximation of $|\nabla I|$ and the degradation of PDE. As shown in Fig. 6.2, O is the target pixel, P_e, P_n, P_w, P_s represents the four midway points which is not directly available from the digital image, while P_E, P_N, P_W, P_S denotes the 4 adjacent pixels of O, and $|\nabla I_{P_e}|$ can be approximated by the following formula

$$|\nabla I_{P_e}| = \frac{1}{h}\sqrt{(I_{P_E} - I_0)^2 + \left[(I_{P_{NE}} + I_{P_N} - I_{P_S} - I_{P_{SE}})/4\right]} \qquad (6.14)$$

A similar discussion applies to the other three directions. After the discretization, it becomes

$$\sum_{p \in \Lambda_o} \frac{1}{|\nabla I_p|}(I_0 - I_p) + \lambda_D(I_o - I_O^0) = 0 \qquad (6.15)$$

where $\Lambda_O = \{P_E, P_N, P_W, P_S\}$ represents the four adjacent pixels of the target pixel O. For any target pixel O, define

$$\omega_P = 1/|\nabla I_P| \, P \in \Lambda_O$$

$$h_P = \omega_P / \left[\sum_{P \in \Lambda_O} \omega_P + \lambda_D(O) \right]$$

$$h_O = \lambda_D(O) / \left[\sum_{P \in \Lambda_O} \omega_P + \lambda_D(O) \right] \qquad (6.16)$$

Here, if $P = P_E$, then p represents the point P_e. Therefore, the formula (6.15) becomes

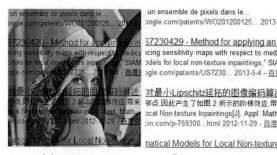 

(a) original image   **(b) text in image**   **(c) results**

Fig. 6.3   Text removal in image

$$I_O = \sum_{P \in \Lambda_O} h_P I_P + h_O I_O^0$$

$$\sum_{P \in \Lambda_O} h_P + h_O = 1 \tag{6.17}$$

In this way, the formula (6.17) can be rewritten into a Gauss-Jacobi iterative form

$$I_O^{(n)} = \sum_{P \in \Lambda_O} h_P^{(n-1)} I_P^{(n-1)} + h_O^{(n-1)} I_O^0 \tag{6.18}$$

When a specific implementation is made, a mask is used to determine the area that needs to be inpainted first (the mask for D needs to be given beforehand), and then according to the information around the area to be inpainted, the image inpainting algorithm is used to restore the information automatically, according to formula (6.14), the algorithm steps are as follows:

(1)   Read the image and the mask information;
(2)   Perform (3), (4), (5) steps for each pixel in the mask;
(3)   Calculate the first derivative value and the modulus value of the gradient of the pixels;
(4)   Set $\lambda_D(O) = 1$, if the pixel is located outside the inpainted area; otherwise, $\lambda_D(O) = 0$;
(5)   By calculating h_P and h_O, the new pixel values are obtained and saved into the new image;
(6)   Calculating the difference between the new image and the old image, if it is less than the given threshold, replace the old image with the new image and exit; otherwise, go to the second step.

The MATLAB code of the algorithm is shown in PROGRAMME 6.1 (Fig. 6.3).

PROGRAMME 6.1: Image inpainting using the total variation model

```
close all;
clear all;
clc;
img=double(imread('lena.jpg'));
mask=rgb2gray(imread('ma.jpg'))>160;
[m n]=size(img);
for i=1:m
    for j=1:n
        if mask(i,j)==0
            img(i,j)=0;
        end
    end
end
imshow(img,[]);
lambda=0.2;
a=0.5;
imgn=img;
for l=1:300
    for i=2:m-1
        for j=2:n-1
            if mask(i,j)==0
                Un=sqrt((img(i,j)-img(i-1,j))^2+((img(i-1,j-1)-img(i-1,j+1))/2)^2);
                Ue=sqrt((img(i,j)-img(i,j+1))^2+((img(i-1,j+1)-img(i+1,j+1))/2)^2);
                Uw=sqrt((img(i,j)-img(i,j-1))^2+((img(i-1,j-1)-img(i+1,j-1))/2)^2);
                Us=sqrt((img(i,j)-img(i+1,j))^2+((img(i+1,j-1)-img(i+1,j+1))/2)^2);
                Wn=1/sqrt(Un^2+a^2);
                We=1/sqrt(Ue^2+a^2);
                Ww=1/sqrt(Uw^2+a^2);
                Ws=1/sqrt(Us^2+a^2);
                Hon=Wn/((Wn+We+Ww+Ws)+lambda);
                Hoe=We/((Wn+We+Ww+Ws)+lambda);
                How=Ww/((Wn+We+Ww+Ws)+lambda);
                Hos=Ws/((Wn+We+Ww+Ws)+lambda);
                Hoo=lambda/((Wn+We+Ww+Ws)+lambda);
```

imgn(i,j)=Hon*img(i-1,j)+Hoe*img(i,j+1)+How*img(i,j-1)+Hos*img(i+1,j)+Hoo*img(i,j);

         end

      end

   end

   img=imgn;

end

figure;

imshow(img,[])

6.3.2 Image Inpainting Based on CDD Model

As shown in Fig. 6.1, the rightmost one is the output from the TV inpainting, in which, the curvature $k = \pm\infty$ at 4 corners a, b, c, and d. However, from the perspective of visual psychology, the curvature of these 4 corners should be zero, that is to say, during the image inpainting process, the curvature k should be as small as possible to obtain the image that conforms to human vision. Based on such analysis, Chan and Shen modified the TV model and proposed the new diffusions Curvature-Driven Diffusions (CDD) inpainting model [25–28].

In the CDD inpainting model, the TV conductivity coefficient is modified to $v = g(|k|/|\nabla I|)$, where g is defined as

$$g(k) = \begin{cases} 0 & k = 0 \\ \infty & k = \infty \\ \text{Finite numbers greater than 0} & 0 < k < \infty \end{cases} \qquad (6.19)$$

Because of this selection, the diffusion will be enhanced where the isophotes have a larger curvature, and the diffusion of small curvature will gradually disappear. Therefore, the CDD inpainting model is

$$\begin{cases} \frac{\partial I}{\partial t} = div\left[\frac{g(|k|)}{|\nabla I|}\nabla I\right], & x \in D \\ I = I^0, & x \in \Omega \backslash D \end{cases} \qquad (6.20)$$

where $k = div\left[\nabla I / |\nabla I|\right]$ is the curvature.

The MATLAB code of image inpainting based on CDD model is shown in PROGRAMME 6.2 (Fig. 6.4).

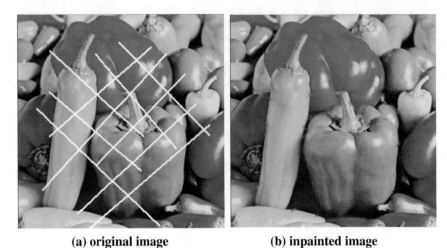

(a) original image   **(b) inpainted image**

Fig. 6.4 CDD inpainting result

PROGRAMME 6.2: Image inpainting based on CDD model

```
close all;
clc;
clear;
imgoriginal=imread('C1001.bmp');
figure(1);
imshow(imgoriginal);
[width,height] = size(imgoriginal);
img= double(imgoriginal);
mask = zeros(width,height/3,3);
for j = 1:height/3
    for i = 1:width
        if ((imgoriginal(i,j,1) >220)&&(imgoriginal(i,j,2) >220)&&(imgoriginal(i,j,3) >220))
            mask(i,j,1) = 255;
            mask(i,j,2) = 255;
            mask(i,j,3) = 255;
```

```
            else
                mask(i,j,1) = 0;
                mask(i,j,2) = 0;
                mask(i,j,3) = 0;
            end
        end
end
figure(2);
imshow(mask);
a=zeros(width,height);
I=cat(3,a,2*a,3*a);
J=cat(3,a,2*a,3*a);
n = 1;
itertimes=1000;
tic;
while n <= itertimes
    for i = 2:width-1
        for j = 2:height/3-1
            if (mask(i,j+1,1) == 255)||(mask(i,j-1,1) == 255)||(mask(i+1,j,1) == 255)||(mask(i-1,j,1) ==
255)
                for k=1:3
                    grid_w(k)    =    (img(i,j,k)-img(i-1,j,k))^2+(1.0/16)*(img(i-1,j+1,k)+img(i,j+1,k)-
img(i-1,j-1,k)-img(i,j-1,k))^2;
                    grid_e(k)    =    (img(i,j,k)-img(i+1,j,k))^2+(1.0/16)*(img(i,j+1,k)+img(i+1,j+1,k)-
img(i,j-1,k)-img(i+1,j-1,k))^2;
                    grid_s(k) = (img(i,j,k)-img(i,j-1,k))^2+(1.0/16)*(img(i+1,j,k)+img(i+1,j-1,k)-
img(i-1,j,k)-img(i-1,j-1,k))^2;
                    grid_n(k)    =    (img(i,j,k)-img(i,j+1,k))^2+(1.0/16)*(img(i+1,j,k)+img(i+1,j+1,k)-
img(i-1,j,k)-img(i-1,j+1,k))^2;
I(i,j,k)=0.5*(img(i+1,j,k)-img(i-1,j,k))/sqrt(0.25*(img(i+1,j,k)-img(i-
1,j,k))^2+0.25*(img(i,j+1,k)-img(i,j-1,k))^2+1);
J(i,j,k)=0.5*(img(i,j+1,k)-img(i,j-1,k))/sqrt(0.25*(img(i+1,j,k)-img(i-1,j,k))^2+0.25*(img(i,j+1,k)-
img(i,j-1,k))^2+1);
Kw(k)=sqrt((I(i,j,k)-I(i-1,j,k)+(I(i-1,j+1,k)+I(i,j+1,k)-I(i-1,j-
1,k)-I(i,j-1,k))/2)^2+(J(i,j,k)-J(i-1,j,k)+(J(i-1,j+1,k)+J(i,j+1,k)-J(i-1,j-1,k)-J(i,j-1,k))/2)^2);
Ke(k)=sqrt((I(i+1,j,k)-I(i,j,k)+(I(i,j+1,k)+I(i+1,j+1,k)-I(i,j-1,k)-I(i+1,j-1,k))/2)^2+(J(i+1,j,k)-
J(i,j,k)+(J(i,j+1,k)+J(i+1,j+1,k)-J(i,j-1,k)-J(i+1,j-1,k))/2)^2);
Ks(k)=sqrt((I(i,j,k)-I(i,j-1,k)+(I(i+1,j,k)+I(i+1,j-1,k)-I(i-1,j,k)-I(i-1,j-1,k))/2)^2+(J(i,j,k)-J(i,j-
1,k)+(J(i+1,j,k)+J(i+1,j-1,k)-J(i-1,j,k)-J(i-1,j-1,k))/2)^2);
```

Kn(k)=sqrt((I(i,j+1,k)-I(i,j,k)+(I(i+1,j,k)+I(i+1,j+1,k)-I(i-1,j,k)-I(i-1,j+1,k))/2)^2+(J(i,j+1,k)-

J(i,j,k)+(I(i+1,j,k)+J(i+1,j+1,k)-J(i-1,j,k)-J(i-1,j+1,k))/2)^2);

```
                        w1(k) = Kw(k)/sqrt(1+grid_w(k))+1;
                        w2(k) = Ke(k)/sqrt(1+grid_e(k))+1;
                        w3(k) = Ks(k)/sqrt(1+grid_s(k))+1;
                        w4(k) = Kn(k)/sqrt(1+grid_n(k))+1;
                        img(i,j,k) =(w1(k)*img(i-1,j,k)+w2(k)*img(i+1,j,k)+w3(k)*img(i,j-1,k)
            +w4(k)*img(i,j+1,k))/(w1(k)+w2(k)+w3(k)+w4(k));
                                    end
                            end
                    end
            end
            n = n+1;
            end
            img = uint8(floor(img));
            toc;
            figure(3);
            imshow(img,[]);
            imwrite(img,'CDD_result.bmp');
```

6.4   Exemplar-Based Image Inpainting Algorithm

The exemplar-based [29] approach is an important class of inpainting algorithms, here, we will describe the method of Criminisi et al. for repairing texture component, which takes isophote into consideration, sampling the best matching patches from the known region, and pastes into the target patches in the missing region.

And the restoration is a sampling process driven by the isophote. As shown in Fig. 6.5, given the input image I, Φ is the source region, the region to be filled is indicated by Ω, and its boundary is denoted $\delta\Omega$, for each point p on the contour $\delta\Omega$, the patch Ψ_p is constructed with p in the center of the patch, n_p is the normal to the contour $\delta\Omega$, besides, $\nabla I_p^{\perp}$ is the direction (the vertical direction of the gradient) and intensity of the isophote at point p.

The core idea of the Criminisi algorithm is to consider the fill order of the target region, that is, when filling the target region, calculating the priority of all the target pixels on the boundary of the inpainting domain, the patch with the highest priority will be filled and updated at first. As we mentioned before, each pixel p on the contour $\delta\Omega$ corresponds to a rectangular patch, which is constructed with p in the centere, and the size of the block is equal to the size of the given module (generally the module is slightly larger than the maximum texture element in the sample area). Select the patch on $\delta\Omega$ with highest priority, filling it with padding. Assuming that Ψ_p is the patch block centered at point $p \in \delta\Omega$, its priority $P(p)$ is defined as the product of two terms:

$$P(p) = C(p) \times D(p) \tag{6.21}$$

In Eq. (6.21), C(p) is the confidence term of Ψ_p, which reflects the number of effective points contained in the small pieces centered on point p, and the larger the value, the greater the effective points around point p, in another word, if we start to restore from Ψ_p, we could have a higher confidence value and reduce the error as far as possible. Here defined C(p) as:

$$C(p) = \frac{\sum C(q)}{|\Psi_p|}, \text{ where } q \in \Psi_p \cap \Phi \tag{6.22}$$

where $|\Psi_p|$ is the area of Ψ_p, here is the size of the module, $C(q)$ is the confidence term of point q which is initialized to:

$$C(q) = \begin{cases} 0 \; \forall q \in \Omega \\ 1 \; \forall q \in \Phi \end{cases} \tag{6.23}$$

From the Eqs. (6.22) and (6.23), we can see that the more pixels in the sample area of the patch, in another word, the more pixels have been filled, the higher confidence item of the patch will be.

In Eq. (6.21), $D(p)$ represents the strength of isophotes hitting the boundary and boosts the priority of a patch that an isophote "flows" into, and is defined as follows

$$D(p) = \frac{|\nabla I_p^{\perp} \cdot n_p|}{\alpha} \tag{6.24}$$

where n_p is the unit vector orthogonal to the front $\delta\Omega$, α is the normalized factor (for 8-bit gray-level image $\alpha = 255$). From the formula (6.24) we can see that, the larger the intensity of the isophote of point p on $\delta\Omega$, and the angle between the

Fig. 6.5 Principle of Criminisi algorithm

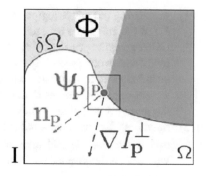

unit vector is smaller, then the calculated value $D(p)$ is larger, which reflects the structural information of the image.

Once the priority of each pixel on edge is calculated, the patch corresponding to the point p with the highest priority is determined, a global search is performed on the whole image to find the most similar example $\Psi_{\hat{q}}$ from the source region Φ to compose the given patch, $\Psi_{\hat{q}}$ satisfying the following conditions:

$$\Psi_{\hat{q}} = \arg \min_{\Psi_q \in \Phi} d\{\Psi_p, \Psi_q\} \tag{6.25}$$

where $d\{\Psi_p, \Psi_q\}$ represents the distance between two generic patches Ψ_p and Ψ_q, using the Sum of Squared Differences (SSD) of pixels as the distance measurement, the definition of SSD is as follows:

$$SSD = \sqrt{\sum_{i=1}^{m} \sum_{j=1}^{n} (p_{ij} - q_{ij})^2} \tag{6.26}$$

where m, n denotes the length and width of the patch, p represents the pixel in the patch to be restored, and q denotes the pixel in the source region Φ. By comparing the corresponding SSD values of each matching pixel, find the exemplar Ψ_q in the source region Φ that minimizes SSD, and then copy image data from $\Psi_{\hat{q}}$ to Ψ_p,which successfully expands the texture and structural information of the image.

In the Criminis algorithm, after the completion of a restoration, the original unknown pixel turned into a known pixel, we can see from the Eq. (6.23) that the confidence terms of these pixels have changed, and the required information for computing filling priorities needs to be updated:

$$C(p) = C(\hat{p}), \forall p \in \Psi_{\hat{p}} \cap \Omega \tag{6.27}$$

Since the contour of the target patch to be restored has changed, a new contour has been formed. If the new restoration area is empty, the inpainting is completed.

Therefore, the steps of exemplar-based Criminisi image inpainting algorithm includes the following four main steps:

(1)   Initializing the target region manually, input the image, extracting the contour $\delta\Omega$ of the area to be restored.
(2)   Calculating the priority of all the pixels according to the formula (6.21), and the block whose corresponding priority is highest is selected. Where the confidence term indicates the proportion of the known pixels in the patch to be restored to the entire pixels, and the data item $D(p)$ is the dot product of the isophote and the unit vector at pixel point p.
(3)   Searching for the most similar image example $\Psi_{\hat{q}}$ from the source region Φ, update the unknown information according to the corresponding information, where $\Psi_{\hat{q}}$ is determined by formula (6.25), whose SSD among the known region is minimum.

(4) Updating the information, in which the boundary $\delta\Omega$ of the target region Ω and the required information for computing filling priorities are updated.

Repeat the above steps until the target area $\Omega = \phi$, then exits the loop and the inpainting is completed.

The MATLAB code of the exemplar-based [29–33] image inpainting is shown in PROGRAMME 6.3.

PROGRAMME 6.3: The exemplar-based image inpainting

```
clc;

clear;

tic

imagepath =imread( 'hua.bmp');         % the original image

maskpath =imread('huamask.bmp');    %the damaged image

fillColor=[255,255,0];         % the RGB color values of the damaged area in the broken image

[Psnr,inpaintedImg] =RGB_Criminisi(imagepath,maskpath,fillColor);

%Criminisi algorithm to repair color images

function [Psnr,inpaintedImg] =RGB_Criminisi(imagepath,maskpath,fillColor)

img0=imagepath;

fillImg=maskpath;

img = double(fillImg);% images that need to be repaired

fillRegion=img(:,:,1)==fillColor(1)&img(:,:,2)==fillColor(2)&img(:,:,3)==fillColor(3);

origImg = img;

ind = img2ind(img);

%-------------------------------------------------------

in=ind;

[A,BB]=find(in);                    % get the coordinates of each point

%-------------------------------------------------------

sz = [size(img,1) size(img,2)];

z1=size(img,1);

z2=size(img,2);

sourceRegion = ~fillRegion;

% Seeking the value of the isophote

[Ix(:,:,3),Iy(:,:,3)] = gradient(img(:,:,3));

[Ix(:,:,2),Iy(:,:,2)] = gradient(img(:,:,2));

[Ix(:,:,1),Iy(:,:,1)] = gradient(img(:,:,1));

Ix = sum(Ix,3)/(3*255); Iy = sum(Iy,3)/(3*255);

temp = Ix; Ix = -Iy; Iy = temp;   % Rotate 90 degrees

%------------------------------------------------------------------------

% to calculate the gradient value
```

```
[ix(:,:,3),iy(:,:,3)] = gradient(img(:,:,3));
[ix(:,:,2),iy(:,:,2)] = gradient(img(:,:,2));
[ix(:,:,1),iy(:,:,1)] = gradient(img(:,:,1));
ix = sum(ix,3)/(3*255); iy = sum(iy,3)/(3*255);
%------------------------------------------------------------------------
% Initializes the confidence term C and the data item D
C = double(sourceRegion);
D = repmat(-.1,sz);
%image inpainting (until all damaged areas have been repaired)
while any(fillRegion(:))
    % seek for the edge
    dR = find(conv2(double(fillRegion),[1,1,1;1,-8,1;1,1,1],'same')>0);
    [Nx,Ny] = gradient(double(~fillRegion));
    N = [Nx(dR(:)) Ny(dR(:))];
    N(~isfinite(N))=0;
    %calculate the confidence value
    for k=dR'
        Hp = qukuai_9(sz,k);
        q = Hp(~(fillRegion(Hp)));
        C(k) = sum(C(q))/numel(Hp);
    end
% calculate the priority
    D(dR) = abs(Ix(dR).*N(:,1)+Iy(dR).*N(:,2)) /255;
    priorities =C(dR).*D(dR);
% Find out the largest priority block Hp
    [unused,ndx] = max(priorities(:));
    p = dR(ndx(1));
    %--------------------
    [Hp,rows,cols] = qukuai_9(sz,p);      %the size is 9x9
    toFill=fillRegion(Hp);
    Wpatch=img(rows,cols,:); % Get the block to be repaired
    %------------------------------------------------------------------------
    % Use the global search to find the best match block
      Hq=whole_match(z1,z2,img,Wpatch,fillColor);
    %------------------------------------------------------------------------
    % Update the fill area
    fillRegion(Hp(toFill)) = false;
```

```
% Update the C (p) value and the value of the isophote
C(Hp(toFill))    = C(p);
Ix(Hp(toFill)) = Ix(Hq(toFill));
Iy(Hp(toFill)) = Iy(Hq(toFill));
%-------------------------------------------------------------------
% Update the gradient value
ix(Hp(toFill)) = ix(Hq(toFill));
iy(Hp(toFill)) = iy(Hq(toFill));
%-------------------------------------------------------------------
% Copy image information from Hq to Hp
ind(Hp(toFill)) = ind(Hq(toFill));
img(rows,cols,:) = ind2img(ind(rows,cols),origImg);
end
inpaintedImg=img;
A=double(img0);
B=double(inpaintedImg);
Psnr=PSNR(A,B);
inpaintedImg=uint8(inpaintedImg);
subplot(1,2,1);
imshow('huamask.bmp');
title(' original image');
subplot(1,2,2);
imshow(inpaintedImg);
title(' the corrected image based on Criminisi');
%%
function [Hp,rows,cols]=qukuai_9(sz,p)
x=floor(rem(p,sz(1)));
y=floor(p/sz(1))+1;% Get the location of the center of the area to be repaired
w=4;% The radius of the block (the radius of the 9*9 size block is 4)
rows=max(1,x-w):min(x+w,sz(1));
cols=max(1,y-w):min(y+w,sz(2));
numhang=length(rows);
numlie=length(cols);
HJ=zeros(numhang,numlie);
LJ=zeros(numhang,numlie);
for ii=1:numlie
    HJ(:,ii)=rows;% the matrix formed by expanding rows
end
```

```
for jj=1:numhang
    LJ(jj,:)=cols;% the matrix formed by expanding lines
end
Hp=(LJ-1)*sz(1)+HJ;% Get the block to be repaired (the specific location in the image)
```
%%***
% Global search to find the best matching block (color image)
%***
```
function Block=whole_match(zx,zy,img,Wpatch,fillColor)
%zx,zy is the size of the image
aa=size(Wpatch,1);
bb=size(Wpatch,2);% The actual size of the block to be repaired
sx=zx-aa+1;
sy=zy-bb+1;% Determines the maximum value of the search area
all=aa*bb;% One-dimensional size
min=1.0000e+10;% Initial value
% Calculate the distance between the found matching block and the block to be repaired at the known
    pixel (just the known pixel)
for i=1:sx
    for j=1:sy
        Mpatch=img(i:i+aa-1,j:j+bb-1,:);% Get the same size as the block to be repaired in the image
posunme=Mpatch(:,:,1)==fillColor(1)&Mpatch(:,:,2)==fillColor(2)&Mpatch(:,:,3)==fillColor(3);
        if any(posunme(:))%It shows that there is a broken point which is not the best matching block
    to be repaired
            continue;
        end
        Err=0;
        for gg=1:aa*bb
            % Traverse each pixel in the block
pf=Wpatch(gg)==fillColor(1)&Wpatch(gg+all)==fillColor(2)&Wpatch(gg+2*all)==fillColor(3);
            if pf    % Indicating that the location of the pixel is broken, do not participate in the
    calculation of distance
                continue;
            end
                cha=Wpatch(gg)-Mpatch(gg);Err=Err+cha*cha;
                 cha=Wpatch(gg+all)-Mpatch(gg+all);Err=Err+cha*cha;
                  cha=Wpatch(gg+2*all)-Mpatch(gg+2*all);Err=Err+cha*cha;
        end
        if Err<min
            min=Err;% Take the blocks that minimize the error and note their starting position in the
```

```
image
            hk=i;
            lk=j;
         end
      end
end
rows=hk:hk+aa-1;
cols=lk:lk+bb-1;% The best matching block is the block with the smallest error
Block=form_patch(rows,cols,zx);
%--------------------------------------------------------------------
% ind2img Convert index images to RGB images
function img2 = ind2img(ind,img)
for i=3:-1:1, temp=img(:,:,i); img2(:,:,i)=temp(ind); end;
%--------------------------------------------------------------------
% Converting RGB images into index images
%--------------------------------------------------------------------
function ind = img2ind(img)
s=size(img); ind=reshape(1:s(1)*s(2),s(1),s(2));
% calculate the PSNR value
%--------------------------------------------------------------------
function PSNR=PSNR(u,v)
[m,n]=size(u(:,:,1));
a=0;
for i=1:m
     for j=1:n
            t(i,j,1)=u(i,j,1)-v(i,j,1);
            a=a+t(i,j,1)^2;
        end
end
mse=a/(m*n);
PSNR_R=10*log10(255^2/mse);
b=0;
for i=1:m
     for j=1:n
            t(i,j,2)=u(i,j,2)-v(i,j,2);
            b=b+t(i,j,2)^2;
        end
end
mse=b/(m*n);
PSNR_G=10*log10(255^2/mse);
c=0;
```

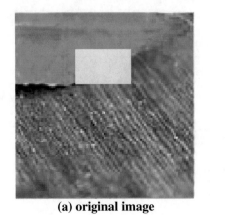
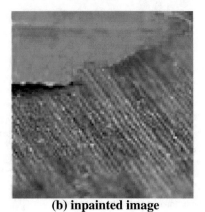

(a) original image   **(b) inpainted image**

Fig. 6.6 Comparison of flowers before and after inpainting

```
for i=1:m
    for j=1:n
        t(i,j,3)=u(i,j,3)-v(i,j,3);
        c=c+t(i,j,3)^2;
    end
end
mse=c/(m*n);
PSNR_B=10*log10(255^2/mse);
PSNR=(PSNR_R+PSNR_G+PSNR_B)/3;
```

In Fig. 6.6, a simulation experiment is conducted. It can be seen from the restored results that the method can obtain accurate texture features for strong directional texture information.

References

1. Shu-gen W, Jing-ling Z (2004) Image inpainting for information lostarea based on the texture matching approach. Bullet Surv Mapp 12:21–23
2. Bertalmio M, Sapiro G, Caselles V, et al (2000) Image in painting. In: Proceedings of international conference on computer graphics and interactive techniques. New Orleans, Louisiana, USA, pp 417–424
3. Chan TF, Shen JH (2001) Non-texture inpainting by curvature-driven diffusions (CDD). J Vis Commun Image Represent 12(4):436–449
4. Chan TF, Shen JH (2001) Mathematical models for local non-texture inpainting. SIAM J Appl Math 62(3):1019–1043
5. Chan TF, Kang SH, Shen JH (2002) Euler's elastica and curvature based inpainting. SIAM J Appl Math 63(2):564–592

6. Tsai A, Yezzi JA, Willsky AS (2001) Curve evolution implementation of the Mumford-Shah functional for image segmentation, denoising, interpolation and magnification. IEEE Trans Image Process 10(8):1169–1186
7. Esedoglu S, Shen JH (2002) Digital inpainting based on the Mum ford-Shah-Euler image model. Eur J Appl Math 13(4):353–370
8. Tang F, Ying YT, Wang J et al (2004) A novel texture synthes is based algorithm for object removal in photographs. In: Proceedings of N in the Asian computing science conference. Chiang Mai, Thailand, pp 248–258
9. Criminisi A, Perez P, Toyama K (2003) Object removal by exemplar-based inpainting. In: Proceedings of IEEE Computer Society Conference on Computer Vision and Pattern Recognition, vol 2. Monona Terrace Convention Center Madison, Wisconsin, USA, pp 18–20
10. Rudin L, Osher S, Faterni E (1992) Nonlinear total variation based noise removal algorithms. Physica D 60(1–4):259–268
11. Bertalmio M, Vese L, Sapiro G, et al Simultaneous texture and structure image in painting. IEEE Trans Image Process 12(8):882–889
12. Efros AA, Leung TK (1999) Texture synthesis by non-parametric sampling. In: Proceedings of the IEEE computer society international conference on computer vision vol 2. Washington DC, USA, pp 1033–1038
13. Harald G. (2004) A combined PDE and texture synthesis approach to in painting. In: Proceedings of 8th European conference on computer vision, vol 2. Prague, Czech Republic, pp 214–224
14. Mumford D, Shah J (1989) Optimal approximations by piecewise smooth functions and associated variational problems. Commun Pure Appl Math 42(5):577–685
15. Shen JH (2004) Bayesian inpainting based on geometric image models [EB/OL]. http://www.math.ucla.edu/~imagers/htmls/inp.html. Accessed 28 Nov 2004
16. Mumford D. Elastica and computer vision. In: Bajaj C (ed) Algebraic geometry and its applications. Springer, New York, pp 491–506
17. Rane SD, Sapiro G, Bertalmio M (2003) Structure and texture filling-in of missing image blocks in wireless transmission and compression applications. IEEE Trans Image Process 12(3):296–303
18. Yamauchi H, Haber J, Seidel HP (2003) Image restoration using multire solution texture synthesis. In: Proceedings of computer graphics international conference (CGI 2003). Tokyo, Japan, pp 1530–1552
19. Drori I, Daniel CO, Hezy Y (2003) Fragment based image completion. ACM Trans Graph 22(3):303–312
20. Zhang YJ, Xiao JG, Shah M (2005) Region completion in a single image [EB/OL]. www.cs.u c.fedu/~vision/papers/zhang_xiao_shah_EG2004.pdf. Accessed 21 April 2005
21. Chan FT, Shen JH (2004) Variational image inpainting [EB/OL]. http://www.math.ucla.edu/~imagers/htmls/inp.html. Accessed 28 April 2005
22. Costanzino N (2004) EN161 project presentation III: structure inpainting via variational methods [EB/OL]. http://mountains.ece.umn.edu/~guille/inpainting.html. Accessed 28 April 2005
23. Xu WW, Pang ZG, Zhang MM (2002) Image inpainting based on total variational model. J Image Graph 7(4):351–355
24. Cohen A, Dahman W, Daubechies I et al (2004) Tree approximation and optimal encoding [EB/OL]. http://www.math.sc.edu/~devore/publications/9909.pdf. Accessed 21 Nov 2004
25. Starck JL, Nguyen MK, Murtagh F. Wavelets and curvelets for image deconvolution: a combined approach. Signal Process 83(10):2279–2283.
26. Donoho DL (2000) Beam lets. In: Invited talk at IMA workshop on image analysis and low level vision. University of Minnesota, Minnesota, USA
27. Chuang YY, Curless B, Salesin DH et al (2001) A bayesian approach to digital matting. In: Proceedings of IEEE computer society's conference on computer vision and pattern recognition. Hawaii, USA, pp 264–271
28. Lin SY, Shi JY (2005) Fast natural image matting in perceptual color space. Comput Graph 29(3):403–414

29. Criminisi A, Perez P, Toyama K (2004) Region filling and object removal by exemplar-based image inpainting. IEEE Trans Image Process 13(9):1200–1212
30. Harrison P (2001) A nonhierarchical procedure for resynthesis of complex texture. In: Proceedings of 9th International conference on central europe computer graphics, visualization, and computer vision [C/OL], Plzen, Czech Republic. Feb. 2001. http://www.csse.monash.edu.au/~pfh/resynthesizer/
31. Borikar S, Biswas KK, Pattanaik S (2205) Fast algorithm for completion of images with natural scenes [EB/OL]. http://www.graphics.cs.uc.fedu/borikar/BorikarPaper.pdf. Accessed 20 April 2005
32. Cheng WH, Hsieh CW, Lin SK et al (2005) Robust algorithm for exemplar-based image inpainting [EB/OL]. http://www.cmlab.csie.ntu.edu.tw/~wisley/publications/CGIV_2005.pdf Accessed 18 April 2005
33. Cheng KY (2005) Research on improving exemplar-based inpainting [EB/OL]. http://graphics.csie.ntu.edu.tw/~kyatapi/Cheng.pdf, Accessed 22 April 2005

Chapter 7
Image Fusion

In this chapter we introduce the image fusion methods, including the wavelet transform based fusion, region based fusion and the fusion method based on fuzzy Dempster-Shafer evidence theory. Moreover, the image quality and fusion evaluations are also introduced. In this chapter we introduce the image fusion methods, including the wavelet transform based fusion, region based fusion and the fusion method based on fuzzy Dempster-Shafer evidence theory. Moreover, the image quality and fusion evaluations are also introduced.

7.1 Introduction

Image fusion [1] is a process of combing multiple input images of the same scene into a single fused image, which synthesizes high quality images from image data collected by multi-source channels with the same target by the image processing and computer technology. In application, the information of multiple images from a single sensor or heterogeneous sensors will be integrated to reduce the uncertainty and redundancy of the output on the basis of the maximum combination of related information, enhance the information transparency of the image so as to form a clear, complete and accurate description of the object, the spatial resolution and spectral resolution of the original image will also be enhanced which is conducive to dynamic monitoring, target identification and decision making. The data form of image fusion is the image containing brightness, color, temperature, distance and other scenery features, which can be given in the form of a picture or a series of images. Generally, a specific algorithm is used to combine relevant information from two or more source images into one single image such that the single image contains most of the information from all the sources images. Image fusion is not a simple overlay, it can produce new and more valuable images. Figure 7.1 shows the general model of image fusion. In this model, spatial registration, information fusion and information representation are the main steps.

© Springer International Publishing AG, part of Springer Nature 2019
S. Gong et al., *Advanced Image and Video Processing Using MATLAB*,
Modeling and Optimization in Science and Technologies 12,
https://doi.org/10.1007/978-3-319-77223-3_7

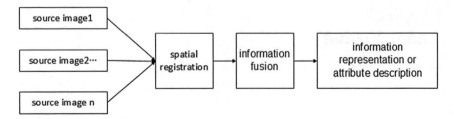

Fig. 7.1 General model of image fusion

7.2   Fusion Categories

7.2.1   Multi-view Fusion

Multiple-view fusion fuses the images obtained from multiple cameras. As an object occluded in one camera view may be visible in other camera views, these images often contain large amounts of complementary and redundant information, thus, the camera views need to be associated, that is to say, the information contained in the original view images needs to be integrated into a new image as complete as possible. By this means, people can have a more vivid and intuitive understanding of the original scene with the help of the newly generated images. The multi-view image fusion plays an irreplaceable role in practical application with its unique advantages. However, because of the characteristics of high resolution and high information quantity, the storage and transportation of multi-view images has become a difficult problem in practical applications.

In order to facilitate the subsequent computation, it is necessary to reduce the dimension of the image in advance. Since the displacement and angle transformation between different images, the image needs to be matched and then fused according to specific conditions at the same time.

Since pixel level fusion is the most common and widely used approach, we often realize the fusion according to the basic characteristics of pixels. Different processing methods are used for different regions. Figure 7.2 shows the schematic diagram of multi-view image fusion.

In Fig. 7.2, I_1 represents the image from the view point 1, and I_2 represents the image from view point 2. Where I_1 and I_2 are equal in size, both of which is $m \times n$. The gray parts in both I_1 and I_2 represent the overlapping regions of the two images. Since image I_1 translated c and f in the horizontal and vertical directions respectively to get I_2, the size of the overlapped area is $(m - f) \times (n - c)$. With the help of the corresponding relationship between I_1 and I_2, the newly fused image I_f can be calculated, whose size is $(m + f) \times (n + c)$, and I_f almost retained most of the information of I_1 and I_2.

For convenience, we divide I_f into five regions, the greyscale area in the middle and four regions labeled with one, two, three, four on the edge. According to the

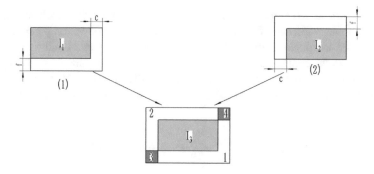

Fig. 7.2 The multi-view image fusion schematic diagram

transformation relationship, region 1 is the remaining part after removing the over-lapping area in I_1, and 2 belongs to the rest of the removed overlapping area in I_2. 3 and 4 are the newly generated parts, while the gray part is the overlapping area of I_1 and I_2. Since 1 and 2 are specific parts of I_1 and I_2, they will be kept directly in the new image. For 3 and 4, as they are adjacent to the part 1 and 2 on the outer boundary, a 3×3 operator S is used to conduct weighted interpolation from their junction place in order to make the pixels of the fused image stable at the boundary, where

$$S = \begin{bmatrix} 1/8 & 1/8 & 1/8 \\ 1/8 & 0 & 1/8 \\ 1/8 & 1/8 & 1/8 \end{bmatrix} \tag{7.1}$$

It is important to note that the index direction of the pixel points is different when the region 3 and 4 are interpolated. For region 3, the positive direction of the index is from right to left, from top to bottom. While for region 4, the positive direction of the index is from left to right, from bottom to top. The process of the grayscale overlapping area is more troublesome. In addition, although the grayscale area in I_1 and I_2 seems alike, their brightness and color information are somewhat different due to the different viewpoints. Therefore, the pixel values of the two parts cannot be simply added and then averaged. In order to improve the fusion quality, wavelet fusion can be adopted.

When the wavelet transform [2–5] is adopted to fuse the overlapping regions, the image will decomposed by N layers wavelet first, then you will get $(3N + 1)$ different frequency bands. These frequency bands includes $3N$ high-frequency sub-images and 1 low-frequency sub-image. At the time of fusion, for the high-frequency part, the value of the absolute maximum value of the corresponding wavelet decomposition coefficient in the two source images is used as the decomposition coefficient of the fusion image. For the low-frequency part, the rules of the processing are relatively complex, and the specific steps are as follows:

(1) Suppose that $C(I)$ represents the coefficient matrix of the wavelet low frequency component of image I, and $p = (m, n)$ represents the space position of the wavelet coefficients, then $C(I, p)$ is the value of the element with the subscript (m, n) of the coefficient matrix of the wavelet low frequency components.

(2) Selecting a small patch Q with p as the center, $u(I, p)$ represents the average value of $C(I)$ in Q with p as the center. $G(I, p)$ is the regional variance significance of $C(I)$ in Q,

$$G(I, p) = \sum_{q \subset Q} w(q) |C(I, q) - u(I, p)|^2 \tag{7.2}$$

where, $w(q)$ is the weight value, the further away from p, the smaller the value is.

(3) Calculating the regional variance significance $G(I_1, p)$ and $G(I_2, p)$ of I_1 and I_2 respectively according to Eq. (7.2), then calculate the region variance matching degree of point p.

$$M_2(p) = \frac{2 \sum_{q \subset Q} w(q) |C(I_1, q) - u(I_1, q)| |C(I_2, q) - u(I_2, q)|}{G(I_1, q) + G(I_2, q)} \tag{7.3}$$

(4) Setting a matching threshold T. When $M_2(p) < T$, the fusion strategy is

$$G(I_f, p) = \begin{cases} C(I_1, p), C(I_1, p) \geq C(I_2, p) \\ C(I_2, p), C(I_1, p) < C(I_2, p) \end{cases} \tag{7.4}$$

While $M_2(p) \geq T$, the fusion strategy is the average strategy.

$$G(I_f, p) = \begin{cases} W_{max} C(I_1, p) + W_{min} C(I_2, p), G(I_1, p) \geq G(I_2, p) \\ W_{min} C(I_1, p) + W_{max} C(I_2, p), G(I_1, p) < G(I_2, p) \end{cases} \tag{7.5}$$

where, $W_{min} = 0.5 - 0.5 \left(\frac{1 - M_2(p)}{1 - T} \right)$, $W_{max} = 1 - W_{min}$. After the above processing, the ideal fused image based on wavelet transform can be obtained after the wavelet reconstruction.

7.2.2   Multimodal Fusion

Multimodal image fusion [6–10] fuse images taken from different modalities of the same scene. A single sensor can only obtain incomplete information of the object being tested. And it can be affected by the environment easily and the stability is not strong enough. Multimodal information fusion can combine information provided by multiple sensors, retain useful information and eliminate error messages. And

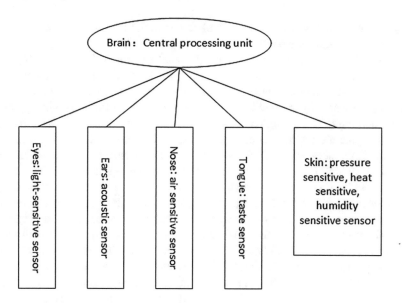

Fig. 7.3 Fusion of human and multimodal data

the reliability of the system can be improved by redundancy information, which can also improve the reliability of the measurement to achieve the final information optimization. This method can be applied to a variety of different environments, which could ensure normal working requirements under bad conditions. Multimodal fusion was first born in the military field and is now developing rapidly in the civilian field.

The most basic multimodal fusion comes from the observation and understanding of the objective things by the organism. As shown in Fig. 7.3, the organism uses a variety of different senses to perceive objects to obtain a large number of different kinds of information. Then send these information back to the central processor of the organism—the brain. Upon receiving these information, the brain will combine and link the information according to the experience that has been summarized and accumulated over a long period of time, and finally get the correct understanding of the observed objects.

The multimodal system will get three kinds of information in the process of information acquisition: (1) Redundant information: redundant information is the multiple duplicate information provided by a variety of sensors to a certain feature of the objective thing, which can improve the reliability of the system. (2) Complementary information: complementary information refers to the independent external characteristic information observed by various sensors, which can extend the performance of the system. (3) Collaborative information: collaborative information refers to the information that a single sensor cannot get and must rely on multiple sensors to cooperate, which can further expand the control scope of the system. Therefore, in the multi-modal fusion research, the key lies in the feature recognition method and the fusion algorithm itself.

Multimodal images are multiple images obtained from different imaging principles and devices, such as the images obtained by different imaging devices in the medical field, like CT, MRI, PET, etc. These images reflect different focus of human tissues and organs, and the clarity is also different. Multimodal image fusion is an emerging topic in image processing in recent years, which merge multisource images to generate new and better quality images with certain algorithms. It can eliminate the differences between information from various sensors by utilizing different spatial resolution, time resolution and spectral resolution images. By the way, the multimodal fusion can also enhance the reliability of the information in the image, improving the accuracy of the image, raising the availability, and obtain a more accurate and clearer description of the target. Minoshiha proposed a fusion method of three different formats to detect Alzheimer's disease, which introduces multimodal data fusion into the medical field. Nowadays, multimodal image fusion method has become an important medical means. In the field of remote sensing, we can get a clearer and more accurate fusion image by integrating high resolution and low resolution images, hyperspectral and low spectral images, multiband image and multi temporal image and so on. Multimodal image fusion technology has high application value in many fields such as video surveillance, medical diagnosis, satellite remote sensing and digital photography, etc.

The classical fusion methods of multimodal image fusion technology include: maximum likelihood estimation, Kalman filter, the least square method, the weighted average, Bayesian estimation, typical inference method and D-S evidence theory method. Modern fusion methods include clustering analysis, fuzzy logic, neural network and so on. The fusion method based on multi-scale transformation is one of the research hotspots of multimodal image fusion, such as pyramid transform, wavelet transform and multiscale geometric transformation. This method usually consists of three steps. First of all, transform the source image into a multiscale space and obtain the low frequency and high frequency conversion coefficients. Then, some certain rules/strategies are used to fuse the low frequency and high frequency transform coefficients respectively so as to obtain the fusion coefficients. Finally, the fusion coefficient is transformed and the fused images are obtained. In the multi-scale transformation based method, the selection of the transformation space and the design of the fusion rule are the two most important factors, most of the research work is carried out around these two elements. Figure 7.4 shows the multimodal image fusion process under wavelet transform.

Then we take Mallat fast algorithm wavelet transform as an example to fuse the two modal images. Firstly, the two images are decomposed by wavelet, and the fusion rule based on combination of selection and weighting factor is adopted for the low-frequency part of the decomposed image, then, for the high frequency part, the fusion rule based on regional energy is adopted. Finally, the fusion image is obtained by inverse wavelet transform.

For the multiscale wavelet decomposition of the image, 4 different subgraphs can be obtained at each decomposition scale. LL is the low-frequency part,

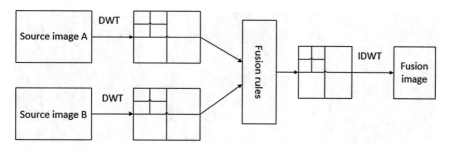

Fig. 7.4 Schematic diagram of multimodal image fusion based on wavelet transform

which represents the main information of the image and concentrates the majority of the energy of the image. While HL, LH and HH are the high frequency parts, which represent the details of the horizontal direction, vertical direction and diagonal direction of the image respectively. In terms of fusion rules, different fusion rules are adopted for high frequency and low frequency components respectively. Since the low-frequency component LL of the multi-scale decomposed image has a great influence on the quality of the restoration image. Thus, the combination of the selection and the weighting factors is used in the fusion of low frequency components.

$$A(i,j) = \alpha(A_1(i,j) + \kappa A_2(i,j)) - \beta(A_1(i,j) - \kappa A_2(i,j)) \tag{7.6}$$

where image size is $M \times N$, $i \in M$, $j \in N \cdot k, \alpha, \beta$ is the weighting factor. By adjusting the parameter k, the dominant proportion of the two images can be adjusted to balance two images with different brightness. If the factor α increases, the image will be brighter, when factor β increases, the edge of the image is enhanced. For different types of images, appropriate adjustment to the factors can reduce the blurred edges and ensure that the edge information of the image is not overly lost.

The result of high frequency component fusion is $HF_k(i,j)$, where $k = HL, LH, HH$. The fusion process first calculates the regional energy centered on the high frequency components at each scale. The size of the region is set to $m \times n$, where $m, n \geq 3$. The following is the calculation of regional energy at the same scale. First, the values of the same region in different directions are added to get the average; then, summing up the high frequency component of the same region and subtracting the corresponding mean value, thus the region energy $VA_1^k(i,j)$ and $VA_2^k(i,j)$ of this high frequency component can be obtained. After calculating the region energy of each high frequency component of the multimodal image, the region matching degree can be calculated by the following formula.

$$Match_k(i,j) = \frac{2VA_1^k(i,j)VA_2^k(i,j)}{[VA_1^k(i,j)]^2 + [VA_2^k(i,j)]^2} \tag{7.7}$$

where, $i \in m, j \in n$. The fusion rule is:

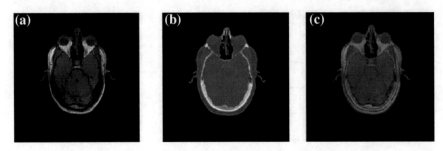

Fig. 7.5 An example of multimodal image fusion in medicine: **a** MRI image; **b** CT image; **c** fusion image

$$HF_k(i,j) = \begin{cases} \frac{1}{2}[HA_1^k(i,j) + HA_2^k(i,j)] & Match_k(i,j) \geq T_k \\ \begin{cases} HA_1^k(i,j) \ VA_1^k(i,j) \geq VA_2^k(i,j) \\ HA_2^k(i,j) \ VA_1^k(i,j) < VA_2^k(i,j) \end{cases} & Match_k(i,j) < T_k \end{cases} \qquad (7.8)$$

T_k is the threshold of the corresponding high frequency component.

Figure 7.5 shows examples of multimodal image fusion of Computed Tomography (CT) and Magnetic Resonance Image (MRI) images in the medical field. CT has the advantage of high spatial resolution, which is imaging based on the principle that various tissues have different degree of absorption of X rays. The bone can be imaged very clearly, which will provide a more accurate reference to the location of the lesion, while the soft tissue is not clearly visible. MRI uses water protons information imaging, whose spatial resolution is not as good as that of CT images, but it is clear for soft tissue imaging and is helpful for defining the range of lesions. However, it lacks rigid bone tissue as reference for location.

Obviously, if the information of medical images provided by different imaging devices is organically combined, the advances in modern medical clinical diagnosis technology will be greatly promoted. Therefore, the new thought of getting more valuable information from the fusion of medical images came into being. In the process of multimodal image fusion, different fusion rules will have a great influence on the fusion result.

7.2.3 Multi-temporal Fusion

Multi-temporal usually refers to the characteristics of a set of images in time series. The maximum characteristic of multi-temporal images can be divided into two aspects. Firstly, the images acquired at different times have different characteristics on the same target. On the other hand, a new targets will appear or some of the existing targets may disappear with the difference in imaging time. After the

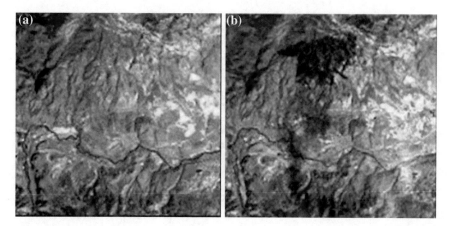

Fig. 7.6 Remote sensing image in Reno area: **a** image on August 5, 1986; **b** image on August 5, 1992

fusion of multi-temporal images of the same scene, we can obtain an image with temporal target distribution information and spatial target distribution information, which can meet the requirements of dynamic analysis, such as studying and tracking the evolution of natural history, monitoring the dynamic change of environment and resources. Therefore, the most important thing of multi-temporal image fusion is to optimize the complementary information at different images in the final fused image. Through detecting the change objects in multi-temporal images, we could get the change characteristics of the objects, including the regional distribution, the size characteristic and the outer edge shape change. Compared with the previous studies on multi-view image fusion and multimodal image fusion, the research on multi-temporal image fusion is relatively small. Figure 7.6 shows an example of the multi-temporal images.

The multi-temporal image fusion can effectively detect the features of the target in the images at different moments. As for high temporal resolution images, the time varying information of target can be extracted from the images, and then fused with the image with high spatial resolution, which can get the fusion image with high resolution of time and space at last. One of the methods to extract the change information of different phase images is the change detection algorithm based on the different images and the transform based methods, such as the principal component analysis (PCA) method, the improved multi-block PCA method, the iterative PCA method and the independent principal component analysis (ICA) method, etc. These methods can detect the strength and weakness of different regions in different phase images. The second is to classify the multi-temporal images respectively, then compare the images after classification, and get the difference of the different phase images from the classification results.

Therefore, it is a common strategy to use the results of the change detection to improve the quality of the final fusion image when different phase images are fused

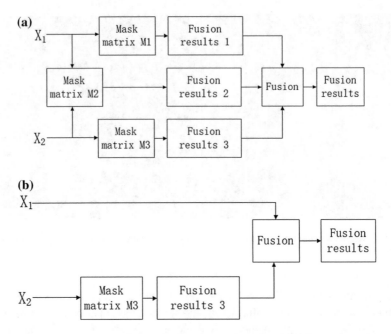

Fig. 7.7 Multi-temporal image fusion scheme: **a** integration scheme 1; **b** integration scheme 2

or compounded. For example, the results of the change detection can be used to extract the template of the target feature change, and the fusion of the phase image can be realized on this basis. The image fusion scheme is as follows:

In Fig. 7.7, M_1, M_2, M_3 are the target feature template, background template of the multi-temporal image X_1 and the target feature template of the multi-temporal image X_2. The Fusion scheme 1 does not distinguish the sequence of images X_1 and X_2. The final fusion image not only shows the distinct characteristics of the target in the two phase-images, but also analyze the difference of the background region. Specifically, the fusion scheme 1 adopts the fusion way with operation to integrate the complementary information of multi-temporal images effectively, which not only reflects the integrity and clarity of the target, but also ensures the smoothness of the background area.

Fusion scheme 2 takes the temporal image in a certain moment as the main image. By combining the salient features of another moment image, the movement information of the same target can be effectively reflected in the final fused image.

7.2.4 Multi-focus Fusion

Multi-focus image fusion is an important branch of multi-source image fusion. In multi focus fusion, the input images will be focusing in different scenes. Some images may focus on the foreground and some on the background. In the application of digital camera, optical lenses suffer from a limited depth of focus, it is often not possible to get an image that contains all relevant objects in focus. Part of the images which is out of focus has less depth of field. One possible solution to overcome this problem is to take several pictures with different focus settings and combing them together into a single frame to get all the information from the less focus area using image fusion method. The goal is to enhance the image quality and information so that it provides more detail information than the information available in single image. This technology can improve the utilization of image information effectively and enhance the reliability of the system. Which lays a good foundation for the subsequent processing of image recognition, edge detection, image segmentation and feature extraction. At present, multi-focus image fusion has been widely used in target recognition, microscopic imaging, military operations, machine vision and other fields.

According to the different stages of multi-focus image fusion in the process of processing, it can be divided into Pixel-level image fusion, Feature-level image fusion, or Decision-level image fusion. No matter which stage the fusion is, the key to multi-focus image fusion is to find a clear area or pixels in the source image. Then a clear fused image of all the scenes can be obtained by reorganizing it.

7.3 Image Fusion Schemes

It is difficult to give an accurate classification to the multi-source image fusion. The actual fusion process can be divided into different levels according to the different forms of information flow. A generally accepted stratification approach is according to the process of fusion, which is similar to the fusion scheme of multi-sensor fusion. The fusion is divided into four levels from low to high: signal level fusion, data level fusion (pixel level fusion), feature level fusion, and decision level fusion, as shown in Fig. 7.8.

(1) Signal level fusion

The signal level fusion is to produce a fused signal by mixing the unprocessed sensors' output at the lowest level in the signal domain. The fused signal is the same as the source signal, but the quality is better. The signal from the sensor can be modeled as a random variable with different related noises. In this case, fusion can be considered as an estimation process. To a large extent, the signal level image fusion

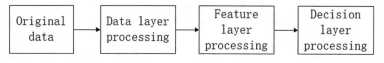

Fig. 7.8 Four layers of image fusion process

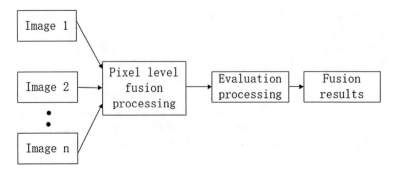

Fig. 7.9 General steps for the fusion of pixel level images

can be regarded as the problem of optimal concentration or distribution detection of signals, and whose registration requirement for the time and space of the signal is the highest.

(2) Pixel level fusion

Data level fusion is also known as pixel level fusion (Fig. 7.9). The narrow sense of image fusion refers to the fusion of pixel level images, which directly processes the data collected by the sensors to obtain the fused images. It is the basis of high level image fusion and one of the key points of the present image fusion research. In other words, this fusion is a fusion of different physical parameters. Therefore, a pixel of a given spatial location of the fused image is derived from each pixel of the source images at that location and their associated neighborhoods. That is to say, there will be more details, such as the extraction of edge and texture, which is helpful to further analysis, processing and understanding of the image. It can also expose potential targets which is benefit to identify potential target pixels. Pixel based methods generally deal with pixel level information directly, they could keep as much information as possible in the source image, providing subtle information that cannot be provided by other fusion levels, and is more suitable for further processing and analysis on the computer. However, the limitation of pixel level image fusion cannot be ignored. As it operates on the lowest level of pixels, these methods are generally time consuming as they require more number of computations. In addition, when data communication is carried out, the amount of information is large, and it is easily affected by the noise. Besides, if the fusion is conduct directly without a strict registration, the contrast of the image will be affected directly by the blurring effect.

In general, the existing pixel level fusion methods can be subdivided into two categories: one is spatial domain based, and the other is transform domain based. In the former, there are many kinds of methods, such as logical filtering method, gray-weighted average method and contrast modulation method, etc. There are also the algorithms of pyramidal decomposition fusion and wavelet transform method in the transformation domain. Where the wavelet transform is the most important and commonly used method at present. Nowadays, there are two main problems in image fusion based on wavelet transform: the selection of optimal wavelet basis functions and the selection of optimal wavelet decomposition layers.

(3) Feature level fusion

As shown in Fig. 7.4, the feature level fusion involves the integration of feature sets corresponding to multiple information sources, where the feature sets extracted from multiple data sources can be fused to create a new feature set to represent the individual. Image fusion based on feature level belongs to the middle level which operates on the characteristics such as size, edge, shape etc., and its advantage is that it achieves certain information compression and is conducive to real-time processing.

Image features include a lot of content, such as physical features (including spectrum, electromagnetic characteristics, etc.), geometric features and mathematical features, etc. It can be the shape, size, texture, contrast, etc. It can also be the observer's target or interest area in the source image, such as the contour, character, building or vehicle, etc. In the field of image recognition, people usually use physical and geometric features to identify objects, as these characteristics are easily discovered by human vision. The geometric feature is the structure description of the visual attribute of a certain aspect of the graph object, which could reflect the characteristics of the target more essentially than the original image. Image geometric features and its extraction technology is a key problem in image information processing. The basic geometric features of the target in the image are the edge points, the line segments and the regions. The edge point is the reflection of the discontinuity of image grayscale. A line segment is the description of a continuous edge point, which often parameterized into a geometric line, such as a line segment and a curve segment. A region is a set of pixels that are connected and have a certain consistent attribute. However, due to the limitation of the physical properties of single image and the influence of various interference factors in imaging process, it is often difficult to obtain the geometric description of the object closely related to the identification purpose. By using multi-source image information, the range and accuracy of the description of the various features of the target and scene are expanded. The characteristics of objects and scenes can be reflected simply and clearly in multi-source image data, so that it is possible to extract the geometric feature closely related to the identification purpose. Obviously, this kind of image geometric feature extraction is more import for image understanding.

At present, there are few studies in this field, and the research content we had seen in the literature including fusion distance and visible light image's edge extraction, fusion of multiband image region extraction, fusion of multiband line extraction, etc. These methods are basically based on the features of single image extraction, and the information between the multi-source images cannot be fully utilized in the feature extraction stage. A better approach is to combine image feature extraction and fusion together. In the process of feature extraction, fused the information of multi-source images effectively. Mining all the information of the image to form a feature description with the fusion nature. This reflects the synthesis of information reflected in each image, which derives from all the information of the image, but it is also different from the individual features extracted from each image.

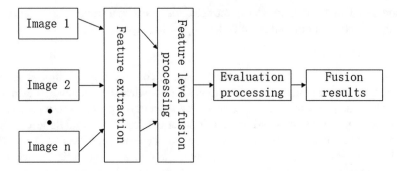

Fig. 7.10 The general steps of feature level image fusion

In the feature level fusion, it is necessary to ensure that different images contain the feature of information. For example, the characterization of the infrared light for the heat of the object, the characterization of the visible light for the brightness of the object and so on.

Feature level fusion compressed the image information in advance, and then use the computer to analyze and process, which will reduce consumed memory and time compared to the pixel level. And the real-time performance of the desired image processing will be improved. Feature level image fusion requires less image matching accuracy than the first layer, and the speed of calculation is also faster than the first one. However, it extracts the image features as fusion information, which will lose a lot of detail characteristics (Fig. 7.10).

(4) Decision level fusion

Decision level fusion is simply a selective voting for obtaining a conclusive decision from the source images, which is a cognitive-based approach. It is not only the highest level of image fusion, but also a higher abstraction of the image information, which is directly aimed at the specific decision making. In the decision level image fusion method, each image data source has been transformed to obtain an independent target attribute estimation, which is already a representative symbol or corresponding decision for information extraction. Then the attribute decision from each data source is fused, as shown in Fig. 7.11. Therefore, its fusion result directly affects the level of decision-making.

The main advantage of decision level fusion is that the computation of the decision level image fusion is the smallest and the fusion center processing cost is low, which requires low demand for information transmission bandwidth. When one or more sensors are wrong, the system can get the correct results through proper fusion. This method has a wide range of applications, and has no special requirements for the original sensor, sensors that provide the original data can be heterogeneous sensors, and can even include information obtained by non image sensors. However, this approach has a strong dependence on the previous level, and the obtained image is not very clear compared with the previous two fusion methods. It is more difficult

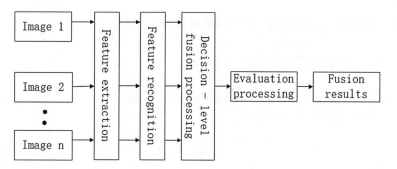

Fig. 7.11 General steps for decision level fusion

to realize the decision level image fusion, but the image transmission noise has the least influence on it.

The decision level image fusion is mainly depending on the subjective requirements, and there are also some rules to make use of the feature information obtained from the feature level images. Then the optimal decision is made directly according to certain criteria and the reliability of each decision (the probability of the existence of the target). The common research contents of multi-source image decision fusion technology can we seen in the current literature include the classification of remote sensing images, the classification of hyperspectral images and object recognition. And the techniques used are voting method, Bayesian method, consensus theory, evidence theory method, neural network and fuzzy integral. In particular, the D-S evidence theory can describe the uncertainty information by "interval estimation" rather than "point estimation" method, which shows great flexibility in distinguishing between what is unknown and what is uncertain as well as accurately reflecting the collection of evidence. Therefore, the D-S evidence theory is a kind of decision fusion method which is suitable for the application of object recognition

(5) Selection of different fusion strategies

The selection of each adaptation level depends on the different factors in the actual situation, such as the image source. At the same time, the selection of different level processing is also related to the result of image preprocessing. Since pixel level fusion is an early stage fusion wherein each pixel of the source images carries similar importance, it becomes the most popular. Most of the proposed image fusion algorithms belong to the fusion at this level. Over the past two decades, multiscale transforms, such as the pyramid transform and discrete wavelet transform (DWT), have been widely used for pixel-level image fusion.

7.4 Image Fusion Using Wavelet Transform

7.4.1 Basis of Wavelet Transform

Let $f(t)$, $\varphi(t)$ be the quadratic integrable function, $\psi(w)$ is the Fourier transform of $\varphi(t)$ and it satisfies the condition:

$$\int_{-\infty}^{\infty} \frac{|\psi(w)|^2}{w} dw < \infty \qquad (7.9)$$

Then,

$$W_f(a, b) = -\frac{1}{\sqrt{a}} \int_{-\infty}^{\infty} f(t)\overline{\varphi(\frac{t-b}{a})}dt, a > 0 \qquad (7.10)$$

is called as continuous wavelet transform (CWT) of $f(t)$, where $\varphi(t)$ is called the wavelet function or wavelet generating function, a is called the scale factor and b is called the translation factor.

In practical applications, especially the implementation on computers, continuous wavelet must be discretized. This discretization here is for the continuous scale parameter a and the continuous translation parameter b, but not the time variable t. If $\varphi(t)$ is the wavelet generating function, the discrete wavelet generating function is:

$$\varphi_{jk} = a_0^{-\frac{1}{2}} \varphi(a_0^{-j}t - kb_0)j, k \in z \qquad (7.11)$$

If $f(t)$ is any quadratic integrable function and satisfies the condition:

$$A\|f\|^2 \le \sum_{jk} |<f, \varphi_{jk}>|^2 \le B\|f\|^2 \qquad (7.12)$$

Then $\{\varphi_{jk}(t), j, k \in z\}$ is called a wavelet framework, where

$$W_j(j, k) = \int_{-\infty}^{\infty} f(t)\overline{\varphi_{jk}(t)}dt \qquad (7.13)$$

is called the discrete wavelet transform of $f(t)$. If the wavelet bases above is orthogonal at the same time, also known as an orthogonal wavelet transform. If $a_0 = 2$, $b_0 = 1$, the wavelet transform described above is dyadic wavelet transform.

7.4.2 Discrete Dyadic Wavelet Transform of Image and Its Mallat Algorithm

Image fusion is to combine two or more images of the same object into an image, making it easier for people to understand than any of the original images. If an image is decomposed by L-layer wavelet, a 3L+1 layer sub-band is obtained, which includes low frequency baseband C_j and the high-frequency sub-band D^h, D^v and D^d of the 3L layer. With $f(x, y)$ on behalf of the original image, recorded as C_0. Let the filter coefficients matrix of the scale coefficients be $\phi(x)$ and the wavelet coefficients $\psi(x)$ be H and G, then the dyadic wavelet decomposition algorithm can be described as

$$\begin{cases} C_{j+1} = HC_jH' \\ D^h_{j+1} = GC_jH' \\ D^v_{j+1} = HC_jG' \\ D^d_{j+1} = GC_jG' \end{cases} \tag{7.14}$$

In this formula, j represents the number of decomposed layers, h, v, d are represented as the horizontal, vertical, diagonal components respectively; H' and G' are conjugate transpose matrixes of H and G. After the two-dimensional image is decomposed by wavelet, low-frequency sub-images and high-frequency sub-images of horizontal, vertical and diagonal directions can be obtained. Low-frequency sub-images can also continue to further decompose. Therefore, if the two-dimensional image is decomposed by the N-layer wavelet, and ultimately there will be $3N + 1$ high-frequency components and a low-frequency component. When $N = 2$, the wavelet decomposition of the image is shown in Fig. 7.12.

The algorithm of wavelet reconstruction is:

$$C_{j-1} = H'C_jH + G'D^h_jH + H'D^v_jG + G'D^d_jG \tag{7.15}$$

Mallat proposed a fast decomposition and reconstruction algorithm for wavelet transform, which uses two one-bit filters to realize the fast wavelet decomposition

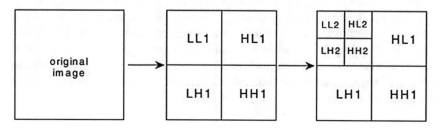

Fig. 7.12 Wavelet decomposition of the image

of two-dimensional images, and then reconstructs the image by using two one-bit reconstruction filters. Let H (low-pass) and G (high-pass) be the two one-bit mirror filtering operators, r and c correspond to the rows and columns of the image respectively. According to Mallat algorithm, there is the following decomposition formula under the scale j:

$$C_{j+1} = H_r H_c C_j$$
$$D_{j+1}^h = G_r H_c C_j$$
$$D_{j+1}^v = H_r G_c C_j$$
$$D_{j+1}^d = G_r G_c C_j \qquad (7.16)$$

In the formula,$C_{j+1}, D_{j+1}^h, D_{j+1}^v, D_{j+1}^d$ correspond to low-frequency components, high-frequency components in the vertical direction, high-frequency components in the horizontal direction, high-frequency components in the diagonal direction of image C_{j+1} respectively. The corresponding Mallat reconstruction algorithm of two-dimensional image is:

$$C_j = H_r^* H_c^* C_{j+1} + H_r^* G_c^* D_{j+1}^2 + G_r^* G_r^* D_{j+1}^3 \qquad (7.17)$$

where H^*, G^* are the conjugate transposed matrices of H, G respectively. The low frequency part reflects the approximate and average characteristics of the original image, and the three high frequency components are the detail parts of the image, reflecting the edge information of the image.

7.4.3   Steps of Implementation

The general structure of image fusion technique based on wavelet transform is shown in Fig. 7.13. Firstly, the original fused image is filtered by the low and high frequency filtering, and the original image is decomposed into 4 sub images with different frequency components. The above process is repeated according to the need of the low frequency sub-images, that is, wavelet tower decomposition of each image is established. And then the fusion layer is merged, and according to different requirement, the different frequencies of layers using different fusion operators for fusion processing. The fusion wavelet pyramid is finally obtained, and the wavelet transform is applied to the reconstructed wavelet pyramid, that is, image reconstructed, the resulting reconstructed image is a fused image. This can combine the details from different images effectively together to meet the actual requirements, which is conducive to human visual effects.

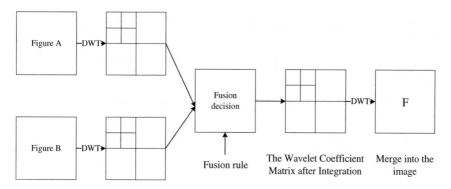

Fig. 7.13 General structure block diagram of image fusion based on discrete dyadic wavelet transform

The specific steps are as follows:

(1) Preprocessing of image

   Image filtering: due to the fusion of distorted image will inevitably lead to image noise into the fusion effect, the original image must be preprocessed to eliminate noise before fusion.

   Image registration: As the information provided by multiple imaging modes or multi-focal sources is often complementary, in order to synthesize multiple imaging modes or multi-focal sources to provide more comprehensive information, it is often necessary to fuse the effective information so that multiple images can be completely matched to the geometric positions in the spatial domain.

(2) The wavelet transform of each original image is carried out, and the wavelet tower decomposition of each image is established to obtain the low and high frequency components of the image.

(3) According to the characteristics of low frequency and high frequency components, each decomposition layer is fused according to their respective fusion algorithms. Different frequency components of each decomposition layer can be fused by different fusion operators, and finally the fused wavelet pyramid is obtained.

(4) The wavelet transform of the wavelet pyramid after fusion is reconstructed, that is to say, the reconstructed image is the fused image.

The image fusion code based on wavelet transform is realized as shown in Programme 7.1

PROGRAMME 7.1: The image fusion code based on wavelet transform

```
clear
[imA,map1] = imread('D:\Program Files\MATLAB\R2010b\workspace\clock1.jpg');
M1 = double(imA) / 256;
[imB,map2] = imread('D:\Program Files\MATLAB\R2010b\workspace\clock2.jpg');
M2 = double(imB) / 256;
zt= 3;
wtype = 'bior6.8';
%       M1 - input image A
%       M2 - input image B
%       wtype    the type of wavelet used
%       Y   - fused image
%%%%%%%%%%%%%%%%%%%%%%%%%%%%%%%%%%%%%%%%%%%%%%%%  %%
    wavelet transform of image fusion
%%%%%%%%%%%%%%%%%%%%%%%%%%%%%%%%%%%%%%%%%%%%%%%%%%%
    the wavelet coefficient with large absolute value in the wavelet transform correspond to significant
    changes in brightness, that is, the significant features in the image.
%%      select the wavelet coefficients with large absolute value as the required wavelet coefficient.
%%      the low-frequency partial coefficient adopts the method of seeking the average of two
%%%%%%%%%%%%%%%%%%%%%%%%%%%%%%%%%%%%%%%%%%%%%%%%%
[c0,s0] = wavedec2(M1, zt, wtype);% decomposition of multi-scale dyadic wavelet
[c1,s1] = wavedec2(M2, zt, wtype);% decomposition of multi-scale dyadic wavelet
%%%%%%%%%%%%%%%%%%%%%%%%%%%%%%%%%%%%%%%%%%%%%%%%  %%
    the wavelet coefficient with large absolute value are taken as the wavelet coefficient after fusion, and
    then the total image effect is reconstructed
%%%%%%%%%%%%%%%%%%%%%%%%%%%%%%%%%%%%%%%%%%%%%%%%%
KK = size(c1);
Coef_Fusion = zeros(1,KK(2));
%%      low-frequency coefficient processing, while also deal with high-frequency coefficient, but the
    results will be covered when the latter part of the processing of high-frequency coefficient.

Coef_Fusion(1:s1(1,1)) = (c0(1:s1(1,1))+c1(1:s1(1,1)))/2;
%%      deal with high-frequency coefficient
    MM1 = c0(s1(1,1)+1:KK(2));
    MM2 = c1(s1(1,1)+1:KK(2));
    mm = (abs(MM1)) > (abs(MM2));
    Y   = (mm.*MM1) + ((~mm).*MM2);
    Coef_Fusion(s1(1,1)+1:KK(2)) = Y;
%%      reconstruction
```

```
Y = waverec2(Coef_Fusion,s0,wtype);
%%      display the image
subplot(2,2,1);imshow(M1);
colormap(gray);
title('input1');
axis square
subplot(2,2,2);imshow(M2);
colormap(gray);
title('input2');
axis square
subplot(223);imshow(Y,[]);
colormap(gray);
title('the fusion image ');
axis square;
%%%%%%%%%%%%%%%%%%%%%%%%%%%%%%%%%%%%%
```

Figure 7.14, taking two multi-focus [11] image fusion as an example, results of the image fusion with being subjected to wavelet transform are given. Figure 7.14a is the right focus image, the right half is clear, the left half is blurred; Figure 7.14b is the left focus image, the left half is clear while the right half is blurred. Figure 7.14c is an image obtained using a wavelet method. Figure 7.14d is a clear image, which is obtained by artificial method. It can be seen from Fig. 7.14 that the fused images are larger than those of the original image (7.14a, b), and the details of the left and right are clear.

7.5 Region-Based Image Fusion

Regional feature refers to the distribution of point [12, 13] or local feature within the object in the image, it also refers to a statistic and a regional geometric feature (area, shape) and so on. The traditional pixel-level method of image fusion separates the connection between pixels. Due to the region can represents the target information, the region-based fusion is more practical than the pixel-based fusion. According to the study in recent years, the method of extracting the region of object and performing the image fusion on the basic of the regional characteristics can achieve more reasonable fusion effect compared with the pixel-level fusion without regional division. The regional fusion takes the correlation between adjacent pixels into account and highlights the characteristics of region, it also reduces the noise sensitivity.

(a)Right focus clock—1   (b)Left focus

clock--2

(c) Fusion image   (d)Clear image of artificial cutting

Fig. 7.14   Results of multi-focus fusion

7.5.1   Basic Framework of Regional Integration

In the process of fusion, the image fusion based on the region take appropriate method
for the two images exactly matched with each other to obtain regional representation
of the image according to the characteristics of the original image, and then take the
results of the two images' regional representation to perform the joint area representation and determines the target and background region of the image on basic of the
joint area. At last, different fusion methods are used to fuse the target and background

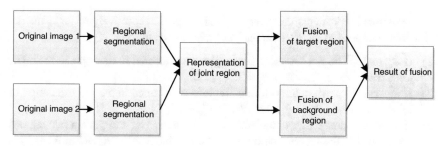

Fig. 7.15 Framework of image fusion based on region representation

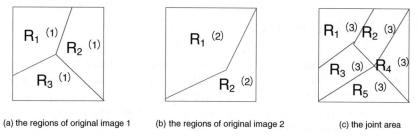

(a) the regions of original image 1   (b) the regions of original image 2   (c) the joint area

Fig. 7.16 Schematic the joint area representation

respectively, then the result of image fusion is available. Figure 7.15 shows the basic fusion framework based on the region representation.

The key of the fusion scheme is that the representation of respective regions of the original image, the representation of the image union region, and the rules of image fusion. There are many ways to carry out the representation of original image effectively. Among them, the method of regional segmentation, the methods of analyzing the characteristic statistic in the region (such as gray statistic, maximum), the methods of regional energy, the method of combining region and transforming are common.

7.5.2   The Strategy of Regional Joint Representation

After the regional representation of the original image is obtained, it is necessary to perform the joint area representation. In general, the information contained in the captured image is different because of the different principles of the components of the images. Therefore, the region composition of the original image will certainly vary widely, which requires a certain algorithm to combine the regional composition so that it can fully include the information contained in the original image, and this joint can be called 'joint area representation'. Figure 7.16 shows a schematic representation of joint area.

The regional representation of the original image is respectively R_m^1, R_m^2, and the joint area is expressed as R_m^u, then the joint rules are as follows:

(1)   If R_m^1 and R_m^2 do not intersect, two regions are formed in the representation of joint area, that is $R_1^3 = R_m^1$, $R_2^3 = R_n^2$;
(2)   If R_m^1 and R_m^2 partly intersect, three regions are formed in the representation of joint area, that is $R_1^3 = R_m^1 \cap R_n^2$, $R_2^3 = R_m^1 - R_1^3$, $R_3^3 = R_n^2 - R_1^3$;
(3)   If there is an area that is completely include, such as $R_m^1 \subset R_n^2$, then two regions are formed in the representation of joint area, that is $R_1^3 = R_n^2$, $R_2^3 = R_n^2 - R_1^3$.

7.5.3   The Rules of Fusion

After defining the joint area representation, according to certain rules, each original image needs to be distinguished between the target region and the background region, that is, the rules of fusion based on the region feature. Currently more common ways are as follows: (1) gradient-based method; (2) regional variance based method; (3) regional energy based method.

Gradient reflects the details of the edge of image, the greater of the gradient value, the more obvious the feature and the greater the degree of change in image information, so that the use of gradient for image fusion can effectively reduce the fuzzy region information on the impact of fusion effects. However, the gradient algorithm only takes the degree of change of the coefficient into account, and it cannot get a good response to the richness of image information, and it is easy to cause the lack of useful information of the high-frequency part of images.

Energy can reflect the richness of the image information well, but it cannot reflect the degree of change of the image information. Therefore, the information of the fuzzy region is introduced to a certain extent, then the effect of fusion is disturbed and ability of representation of the image is weakened.

Variance describes the degree of variation and the dispersion of the pixels in the region. The larger the variance value, the more dramatic the pixel changes in the region and the more scattered the gray scale.

7.5.4   Wavelet Fusion of Regional Variance

Firstly, the original image is decomposed by discrete wavelet frame, and different rules of image fusion are used to fuse the low and high frequency images. The low frequency sub band is generally fused by the weighted average operator, and the fusion rule of high frequency sub band coefficient take a local widow as the object of study to calculate the statistical characteristics in the local area. Because there is an intense correlation among the pixels of the image, it is more likely to reflect the characteristics and trends of the image in a region than in a signal pixel. In a local window, the statistical features are more obvious, indicating that the greater the gray scale changes, the richer the details contain. There are many statistical feature of a local window, such as variance, gradient, energy and so on. Therefore, fusion rules based on local window have different forms according to the use of the different statistical features.

The process of fusion is as follows:

(1) The original image is transformed into high and low frequency sub-image by wavelet transform.
(2) The energy of the image is dispersed on the low and high frequency component. For low frequency components, the averaging method is used to obtain the low frequency components required for reconstruction. For high frequency components, a 3×3 sliding window is used to find its variance, and the high frequency coefficient with large variance is chosen as the high frequency component needed for reconstruction.
(3) Finally, the new image is reconstructed by the new coefficient by inverse transformation of discrete wavelet framework to obtain the fused image.

Figure 7.17 shows the results of multi-focus image fusion based on regional variance wavelet fusion and several other fusion methods. From the comparison of fusion results of the different methods given in Fig. 7.17, it is can be found that the simplest image fusion strategy with the highest absolute value of pixel grayscale value makes the whole image not very clear, and the effect of the image fusion strategy is not very good that taking a large absolute value of low or high frequency in the low-frequency part of the image fusion. Due to the low frequency coefficients represent the overall contour of the image, the majority of the information of the original image is concentrated, which reflects the original image's profile at that resolution and the high frequency information reflects the brightness mutation characteristic of the original image, that is, the edge of the original image, regional boundary characteristics, the result of the fusion of high frequency coefficient affects the details of the image information, that is the key to fusion, the simple maximum value cannot retain most of the image information before the fusion; the method of regional variance is used to make a plurality of pixels constituting a local area participate as a whole in the fusion process, this integrated visual effect of the fusion image is better, and the fusion trace can be effectively suppressed, and the resulting fusion image is clearer than the other methods.

The wavelet fusion code based on the regional variance as shown in Programme 7.2.

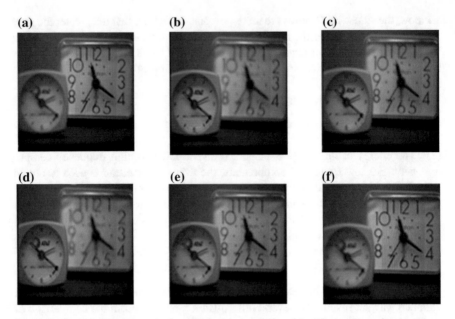

Fig. 7.17 Several different methods **a** original image 1; **b** original image 2; **c** take the largest regional variance; **d** take the largest absolute value; **e** take the larger absolute value of fusion strategy of low and high frequency; **f** average value of low frequency and largest value of high frequency

PROGRAMME 7.2: The wavelet fusion based on the regional variance

```
Clear all;
close all;
%    read the original image
x=imread('A.jpg');
figure;
imshow(x);
title('original image 1')
y=imread('B.jpg');
figure;
imshow(y);
title('original image 2')
a=x;
```

```
b=y;
a=double(a);
b=double(b);
%     perform 2D wavelet transform
[aA,aH,aV,aD]=dwt2(a,'bior2.4');
[bA,bH,bV,bD]=dwt2(b,'bior2.4');
newA=zeros(size(aA));
newH=zeros(size(aH));
newV=zeros(size(aV));
newD=zeros(size(aD));
%     take average of the low frequency coefficient
[m,n]=size(aA);
for i=1:m
    for j=1:n
        newA(i,j)=(aA(i,j)+bA(i,j))/2;
    end;
end;
%   with 3*3 sliding window foe variance
fun=inline('var(x(:))'');
Var_aH=nlfilter(aH,[3 3],fun);
Var_aV=nlfilter(aV,[3 3],fun);
Var_aD=nlfilter(aD,[3 3],fun);
Var_bH=nlfilter(bH,[3 3],fun);
Var_bV=nlfilter(bV,[3 3],fun);
Var_bD=nlfilter(bD,[3 3],fun);
%   select the corresponding coefficient of larger variance as the high frequency coefficient
for i=1:m
    for j=1:n
        if Var_aH(i,j)>=Var_bH(i,j);
            newH(i,j)=aH(i,j);
        else
            newH(i,j)=bH(i,j);
        end
    end
end
for i=1:m
    for j=1:n
```

```
            if Var_aV(i,j)>=Var_bV(i,j);
                newV(i,j)=aV(i,j);
        else
                newV(i,j)=bV(i,j);
        end
    end
end
for i=1:m
    for j=1:n
        if Var_aD(i,j)>=Var_bD(i,j);
                newD(i,j)=aD(i,j);
        else
                newD(i,j)=bD(i,j);
        end
    end
end
%     reconstruction of image
new=idwt2(newA,newH,newV,newD,'bior2.4');
new=uint8(new);
figure;
imshow(new);
title('take large regional variance ')
```

7.6   Image Fusion Using Fuzzy Dempster-Shafer Evidence Theory

The process of image fusion using fuzzy Dempster-Shafer evidence theory is shown as Fig. 7.18. First, fuzzy C-Means clustering is operated on two source images to get the fuzzy membership degree of each point in each image. Second, the simple hypothesis and compound hypothesis are determined according to the fuzzy category. Third, the single and compound basic probability assignment mass function values of each pixel in the two images are determined by the heuristic least squares algorithm. Finally, the basic probability assignment of two images is fused with the Dempster criterion in D-S evidential theory and the final fusion result is obtained by decision.

Choosing two images that having the same object to do the fusion, shown as Fig. 7.19, it can fuse the information of two images and obtain a better image with more information.

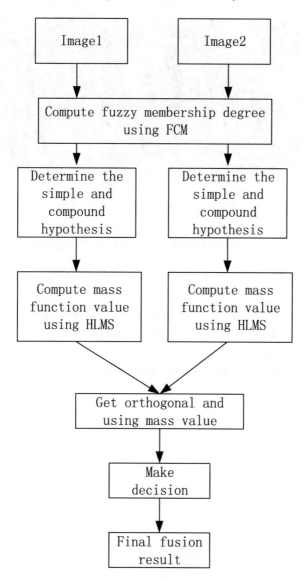

Fig. 7.18 Flow chart of algorithm

The MATLAB code of image fusion which use Fuzzy Dempster-Shafer evidence theory is shown as Programme 7.3.

Fig. 7.19 Fuzzy evidence fusion example

PROGRAMME 7.3: Image fusion using Fuzzy Dempster-Shafer evidence theory

```
% DS-fusion theory function
function x=DS_fusion(x,y)
% function:fuse x, y two vectors(classic Dempster-Shafer Combination formula)
% x,y s' format looks like[m1 m2 m3, ... , mk, m(complete set), m(null set)]
% Require m1 m2 m3 ...don't have Intersection
% m(complete set)could be zero or not, representing Uncertainty
% m(null se)must be 0
[nx,mx]=size(x);
if 1~=nx
    disp('x should be row vector');
    return;
end
[ny,my]=size(y);
if 1~=ny
    disp('y should be row vector);
    return;
end
if mx~=my
    disp('x,y should have equal cols');
    return;
end
temp=0;
for i=1:mx-1
    if i==mx-1
        x(1,i)=x(1,i)*y(1,i);    %special operation on complete set
    else
        x(1,i)=x(1,i)*y(1,i)+x(1,i)*y(1,mx-1)+y(1,i)*x(1,mx-1);
```

```
        end
            temp=temp+x(1,i);
    end
    for i=1:mx-1
        x(1,i)=x(1,i)/temp;
    end
    x(1,mx)=0;

    % Image fusion function
    subplot(1,3,1);
    imshow(g);
    title('Fist image'');
    subplot(1,3,2);
    imshow(h);
    Title  ('Second image');
    g1=g(:);
    h1=h(:);
    g2=g1';
    h2=h1';
    a(1,1:65536)=DS_fusion(g2,h2);
    b=a;
    k=6500000.*b;
    l=uint8(k);
    l2=reshape(l,256,256);
    subplot(1,3,3);
    l3=imadjust(l2,[],[],1);
    imshow(l3);
    title('fusion result');
    %%%%%%%%%%%%%%%%%%%%%%%%%%%%%%%%%%%%%%%%%%%%%%%%%%%%
```

7.7 Image Quality and Fusion Evaluations

Although the image fusion method is numerous, the technology is also endless, the purpose is nothing more than to improve the picture quality or increase the content of the image information, which is the effect of the evaluation of the fundamental starting point. For different levels of fusion, the evaluation of the effect of indicators is not the same. In terms of the underlying fusion, generally the visual effects can be compared and analyzed, and the higher the level, the greater the degree of demand satisfaction. In theory, the fusion of image should be to preserve the effective information in two or more images and to synthesize them into an image. Therefore, the evaluation of the fusion effect should include two aspects: the improvement level and the continuation level [14–16].

For image observers, the meaning of the image mainly includes two aspects: one is the fidelity of the image, the other is the image of the comprehensibility. The existing methods of image fusion performance evaluation can be divided into: objective and subjective evaluation of fusion quality. The former by virtue of observation, which depends largely on the observer's subjective consciousness, with as well as the difference and variation, will change with the application area, where the situation, personal preferences and other changes, the latter is a quantitative calculation, through the value to judge, in general, it has a certain relevance to subjective evaluation.

7.7.1   Subjective Evaluation of Image Fusion

In the evaluation of image fusion effect, subjective evaluation mainly from the following aspects:

(1)   Registration accuracy evaluation. If the degree of registration deviation is small, ghosting will occur, if the deviation is large, there will be serious dislocation.
(2)   Color distribution evaluation. If the color distribution is reasonable, the naked eye will feel comfortable; if the distribution is unreasonable, the whole image color distribution is uneven, visual impact will increase.
(3)   Sharpness evaluation. If the sharpness is close to or improved with the original image, the fused image is clear; if the sharpness is reduced, the fused image will appear to a certain extent blurred.
(4)   Brightness and contrast evaluation. If this two are inappropriate, the fused image will have patches or fog and other noise-like parts.
(5)   Texture information evaluation. If the texture information is sufficient, the fused image will look plumper, if there a loss in the fusion process, it will become dull and lack of hierarchy.

In term of this aspect of evaluation, there are common international 5-point evaluation criteria, see Sect. 5.2.

7.7.2   Objective Evaluation of Image Fusion

For the subjective evaluation, the human eye can only see the obvious changes, the small differences are not sensitive, and subjective judgments will be affected by many factors and always vary. Therefore, a quantitative evaluation method with a uniform standard is indispensable. Now according to the evaluation principle, the objective evaluation method can be divided into statistical characteristics evaluation, information content evaluation, sharpness evaluation, signal to noise ratio (SNR) evaluation and spectral information evaluation. The following is a brief introduction to the main method. In the following evaluation indicators, the original image is $f(x, y)$, the fused image is $g(x, y)$, the ideal image is $i(x, y)$ and the size of image is $N \times N$.

1. Evaluation based on statistical characteristics
   There is no ideal standard reference image, so the fusion effect of image is objectively evaluated based on the statistical characteristics of the fusion image and the performance index which reflects the relationship between the fusion image and original image.

   (1) Average Value (AV) of image
       The size of the mean represents the average size of the image pixel values, which is an evaluation index that belongs to the statistical characteristics. The brightness that human eye can be perceived in grayscale images in the form of grayscale, so the average value of the gray scale has a greater effect on the visual effect of the image. If the average value of the image is appropriate, the result of fusion is better. The average value of the image is defined as:

$$\mu = \frac{1}{N \times N} \sum_{x=0}^{N-1} \sum_{y=0}^{N-1} g(x, y) \tag{7.18}$$

   (2) Standard Deviation
       The centrality or discretization of the image gray value relative to the gray scale is generally reflected by the standard deviation which reflecting the distribution of the pixel values and showing the contrast of the image. If the standard deviation of the fusion image is small, the contrast is small, that is, the amount of information contained therein is smaller. The larger the standard deviation, the more gray-scale distribution, the better the visual effect. The standard deviation is obtained indirectly by the average value, and the standard deviation of the image is defined as:

$$\sigma = \frac{1}{N \times N} \sqrt{\sum_{x=0}^{N-1} \sum_{y=0}^{N-1} [g(x, y) - \mu]} \tag{7.19}$$

   (3) Root mean square error
       The root mean square error can be used to detect the degree of deviation between the image to be detected and the ideal image, which can be evaluated using the known ideal image. The smaller the deviation between the fusion result and ideal image, the better the fusion effect. It is defined as:

$$RMSE = \sqrt{\frac{1}{N \times N} \sum_{x=0}^{N-1} \sum_{y=0}^{N-1} [g(x, y) - i(x, y)]^2} \tag{7.20}$$

2. Objective evaluation based on information content

    (1) Information entropy

        Information entropy is an important indicator of the degree of abundance
of image information. It is reflected degree of deviation in the image range
from the peak area of the gray histogram. The larger the entropy of the fused
image, the more the information volume of the fused image increases, the
richer the image is, the better effect of the image fusion. it is defined as:

$$H = - \sum_{l=0}^{L} h(l) \log[h(l)] \tag{7.21}$$

        where L represents the total gray level of the fused image F, $h(l)$ represents
the ratio of the number of pixels n_l of the gray scale value l to the total number
N of images, that is $h(l) = \frac{n_l}{N}$, which reflects the probability distribution
of the pixel with the gray value of l in the image can be regarded as the
normalized histogram of the image.

    (2) Joint entropy

        Joint entropy is also a parameter that reflects the amount of information
contained in the image. On this basis, it reflects the correlation between
the original image and the fusion result, and quantitatively measures the
correlation between them. Similarly, the greater the joint entropy of the
fusion result, the larger the amount of information carried, and the better
effect. It is defined as:

$$C(f : g) = - \sum_{l_2=0}^{L} \sum_{l_1=0}^{L} P_{fg}(l_1, l_2) \log P_{fg}(l_1, l_2) \tag{7.22}$$

3. Objective evaluation based on the sharpness

    (1) Average gradient

        The average gradient is also called sharpness, which reflects the small detail
contrast and texture change in the image, and also reflects the sharpness
of the image, which can be used as an index to judge the sharpness of the
fusion result. It is defined as:

$$A = \frac{1}{N \times N} \sqrt{\sum_{x=0}^{N-1} \sum_{y=0}^{N-1} [G_X^2(x, y) + G_Y^2(x, y)]^{1/2}} \tag{7.23}$$

        where $G_X^2(x, y)$ and $G_Y^2(x, y)$ are the difference in x and y direction, respec-
tively. In general, the larger the average gradient of the image, the greater
the clarity of the image, the better the fusion effect.

4. Objective evaluation based on spectral information

    (1) Correlation Coefficient

The correlation coefficient reflects the correlation degree of the two spectral features of the image. Generally speaking, the closer the correlation coefficient of the image is to 1, the better the proximity of the image is, the more information is obtained from the original image, the less information, the better the fusion effect. It is defined as:

$$CC = \frac{\sum_{x=0}^{N-1} \sum_{y=0}^{N-1} (f(x,y) - \mu_f)(g(x,y) - \mu_g)}{\sqrt{\sum_{x=0}^{N-1} \sum_{y=0}^{N-1} (f(x,y) - \mu_f)^2 (g(x,y) - \mu_g)^2}} \qquad (7.24)$$

where μ_f, μ_g are the average of the original image $f(x,y)$ and the fused image $g(x,y)$ respectively.

(2) Structure Similarity

The structural similarity is calculated as (7.25).

$$SSIM(f,g) = l(f,g)c(f,g)s(f,g) = \frac{(2\mu_f \mu_g + c_1)(2\sigma_f \sigma_g + c_2)(\sigma_{fg} + c_3)}{(\mu_f^2 + \mu_g^2 + c_1)(\sigma_f^2 + \sigma_g^2 + c_2)(\sigma_f \sigma_g + c_3)} \qquad (7.25)$$

where $l(f,g) = \frac{2\mu_f \mu_g + c_1}{\mu_f^2 + \mu_g^2 + c_1}$, $c(f,g) = \frac{2\sigma_f \sigma_g + c_2}{\sigma_f^2 + \sigma_g^2 + c_2}$, $s(f,g) = \frac{\sigma_{fg} + c_3}{\sigma_f \sigma_g + c_3}$ is brightness comparison, contrast comparison and structural comparison, respectively; $\mu_f, \mu_g, \sigma_f^2, \sigma_g^2, \sigma_{fg}$ represent the average, variance and covariance of the original image and the fusion image, respectively.

5. Objectively evaluation based on signal to noise ratio (SNR)

(1) Signal to noise ratio (SNR)

In the process of image fusion, the noise from the sensor that acquires the image is also a key factor to consider. Therefore, the signal to noise ratio has been applied, the greater the value and the better fusion effect. It is defined as:

$$SNR = 10 \lg \frac{\sum_{x=0}^{N-1} \sum_{y=0}^{N-1} (g(x,y))^2}{\sum_{x=0}^{N-1} \sum_{y=0}^{N-1} (f(x,y) - g(x,y))^2} \qquad (7.26)$$

(2) Difference Index (DI)

The average of the ratio of the absolute value of the difference between the fusion image and the original image and the original image value is called difference index. In general, the smaller the difference index, the smaller the degree of fusion image deviation from the original, the more the original grayscale information remains. It is defined as:

$$D = \frac{1}{N \times N} \sum_{x=0}^{N-1} \sum_{y=0}^{N-1} \frac{|f(x,y) - g(x,y)|}{f(x,y)} \qquad (7.27)$$

Ideally $DI = 0$.

(1)   Peak Signal to Noise Ratio (PSNR)

PSNR is achieved by assuming that the difference between the fused and original image is caused by noise, and the original image is treated as useful information to evaluate quality of the fused image. The larger its value, the closer relation between fusion image and the original image. It is defined as:

$$PSNR = 10 \lg \frac{N \times N \times [\max g(x, y) - \min(g(x, y))]}{\sum_{x=0}^{N-1} \sum_{y=0}^{N-1} [f(x, y) - g(x, y)]^2} \qquad (7.28)$$

(2)   Degree of Distortion (DD)

DD reflects the degree of distortion of the fused image relative to the original image, the smaller the value, the better the effect of fusing the image. It is defined as:

$$DD = \frac{1}{N \times N} \sum_{x=0}^{N-1} \sum_{y=0}^{N-1} |f(x, y) - g(x, y)| \qquad (7.29)$$

In addition to the above several indicators of image quality evaluation, there are some other evaluation indicators, such as general indictors of image quality evaluation, indictors of weighted fusion evaluation. Although the above list of indicators in most cases can accurately evaluate the quality of the image, but the exception of events has also occurred, that is why subjective evaluation is the main evaluation and objective evaluation is auxiliary in the practical application. Therefore, it is one of the hot issues in the study to process a general objective evaluation index which can accurately reflect the quality of the image.

References

1.   Waltz EL, Buede DM (1986) Data fusion and decision support for command and contral. IEEE Trans SMC. 16(16):865–879
2.   Yankui S (2012) Wavelet transform and image, graphics processing technology. Tsinghua University Press, Beijing, pp 6–8
3.   Gonzalez RC, Woods RE (2010) Digital image processing. Electronic Industry Press,p 491
4.   Lin N (2010) Wavelet transform and image processing. China Science and Technology University Press, China, p 19
5.   Defeng Z (2012) MATLAB wavelet analysis. Mechanical Industry Press, p 54–55
6.   Linas J, Waltz E (1990) Multisensor data fusion. Anech House, Norwood, Massachusetts
7.   Gonzalez JP, Ozguner U (2000) Lane detection using histogram—based segmentation and decision trees. In: Proceedings of the intelligent transportation systems, pp 346–351. IEEE, Dearborn, USA
8.   Jiallg H et al (1993) High-speed dual-spectral infrared imaging. Opt Eng 6:1281–1283
9.   Shiyi M, Wei Z (2002) Multi-sensor image fusion technology review. Beijing Univ Aeronaut Astronaut J 28(5):512–518

10. Xi G, Shuguang Z (2001) Based on the multi-sensor image fusion of gradient tower decomposition. Photoelectron Laser 12(3):293–296
11. Hui Y (2006) Research of multi-focus image fusion algorithm
12. Yinglei C, Chunrong Z, Weihua L et al Based on pixel level image fusion method. Comput Appl Res 2021(2):169–172
13. mingqi X, friend H, wen o et al (2003) Based on wavelet analysis. Infrared Laser Eng 32(2):177–181
14. Gonzalez RC, Woods RE, Eddins SL (2009) Digital image processing using MATLAB, vol 2. Gatesmark Publishing, Tennessee
15. Pajares G, Manuel de la Cruz. J (2004) A wavelet-based image fusion tutorial. Pattern Recogn 37(9):1855–1872
16. Yanxing C (2010) Research and implementation of splicing technology based on video image. Northeastern University

Chapter 8
Image Stitching

Abstract In this chapter we firstly introduce the application background and basic process of image stitching, then depict several image stitching methods based on region, image stitching methods based on feature points, and panoramic video image stitching techniques.

8.1 Introduction

In practice, it often needs the wide-view and high-resolution panoramic images, but the size of the image depends on the performance of the camera due to the limitation of the imaging device. Therefore, an approach which utilizes the computer software to stitch images was then raised for making panoramas. Image stitching refers to put several images with overlapping parts together into a large, seamless and high-resolution image. Figure 8.1 shows the sketch map of image stitching.

Generally, image stitching mainly includes the following five steps:

(1) Image preprocessing. It contains basic operations of digital image processing (such as denoising, edge extraction and histogram processing), establishment of image matching templates, image transforms (FT, WT, etc.) and other operations.
(2) Image registration. It adopts some kinds of matching algorithms to find the corresponding positions of the templates or feature points in stitching images so as to determine the transformation relation between two images.
(3) Build the transform model. A mathematical transform model can be built between two images by calculating parameters of the model based on the correspondences of image templates or features.
(4) Unified coordinate transformation. In accordance with the mathematical transform model built in step 3, the image to be stitched will be transferred into the coordinate system of the reference image in order to accomplish the unified coordinate transformation.

© Springer International Publishing AG, part of Springer Nature 2019   271
S. Gong et al., *Advanced Image and Video Processing Using MATLAB*,
Modeling and Optimization in Science and Technologies 12,
https://doi.org/10.1007/978-3-319-77223-3_8

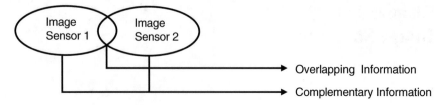

Fig. 8.1 Sketch map of image stitching

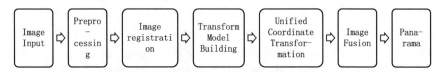

Fig. 8.2 Flowchart of image stitching

(5) Image fusion and reconstruction. Merging the overlapping portions of the images to be stitched to a smooth and seamless reconstructed panorama.

Figure 8.2 shows the basic flowchart of image stitching.

Image registration is the key to image stitching algorithms. According to different image registration methods, the image stitching algorithms can be classified into two categories: image stitching based on region and image stitching based on feature points.

8.2   Image Stitching Based on Region

Image stitching based on region starts from comparing the grayscale values of an area in image to be stitched with the area in referenced image which have the same size by the least squares methods and other mathematical methods. From the comparisons we can measure the similarity between the overlapping areas in images to be stitched, and get the range and position of the overlapping area in the image to be stitched to accomplish the image stitching task. We can also transform the images from spatial domain into frequency domain with FFT and operate the image registration later. To the images with large displacement, we can correct the rotation of the image and then establish the mapping between two images. When take the difference between the grayscale values of pixels in two regions as criterion, the simplest approach is directly adding up the differences pixel by pixel. Another way is to calculate the correlation coefficient between the pixel grayscale values of the two areas. The larger the correlation coefficient is, the higher the matching degree of the two images will be, and this way shows better performances as well as a higher success rate. Nowadays, the commonly used image stitching algorithms based on region include Ratio Matching, Block-based Matching, Line Matching and Grid Matching.

8.2.1 Image Stitching Based on Ratio Matching

Image stitching based on ratio matching first selects the ration of two columns of pixels with a certain distance between the overlapped parts of the image as a template [1]. Then search the best match for the overlapped region in second image and find the two columns corresponding to the template taken from the first image to complete the image stitching. Figure 8.3 is a sketch map of the algorithm. Picture 1 stands for an $(W_1 \times H)$ image in pixels and Picture 2 is a $(W_2 \times H)$ one. W_1 and W_2 may be equal or not. Picture 1 is on the left of Picture 2. Another situation that images are vertical overlapped will not be discussed in this chapter since we can handle it in a similar way.

Following are the steps of this algorithm:

(1) Select two columns of pixels with the interval *span* from overlapped area of Picture 1, calculate the corresponding pixel ratio as template *a*.

$$a(i,j) = \frac{P_1(i,j)}{P_1(i,j+span)}, \quad i \in (1,H) \tag{8.1}$$

(2) In Picture 2, each two columns with the interval of span are selected in turn from the first column, the ratio of its corresponding pixels is calculated as template *b*.

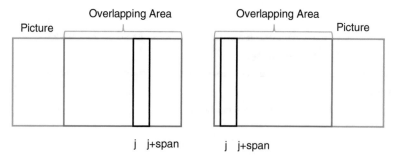

Fig. 8.3 Sketch map of template choosing

$$b(i, j) = \frac{P_{21}(i, j)}{P_{22}(i, j)} \tag{8.2}$$

$$P_{21}(i, j) = P_2(i, j), (i \in (1, H), j \in (1, W_2 - span))$$
$$P_{22}(i, j) = P_2(i, j), (i \in (1, H), j \in (1, W_2 - span))$$

(3) Calculate the differences between template a and b as template c.

$$c(i, j) = (a(i, j) - b(i, j))^2, \quad i \in (1, H), j \in (1, W_2 - span) \tag{8.3}$$

(4) c is a two-dimensional array. Add each column vector up into another array called sum: $sum(j) = \sum_{i=1}^{H} c(i, j)$. The value of $sum(j)$ reflects the difference of selected columns in two images. The column coordinates of $sum(j)$'s minimum value sum_{min} are the best match.

PROGRAMME 8.1 is the code of image stitching based on ratio matching.

PROGRAMME 8.1: Image stitching based on ratio matching

```
clear;
clc;
A=imread('lenna_left.jpg');% read Picture 1, A represents the pixel array of Picture 1
B=imread('lenna_right.jpg');% read Picture 2
[x1,y1]=size(A(:,:,1));
% transform into greyscale images, calculate the length and height of Picture 1
[x2,y2]=size(B(:,:,1));
A1=double(A);
B1=double(B);
sub_A=A1(:,end-1)./A1(:,end);%calculate the ratio of last two columns in Picture 1
sub_D = zeros(size(B,2)-1,2);%define sub_D
for y=1:y2-1
    sub_B=B1(:,y)./B1(:,y+1);% calculate the ratio of two adjacent columns in Picture 2
sub_C=(sub_A-sub_B)'*(sub_A-sub_B);
%calculate the difference between template a and b, calculate the sum of column vector
    sub_D(y,1)=y;
    sub_D(y,2) =sub_C;%sub_D is a two-dimensional array
end
```

```
[a b]= sort(sub_D(:,2));% ascending sort
row = b(1,:);% the coordinate of first element is the best match
x3=x1;
y3 = y1-1+y2-row;%length and height of image to be stitched
C=zeros(x3,y3);
for i=1:x3
    for j=1:y3
      if j<y1
          C(i,j)=A(i,j);
      else
          C(i,j)=B(i,row+j-y1);
      end
    end
end% image stitching
imwrite(C,'picture3.bmp');
imshow(mat2gray(C));
```

In order to confirm the effectiveness of image stitching based on ratio matching, simulation experiments are carried out for two images with overlapped regions. Figure 8.4 shows the result.

    (a) Lenna_left   **(b) Lenna_right**   **(c) Output**

Fig. 8.4   Input and output of experiment

8.2.2   *Image Stitching Based on Line and Plane Feature*

Image stitching based on line and plane feature mainly includes: image preprocessing, feature block searching, image stitching and image fusion. Figure 8.5 shows the flowchart.

(1)   Image preprocessing. Because of the different illumination, it is easy to make stitching errors if the raw images obtained directly from the camera are stitched. Histogram equalization is an effective way to alleviate the effects of illumination. After applying histogram equalization to the two images to be stitched, the grayscale histograms of two images are spread into all gray level ranges and the difference of illumination in adjacent images is reduced efficiently, which will make the image stitching easy to realize.

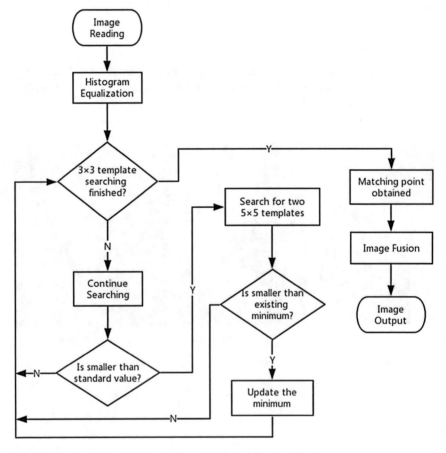

Fig. 8.5   The flowchart of image stitching using line and surface feature

(2) Feature area searching. We take P_1 as the referenced image and P_2 as the image
     to be stitched. The size of P_1 is $(M_1 \times N_1)$ and Picture 2 is $(M_2 \times N_2)$. This
     algorithm will select 3 tiny feature templates in P_1 for matching. First of all,
     limit the area used to select the template is from line 1 to line M_1 and column
     $N_{1/2}$ to column N_1 in P_1. We will first select a 3×3 small template named H_1 in
     this area. Then, according to the image features, we select the other two 5×5
     templates named H_2 and H_3 respectively in the area. As shown in Fig. 8.6, a
     feature template group consisting of 3 tiny templates is made up. (Note: We
     suppose that H_2 and H_1 are at a same level in horizonal direction and so as H_3
     and H_1 are in vertical, and the distance from Hl is Ll and L2, respectively.)
     We adopt the method which calculates the variance values of pixels in templates
     when selecting the feature template. We select the template with maximum sum
     of variance as the standard template because the detail features in images are
     determined by edge features or inflection points of the grayscale value. Where
     the maximum sum of variance is equals to the position of edges or inflection
     points that fluctuate the most in the curves of gray levels. We can measure how
     many detail features there are in a template with the sum of pixel variance values.
     The more details and texture information P_1 contains, the easier to find similar
     areas in P_2. Here are the equations for selecting the feature template.

$$S(x, y) = \sum_{i=1}^{M} \sum_{j=1}^{M} |I(i, j) - \omega|^2 \tag{8.4}$$

$$S_{result}(x, y) = MAX(S(x, y)) \tag{8.5}$$

In Eq. (8.4), $I(i, j)$ represents the pixel value and ω stands for the average gray
value of the template. When we set M as 3, we can obtain the best 3×3 feature

Fig. 8.6 Extracting the
group of feature template

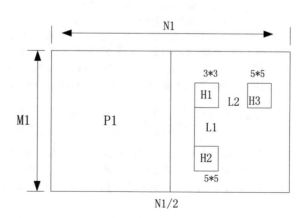

$N1/2$

template through Eqs. 8.4 and 8.5. Repeat the calculation twice to get the other two 5×5 templates. In this way, 3 feature templates with best details are extracted. Record the distance information around them to compose the feature template group.

After extracting appropriate feature template groups, searching from top to bottom and left to right with 3×3 template Fl in P_2, and calculate the pixel difference between F_1 and H_1 one by one. MSE function is used to define the difference function. Here's the definition:

$$S_{3 \times 3}(x, y) = \frac{1}{9} \sum_{i=1}^{3} \sum_{j=1}^{3} [F_1(i, j) - H_1(i, j)]^2 \qquad (8.6)$$

where $F_1(i, j)$ stands for the grey value of corresponding pixels in F_1 and so as $H_1(i, j)$ for H_1.

The experimental results indicate that when the difference of the light intensity in images is small, setting the standard value of the difference function to 30 is an appropriate value if the selected feature template is 3×3. When the selected standard value is greater than 30, the number of templates to meet the conditions is increasing rapidly, which will lead to a longer time spent in computation. If the chosen standard value is less than 30, it is difficult to find the templates meeting the criteria when there is high interference between two images and cause the failure of the algorithm.

When the difference calculated by the Eq. 8.6 is greater than the standard value 30, the difference between the two templates is considered to be very small and it is almost impossible to be a matched area. Hence, we should discard the template and go on with next calculation. When the difference is less than 30, it is considered that the template is very likely to be a matched template. However, the 3×3 template is too small to locate accurately. So according to the recorded distance information between the templates 3×3 and 5×5 in a template group, we can find the corresponding two 5×5 templates in the location of the same distance information around the 3×3 interest template in P_2. Then calculate the difference between the 5×5 templates corresponding to the feature template group. The function for the sum of differences is defined as below.

$$S_{5 \times 5}(x, y) = \frac{1}{25} \sum_{i=1}^{5} \sum_{j=1}^{5} [F_2(i, j) - H_2(i, j)]^2 + \frac{1}{25} \sum_{i=1}^{5} \sum_{j=1}^{5} [F_3(i, j) - H_3(i, j)]^2 \quad (8.7)$$

Calculate all F_1 whose difference is less than 30 through Eq. 8.7 and save every result. Besides, saving the transverse and ordinate values of the most left upper corner pixel points of the template Fl at the same time. Finally, using Eq. 8.8

Fig. 8.7 Traversal matching of the feature template

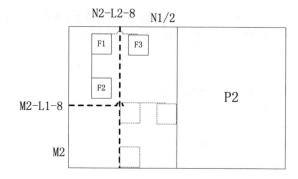

to get the minimum difference sum, the transverse and ordinate values of the upper left corner of the Fl template corresponding to the minimum difference sum are the coordinates of the matching points obtained.

$$S_{find}(x, y) = MIN(S_{5\times5}(x, y)) \tag{8.8}$$

This algorithm actually obtains the best matching template by filtering templates twice. First measure the relevance of the small 3×3 template and save each template with high correlation. Then calculate the relevant of two 5×5 templates that correspond to saved templates and collect the 5×5 template with highest relevance. This means to filter the templates obtained in the first step again for the best matching template (Fig. 8.7).

(3) Image stitching and image fusion. After finding the matching point, simple superposition will cause obvious borders in the picture which is undesirable. A smooth transition for image stitching is required to eliminate such undue influences. Gradated in-and-out algorithm can gain seamless images, but during the image fusion period, the overlapping areas of two images are superimposed by linear weighting and this certainly makes the overlapping areas more blurred than the original image. Hence, we use Gaussian fusion instead. By making the change of gradient factor from 0 to 1 follows the distribution characteristic of Gauss curve approximately, and achieves quick transition between two images. The overlapping area of the stitching result is clearer than that of gradated in-and-out approach.

Matlab programme of the algorithm mentioned above is shown as follows:

PROGRAMME 8.2: Image stitching using line and surface feature

```
%%%%%%%%%%%%%%%%%%%%%%%%%%%
A=imread('1.bmp');
subplot(1,2,1),imshow(A)
title(' Source Image A')
B=imread('2.bmp');
subplot(1,2,2),imshow(B)
title('Source Image B')
%%%%%%%%%%%%%%%%%%%%%%%%%%%
[high,wid]=size(A);
A1=double(A);
B1=double(B);
sub_A=A1(high/2-39:high/2,3*wid/4:3*wid/4+39);
% sub_A=A1(high/2-39:high/2,end-39:end);
sub_B1=B1(11:50,11:50);
mod1=sub_A-sub_B1;
mat1=sum(sum(mod1.*mod1));
mat_best=mat1;
for x1=1:40:wid-40
    for y1=1:40:high-40
        sub_B=B1(y1:y1+39,x1:x1+39);
        mod=sub_A-sub_B;
        mat=sum(sum(mod.*mod));
        if   mat<=mat_best
            mat_best=mat;
            xx=x1;
            yy=y1;
        end
    end
end
x=xx;        % custom settings
y=yy;
for x2=xx-40:xx
    for y2=yy-20:yy+80
        sub_B2=B1(y2:y2+39,x2:x2+39);
        mod2=sub_A-sub_B2;
        mat2=sum(sum(mod2.*mod2));
        if mat2<=mat_best
            mat_best=mat2;
            x=x2;
            y=y2;
        end
    end
```

```
        end
   end
% figure
% colormap(gray);
% subplot(2,1,1);imagesc(sub_A)
% subplot(2,1,2);imagesc(sub_B2)
%%%%%%%%%%%%%%%%%%%%%%%%%%%%%%%%%
x=55;y=1;
%***********************
  if y==0
     AA=A(1:high-1,1:3*wid/4);
     BB=B(1:high-1,1:wid-1);
else if y>=high/2
          AA=A(1:high-y,1:3*wid/4);
          BB=B(y:high-1,x:wid-1);
          else
          AA=A(y:high-1,1:3*wid/4);
          BB=B(1:high-y,x:wid-20);
          end
end
C=[AA BB];
%imwrite(C,'Directly stitched image 34.bmp');
figure,imshow(C)
title(' Directly stitched image')
if y==0
     A2=A(1:high-1,:);
     B2=B(1:high,:);
else if y>=50
          A2=A(1:high-y,:);
          B2=B(y:high-1,:);
     else
          A2=A(y:high-1,:);
          B2=B(1:high-y,:);
     end
end
x=140;%**********************
[high2,wid2]=size(A2);
a1=A2(1:high2,wid-x+1:wid);
b1=B2(1:high2,1:x);
a=double(a1);
```

```
b=double(b1);
d1O=linspace(1,0,x);
d=1:high2;
d1=d1O';
[X1,y1]=meshgrid(d1,d);
im1=a.*X1;%***********************
d20=linspace(0,1,x);
d2=d20';
[X2,y2]=meshgrid(d2,d);
im2=b.*X2;
im11=uint8(im1);
im22=uint8(im2);
im3=imadd(im11,im22);
figure,imshow(im3)
title('Fusion zone with gradated')
a_b=imadd(im11,im22);
aa=A2(1:high2,1:wid-x);
bb=B2(1:high2,x:wid);
%D2=[aa a_b bb];
D2=[aa(:,1:wid2-x) a_b bb];%*********************
imwrite(D2,'6.bmp');
figure,imshow(D2)
title('Image with gradated');
```

Figure 8.8 shows the result.

(a) Source image 1   (b) Source image 2   (c) Final output

Fig. 8.8   Result of stitching

8.2.3 Image Stitching Based on FFT

Panorama refers to the formation of a full view, high resolution 360° image through image processing. It is an integrated reproduction of the view which observers looking around, and it can show better integral information of the surroundings.

Image stitching based on FFT first converts the image to the frequency domain and calculates the rotation amounts and offsets according to its phase cross power spectrum. Then reset the coordinate of the image and apply the movement. At last, the images are stitched together. When stitching a 360° panorama, conversions of the focal length and projections are needed before calculating with phases. Figure 8.9 presents the flowchart of image stitching based on FFT.

The approach applied for stitching cylindrical panoramic image can be divide into 3 parts:

(1) Construct a function with the phase correlation of frequency domain. This function will carry out the 2-D Fourier transformation on two input images and return the offset values between two adjacent images.
(2) Calculate the focal length values of a set of 360° live-action photos and apply the cylindrical projection to the image sequence.
(3) Call the function in part 1 one by one to stitch the images after the projection, and the lighting is processed to generate the cylindrical panoramic image. is generated.

Focal length f is a significant parameter when using the cylindrical projection formula for projection transformation. We set the translations between every two

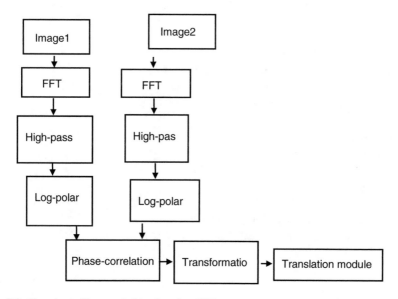

Fig. 8.9 Flowchart of image stitching based on FFT

adjacent images in the image sequence before projection as $C_1, C_2, \ldots, C_n$ respectively, where C_k represents the horizontal translation between the image k and the image $k + 1$. The initial value of focal length named f_0 can be calculated through Formula 8.9:

$$f_0 = \sum_{k=1}^{k=n} \frac{C_k}{2\pi} \tag{8.9}$$

The source code of image stitching based on FFT is shown as PROGRAMME 8.3.

PROGRAMME 8.3: Image stitching based on FFT

```
Function main()
clc;clear;
image1=(imread('D1.bmp'));[h,w,r]=size(image1);
image2=(imread('D2.bmp'));T0(:,:,:,1)=image1;T0(:,:,:,2)=image2;
subplot(121),imshow(image1);subplot(122),imshow(image2);
image11=multi_resolution(image1,2);
image22=multi_resolution(image2,2);
[r1,c1,d1]=size(image11);
[r2,c2,d2]=size(image22);
% calculate phase correlation offsets
tic
fprintf(' calculate phase correlation offsets...');
[i,j]=poc_2pow(image11,image22);
coor_shift(1,1)=i;
coor_shift(1,2)=j;
coor_shift(2,1)=0;coor_shift(2,2)=0;
coor_shift=coor_shift*2^2;%%% convert to offsets of original image
toc
%transform into cylindrical coordinate system
tic
f=sqrt(h^2+w^2);
[T1,coor_shift02]=coortransf(T0,f,coor_shift);
toc
%fuse overlapping areas
tic
fprintf('Fusing overlapping areas and stitching the image...');
panorama1=mosaic(T1(:,:,:,1),T1(:,:,:,2),coor_shift02(1,1),coor_shift02(1,2));
toc
```

```
%image reconstruction
tic
fprintf('Saving and displaying the result...');
imwrite(panorama1,'pic2.jpg','jpg');
imshow(panorama1,[]);
Toc


function T=multi_resolution(Xb,n)
% multiresolution decomposition
[r1,c1,d1,N]=size(Xb);
for i=1:N
    Xb(:,:,1,N)=filter2(fspecial('gaussian'),Xb(:,:,1,N));%%Default parameters of Gaussian filter [3 3],
  sigma=0.5
    Xb(:,:,2,N)=filter2(fspecial('gaussian'),Xb(:,:,2,N));
    Xb(:,:,3,N)=filter2(fspecial('gaussian'),Xb(:,:,3,N));
    end
    step=2^n;
    for i=1:step:r1
        for j=1:step:c1
            T((i+step-1)/step,(j+step-1)/step,:,:)=Xb(i,j,:,:);
        end
    end


    function [dis,dm]=poc_2pow(imageL,imageR);
    %%% phase correlation algorithm
    % imageL=image1;
    % imageR=image2;
    [H1,W1,d1]=size(imageL);
    [H2,W2,d2]=size(imageR);
    if d1==3 imageL=rgb2gray(imageL);end%%%grayscale
    if d2==3 imageR=rgb2gray(imageR);end
    % extract the binary outlines of the 2-exponential histogram
    [imageL,t1]=edge(imageL,'canny',[],1.2);%%%%sigma=1.2(Default 1)
    [imageR,t2]=edge(imageR,'canny',[],1.2);%%%% auto select the threshold value
    Xb=imageL;Yb=imageR;
    %2-exponential histogram
    for i=5:11
        index2=2^i;
        if index2<=H1 && index2<=W1 h1=index2;end
        if index2<=H2 && index2<=W2 h2=index2;end
```

```
end
% minhw1=min(h1,w1);
% minhw2=min(h2,w2);
minhw1=h1;minhw2=h2;
offset1=round((H1-minhw1)/2);
offset2=round((H2-minhw2)/2);
imageL=imageL(offset1:offset1+minhw1-1,W1-minhw1+1:W1);%%%choose right center in the left
    image
imageR=imageR(offset2:offset2+minhw2-1,1:minhw2);%%%% choose left center in the right image
% phase correlation algorithm for measuring offsets
A=fft2(im2double(imageL));%FFT in frequency domain
B=fft2(im2double(imageR));
AB=conj(A).*(B);%%% conjugated convolution, equals to phase transformation
modAB=abs(AB);
%peak value，（I，J）save peak coordinates which are offsets
COR=ifft2(AB);%%%unnormalize, reverse transformation for coeerlation
emin=100000;
% for i=1:10
    [maxC,sorti]=max(COR);
    [C,J]=max(maxC);
    I=sorti(J);
    if I<20 dis=I;
    elseif H2-I<20 dis=(I-H2);
    else dis=0;
    end
    dm=J;


function [T1,coor_shift02]=coortransf(T0,f,coor_shift)
%%transformation from image coordinate to cylindrical coordinate
%transform input image sequence T0 with focal length f
coor_shift02=coor_shift;%%%% the first dimension (row values) stay unchanged and the second
    dimension (column values) update after mapping
[H,W,r,N]=size(T0);
w2f=W/2/f;
h2=H/2;
constant2=f*atan(W/(2*f));
constant1=h2;
```

```
for y=1:W         %%%%%      columns
    angle=atan(y/f-w2f);%%%%atan((y-W/2)/f);
     y1=uint16(f*angle+constant2);
     if  y1==0   y1=1;   end
    for x=1:H   %%%%%%%%%%%%     rows
    x1=uint16((x-h2)*cos(angle)+constant1);
    if   x1==0    x1=1;end
    if r==3     %%%%%%%%%%%%%%color image
        for n=1:N    %%%%%%%%%
            if (y==coor_shift(n,2)) coor_shift02(n,2)=y1; end%%%corresponding offsets
            T1(x1,y1,:,n)=T0(x,y,:,n);%%%% mapping of points
        end
    elseif r==1
    end
    end
end
[h,w,a,N]=size(T1);
for i=1:60
    for j=1:w
        if (T1(i,j,:,:)==0)
            T1(i,j,:,:)=255;
        end
        if (T1(h-i,j,:,:)==0)
            T1(i,j,:,:)=255;
        end
    end
end


function D=mosaic(image1,image2,i,j)
[ra,ca,a]=size(image1);
[rb,cb,b]=size(image2);
Xa=image1;Ya=image2;
% dis=i;%%% top and bottom offsets
dis=i;
EXa=zeros(abs(dis),ca,3)+255;
EXb=zeros(abs(dis),cb,3)+255;
if dis>1
   Xa=[EXa;Xa];
   Ya=[Ya;EXb];
elseif dis< -1
    Xa=[Xa;EXa];
    Ya=[EXb;Ya];
```

```
end
dm=j;%%% stitching crack width ,limited no more than 50 pel
    A=Xa(:,1:(ca-dm-1),:);
    B1=Xa(:,(ca-dm):ca,:);
    B2=Ya(:,1:dm,:);
    B=imagefusion02(B1,B2);%%partial overlapping(fusion)
    C=Ya(:,(dm+1):cb,:);%% cut out the rest part of the second image
    D=[A,B,C];%%merge and complete stitching
    %%% eliminate accumulative errors
    [r,c]=size(D);
    if dis>1
        D=D(1:(r-dis),:,:);
    elseif dis<-1
        D=D((abs(dis)+1):r,:,:);
    end
    function [dis,dm]=phase_correlation(image1,image2);
    % phase correlation algorithm
    %%% C=phase_correlation(imagea,imageb)    input two images
    [H1,W1,d1]=size(image1);
    [H2,W2,d2]=size(image2);
    if d1==3 image1=rgb2gray(image1);end%%%grayscale
    if d2==3 image2=rgb2gray(image2);end
    Xb=image1;Yb=image2;
    [image1,t1]=edge(image1,'canny');
      [image2,t2]=edge(image2,'canny');
    A=fft2(im2double(image1));%FFT in frequency domain
    B=fft2(im2double(image2));
    AB=conj(A).*(B);%%% conjugated convolution, equals to phase transformation
    COR=ifft2(AB);%%% unnormalize, reverse transformation for coeerlation
    [C,i]=max(COR);[C,J]=max(C);I=i(J);%%% peak value，（I，J）save peak coordinates which are offsets
    if I<15 dis=I;
    elseif H2-I<15 dis=I-H2;
    else dis=0;
    end
    dm=J;
    function C=imagefusion02(A,B)
    %%%image fusion
    [M,N,D]=size(A);
```

```
if D==3
for i=1:(N-1)
C(:,i,:)=round((double(A(:,i,:))*(N-i)+double(B(:,i,:))*i)/N);
end
elseif D==1
for i=1:(N-1)
C(:,i)=round((double(A(:,i))*(N-i)+double(B(:,i))*i)/N);
end
end
% figure,imshow(C/max(max(max(C))));
```

Figure 8.10 is the result of image stitching.

Image stitching based on FFT demands images to have the same size and more than 30% overlapping areas. Moreover, it is only applicable to image registration with translation, rotation and scaling. Non-linear distortions as tangential transformation is not applicable. This algorithm only utilizes the phase information in cross power spectrum for image registration hence it is insensitive to the changes of brightness among images.

(a) The source image A   **(b) The source image B**

(c) The output panorama

Fig. 8.10 The sketch map for image stitching

8.3   Images Stitching Based on Feature Points

Instead of using pixel values of the images, image stitching [2] based on feature points [3] calculates features such as texture, edges, objects and etc., from pixels, and then uses features as standards and search matches for the corresponding feature areas of overlapping images. This kind of approaches are more robust. There are two processes for image stitching based on feature points: feature extraction and feature registration. First, extract points, lines and regions where gray scale changes obviously to form a feature set. Second, try to choose the paired features using feature matching algorithms between two feature sets. A series of image segment approaches have been applied to feature extraction and edge detection, such as canny descriptor, Laplace-gauss descriptor and region seeds growing (RSD). The extracted spatial features include closed edges, open edges, crossed lines and other features. Feature matching algorithms include cross correlation, distance transformation, dynamic programming, structure matching and chain code correlation algorithms.

8.3.1   SIFT Feature Points Detection

The process of image stitching based on SIFT feature points include: image acquisition, feature extraction and matching, image registration (calculating H) and finally image stitching.

(1) Image acquisition. Image acquisition is the precondition for image stitching. Different image acquisition methods can obtain different input image sequences and produce different image stitching effects. Currently, there are three different methods to obtain image sequences: (1) fix the camera to the tripod and rotate it to get the image data; (2) fix the camera on a movable platform, and the image data is obtained by parallel moving it; (3) Hand-held the camera for capturing image data by a fixed-point rotating or moving in the direction perpendicular to the camera's optical axis. This process utilizes given images.

(2) Feature extraction and matching. Extract SIFT feature points from the input image sequences. The algorithm calculates and extracts the feature points simultaneously in the spatial domain and the scale domain. Therefore, the obtained feature points have the scale invariance, which can correctly extract the feature points that exist in the image sequences with large scale and angle change. Euclidean distance is used to calculate the distance between two SIFT feature point descriptors.

(3) Image registration with calculation of H. Image registration based on feature points means that the transformation matrix between image sequences is constructed by matching points to complete the stitching of panoramic images. To improve the precision of image registration, RANSAC algorithm [4] is used to calculate and refine the transformation matrix. H algorithm to automatically calculate the transformation matrix: calculate feature points in each image;

Fig. 8.11 The process of image stitching based on SIFT

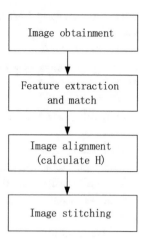

match feature points; calculate initial value of the matrix; use iteration to refine H transformation matrix; boot matching; repeat iterations until the number of corresponding points is stable.

(4)   Image fusion. According to the transformation matrix H of two images, the corresponding images can be transformed to determine the overlapping region of the images, and register the images to be merged into a new blank image to form a mosaic diagram. A quick and simple weighted smoothing algorithm is used to deal with the stitching seam problem.

The process of image stitching algorithm based on SIFT feature points is shown in Fig. 8.11:

The MATLAB source programme (main code) of image stitching algorithm based on SIFT feature points is shown in PROGRAMME 8.4. SIFT feature detection programme is shown in Chap. 4 PROGRAMME 4.9.

PROGRAMME 8.4: Image stitching based on SIFT feature points

```
function [ imgout ] = imMosaic( img1,img2,adjColor )
% use SIFT to find corresponding points
[matchLoc1 matchLoc2] = siftMatch(img1, img2);
% use RANSAC to find homography matrix
[H corrPtIdx] = findHomography(matchLoc2',matchLoc1');
H    %#ok
tform = maketform('projective',H');
```

```
img21 = imtransform(img2,tform); % reproject img2
%fis(img1)
%fis(img21)
% adjust color or grayscale linearly, using corresponding infomation
[M1 N1 dim] = size(img1);
[M2 N2 ~] = size(img2);
if exist('adjColor','var') && adjColor == 1
    radius = 2;
    x1ctrl = matchLoc1(corrPtIdx,1);
    y1ctrl = matchLoc1(corrPtIdx,2);
    x2ctrl = matchLoc2(corrPtIdx,1);
    y2ctrl = matchLoc2(corrPtIdx,2);
    ctrlLen = length(corrPtIdx);
    s1 = zeros(1,ctrlLen);
    s2 = zeros(1,ctrlLen);
    for color = 1:dim
      for p = 1:ctrlLen
          left = round(max(1,x1ctrl(p)-radius));
          right = round(min(N1,left+radius+1));
          up = round(max(1,y1ctrl(p)-radius));
          down = round(min(M1,up+radius+1));
          s1(p) = sum(sum(img1(up:down,left:right,color))); % calculate chroma of the around points
      end
      for p = 1:ctrlLen
          left = round(max(1,x2ctrl(p)-radius));
          right = round(min(N2,left+radius+1));
          up = round(max(1,y2ctrl(p)-radius));
          down = round(min(M2,up+radius+1));
          s2(p) = sum(sum(img2(up:down,left:right,color)));
      end
      sc = (radius*2+1)^2*ctrlLen;
      adjcoef = polyfit(s1/sc,s2/sc,1);
      img1(:,:,color) = img1(:,:,color)*adjcoef(1)+adjcoef(2);
    end
end
% do the mosaic
pt = zeros(3,4);
pt(:,1) = H*[1;1;1];
pt(:,2) = H*[N2;1;1];
pt(:,3) = H*[N2;M2;1];
```

```
pt(:,4) = H*[1;M2;1];
x2 = pt(1,:)./pt(3,:);
y2 = pt(2,:)./pt(3,:);
up = round(min(y2));
Yoffset = 0;
if up <= 0
    Yoffset = -up+1;
    up = 1;
end
left = round(min(x2));
Xoffset = 0;
if left<=0
    Xoffset = -left+1;
    left = 1;
end
[M3 N3 ~] = size(img21);
imgout(up:up+M3-1,left:left+N3-1,:) = img21;
    % img1 is above img21
imgout(Yoffset+1:Yoffset+M1,Xoffset+1:Xoffset+N1,:) = img1;
end
% [matchLoc1 matchLoc2] = siftMatch(img1, img2)
% This function reads two images, finds their SIFT features, and
%     displays lines connecting the matched keypoints.   A match is accepted
%     only if its distance is less than distRatio times the distance to the
%     second closest match.
% It returns the matched points of both images, matchLoc1 = [x1,y1;x2,y2;...]
% Example: match('scene.pgm','book.pgm');
function [matchLoc1 matchLoc2] = siftMatch(img1, img2)
% load matchdata
% load img1data
% load img2data
%{,
% Find SIFT keypoints for each image
[des1, loc1] = sift(img1);
[des2, loc2] = sift(img2);
% save img1data des1 loc1
% save img2data des2 loc2
% For efficiency in Matlab, it is cheaper to compute dot products between
%   unit vectors rather than Euclidean distances.   Note that the ratio of
%   angles (acos of dot products of unit vectors) is a close approximation
```

```
%    to the ratio of Euclidean distances for small angles.
% distRatio: Only keep matches in which the ratio of vector angles from the
%    nearest to second nearest neighbor is less than distRatio.
distRatio = 0.6;
% For each descriptor in the first image, select its match to second image.
des2t = des2';                                      % Precompute matrix transpose
matchTable = zeros(1,size(des1,1));
for i = 1 : size(des1,1)
    dotprods = des1(i,:) * des2t;            % Computes vector of dot products
    [vals,indx] = sort(acos(dotprods));     % Take inverse cosine and sort results
% Check if nearest neighbor has angle less than distRatio times 2nd.
    if (vals(1) < distRatio * vals(2))
        matchTable(i) = indx(1);
    else
        matchTable(i) = 0;
    end
end
% save matchdata matchTable
%}
% Create a new image showing the two images side by side.
img3 = appendimages(img1,img2);
% Show a figure with lines joining the accepted matches.
figure('Position', [100 100 size(img3,2) size(img3,1)]);
colormap('gray');
imagesc(img3);
hold on;
cols1 = size(img1,2);
for i = 1: size(des1,1)
    if (matchTable(i) > 0)
        line([loc1(i,2) loc2(matchTable(i),2)+cols1], ...
                [loc1(i,1) loc2(matchTable(i),1)], 'Color', 'c');
    end
end
hold off;
num = sum(matchTable > 0);
fprintf('Found %d matches.\n', num);
idx1 = find(matchTable);
idx2 = matchTable(idx1);
x1 = loc1(idx1,2);
x2 = loc2(idx2,2);
```

```
y1 = loc1(idx1,1);
y2 = loc2(idx2,1);
matchLoc1 = [x1,y1];
matchLoc2 = [x2,y2];
end
% [descriptors, locs] = sift(img)
function [f inlierIdx] = ransac1( x,y,ransacCoef,funcFindF,funcDist )
%[f inlierIdx] = ransac1( x,y,ransacCoef,funcFindF,funcDist )
%   Use RANdom SAmple Consensus to find a fit from X to Y.
%   X is M*n matrix including n points with dim M, Y is N*n;
%   The fit, f, and the indices of inliers, are returned.
%
%   RANSACCOEF is a struct with following fields:
%   minPtNum,iterNum,thDist,thInlrRatio
%   MINPTNUM is the minimum number of points with whom can we
%   find a fit. For line fitting, it's 2. For homography, it's 4.
%   ITERNUM is the number of iteration, THDIST is the inlier
%   distance threshold and ROUND(THINLRRATIO*n) is the inlier number threshold.
%
%   FUNCFINDF is a func handle, f1 = funcFindF(x1,y1)
%   x1 is M*n1 and y1 is N*n1, n1 >= ransacCoef.minPtNum
%   f1 can be of any type.
%   FUNCDIST is a func handle, d = funcDist(f,x1,y1)
%   It uses f returned by FUNCFINDF, and return the distance
%   between f and the points, d is 1*n1.
%   For line fitting, it should calculate the dist between the line and the
%   points [x1;y1]; for homography, it should project x1 to y2 then
%   calculate the dist between y1 and y2.
minPtNum = ransacCoef.minPtNum;
iterNum = ransacCoef.iterNum;
thInlrRatio = ransacCoef.thInlrRatio;
thDist = ransacCoef.thDist;
ptNum = size(x,2);
thInlr = round(thInlrRatio*ptNum);
inlrNum = zeros(1,iterNum);
fLib = cell(1,iterNum);
for p = 1:iterNum
    % 1. fit using   random points
    sampleIdx = randIndex(ptNum,minPtNum);
    f1 = funcFindF(x(:,sampleIdx),y(:,sampleIdx));
```

```
    % 2. count the inliers, if more than thInlr, refit; else iterate
    dist = funcDist(f1,x,y);
    inlier1 = find(dist < thDist);
    inlrNum(p) = length(inlier1);
    if length(inlier1) < thInlr, continue; end
    fLib{p} = funcFindF(x(:,inlier1),y(:,inlier1));
end
% 3. choose the coef with the most inliers
[~,idx] = max(inlrNum);
f = fLib{idx};
dist = funcDist(f,x,y);
inlierIdx = find(dist < thDist);
end
clear
close all
f = 'a';
ext = 'jpg';
img1 = imread([f '1.' ext]);
img2 = imread([f '2.' ext]);
% img3 = imread('b3.jpg');
img0 = imMosaic(img2,img1,1);
% img0 = imMosaic(img1,img0,1);
fis(img0)
imwrite(img0,['mosaic_' f '.' ext],ext)
```

The result of image stitching is shown in Fig. 8.12:

(a)original image A (b)original image B (c)stitching iamge

Fig. 8.12 Image stitching result

8.3.2   *Image Stitching Based on Harris Feature Points*

The process of image stitching based on Harris feature points is as follows.

(1)   Detect Harris feature points of images;
(2)   Connect the feature points between two images and complete image matching;
(3)   Filter all matching points and obtain points which are needed for image stitching;
(4)   Calculate the distance between feature points of two images and smooth the overlapping parts of the images.

The core code is shown in PROGRAMME 8.5.

PROGRAMME 8.5: Image stitching based on Harris feature points

```
clc
clear all
% read image
pic1=imread('lena1.jpg');
pic2=imread('lena2.jpg');
% Harris feature points detection
points1=myHarris(pic1);
points2=myHarris(pic2);
% draw Harris feature points
figure(1)
drawHarrisCorner(pic1,points1,pic2,points2);
% describe Harris feature
des1=myHarrisCornerDescription(pic1,points1);
des2=myHarrisCornerDescription(pic2,points2);
% coarse match
matchs=myMatch(des1,des2);
% obtain position of match points
matchedPoints1=points1(matchs(:,1),:);
matchedPoints2=points2(matchs(:,2),:);
% line coarse match points
figure(2)
drawLinedCorner(pic1,matchedPoints1,pic2,matchedPoints2);
% Harris feature points fine matching
[newLoc1,newLoc2]=pointsSelect(matchedPoints1,matchedPoints2);
% line fine match points
figure(3)
drawLinedCorner(pic1,newLoc1,pic2,newLoc2);
% stitch images
```

```
im=picMatched(pic1,newLoc1,pic2,newLoc2);
% show the stitching image
figure(4)
imshow(im);
set(gcf,'Color','w');
function points=myHarris(pic)
% function:Harris feature points detection
% input: RGB image or gray scale image
% output:the row and col N×2 matrix of the feature point
if length(size(pic))==3
pic=rgb2gray(pic);
end
pic=double(pic);
hx=[-1 0 1];
Ix=filter2(hx,pic);
hy=[-1;0;1];
Iy=filter2(hy,pic);
Ix2=Ix.*Ix;
Iy2=Iy.*Iy;
Ixy=Ix.*Iy;
h=fspecial('gaussian',[7 7],2);
Ix2=filter2(h,Ix2);
Iy2=filter2(h,Iy2);
Ixy=filter2(h,Ixy);
[heigth,width]=size(pic);
alpha=0.06;
R=zeros(heigth,width);
for i=1:heigth
for j=1:width
M=[Ix2(i,j) Ixy(i,j);Ixy(i,j) Iy2(i,j)];
R(i,j)=det(M)-alpha*(trace(M)^2);
end
end
Rmax=max(max(R));
pMap=zeros(heigth,width);
for i=2:heigth-1
for j=2:width-1
if R(i,j)>0.01*Rmax
tm=R(i-1:i+1,j-1:j+1);
tm(2,2)=0;
```

```
if R(i,j)>tm
pMap(i,j)=1;
end
end
end
end
[row,col]=find(pMap==1);
points=[row,col];
function drawHarrisCorner(pic1,points1,pic2,points2)
% function:draw Harris feature points' match connection
% input:
% pic1  pic2:the images need stitching
% points1  points2: Harris feature points position
X1=points1(:,2);
Y1=points1(:,1);
X2=points2(:,2);
Y2=points2(:,1);
dif=size(pic1,2);
imshowpair(pic1,pic2,'montage');
hold on
plot(X1,Y1,'b*');
plot(X2+dif,Y2,'b*');
set(gcf,'Color','w');
function des=myHarrisCornerDescription(pic,points)
% Function: describe Harris feature points
% Input:
% pic:source image
% points:feature points' position
% Output:
% des: 8×N matrix describing Harris feature points
if length(size(pic))==3
pic=rgb2gray(pic);
end
len=length(points);
des=zeros(8,len);
for k=1:len
p=points(k,:);
pc=pic(p(1),p(2));
des(1,k)=pic(p(1)-1,p(2)-1)-pc;
des(2,k)=pic(p(1),p(2)-1)-pc;
```

```
des(3,k)=pic(p(1)+1,p(2)-1)-pc;
des(4,k)=pic(p(1)+1,p(2))-pc;
des(5,k)=pic(p(1)+1,p(2)+1)-pc;
des(6,k)=pic(p(1),p(2)+1)-pc;
des(7,k)=pic(p(1)-1,p(2)+1)-pc;
des(8,k)=pic(p(1)-1,p(2))-pc;
des(:,k)=des(:,k)/sum(des(:,k));
end
function matchs=myMatch(des1,des2)
% Function:feature point bidirectional match
% input:
% des1、 des2:feature point description matrix
% Output:
% matchs:correspondence relation of match points
len1=length(des1);
len2=length(des2);
match1=zeros(len1,2);
cor1=zeros(1,len2);
for i=1:len1
d1=des1(:,i);
for j=1:len2
d2=des2(:,j);
cor1(j)=(d1'*d2)/sqrt((d1'*d1)*(d2'*d2));
end
[~,indx]=max(cor1);
match1(i,:)=[i,indx];
end
match2=zeros(len2,2);
cor2=zeros(1,len1);
for i=1:len2
d2=des2(:,i);
for j=1:len1
d1=des1(:,j);
cor2(j)=(d1'*d2)/sqrt((d1'*d1)*(d2'*d2));
end
[~,indx]=max(cor2);
match2(i,:)=[indx,i];
end
```

```
matchs=[];
for i=1:length(match1)
for j=1:length(match2)
if match1(i,:)==match2(j,:)
matchs=[matchs;match1(i,:)];
end
end
end
function drawLinedCorner(pic1,loc1,pic2,loc2)
% Function:draw connections between match points
% Input:
% pic1、 pic2:image to need stitching
% loc1、 loc2:position of the paired points
X1=loc1(:,2);
Y1=loc1(:,1);
X2=loc2(:,2);
Y2=loc2(:,1);
dif=size(pic1,2);
imshowpair(pic1,pic2,'montage');
hold on
for k=1:length(X1)
plot(X1(k),Y1(k),'b*');
plot(X2(k)+dif,Y2(k),'b*');
line([X1(k),X2(k)+dif],[Y1(k),Y2(k)],'Color','r');
end
set(gcf,'Color','w');
function [newLoc1,newLoc2]=pointsSelect(loc1,loc2)
% Filter:filter the paired match points and obtain the fine match points
% Input:
% loc1、 loc2:position of coarse match points
% Output:
% newLoc1、 newLoc2:position of fine match points
slope=(loc2(:,1)-loc1(:,1))./(loc2(:,2)-loc1(:,2));
for k=1:3
slope=slope-mean(slope);
len=length(slope);
t=sort(abs(slope));
thresh=t(round(0.5*len));
ind=abs(slope)<=thresh;
slope=slope(ind);
loc1=loc1(ind,:);
```

```
loc2=loc2(ind,:);
end
newLoc1=loc1;
newLoc2=loc2;
function im=picMatched(pic1,newLoc1,pic2,newLoc2)
% Function: obtain the stitching image
% Input:
% pic1、 pic2: images need stitching
% newLoc1、 newLoc2:new position of feature points
% Output:
% im: the stitching image
if length(size(pic1))==2
pic1=cat(3,pic1,pic1,pic1);
end
if length(size(pic2))==2
pic2=cat(3,pic2,pic2,pic2);
end
SZ=2000;
X1=newLoc1(:,2);
Y1=newLoc1(:,1);
X2=newLoc2(:,2);
Y2=newLoc2(:,1);
sel=randperm(length(newLoc1),3);
x=X2(sel)';
y=Y2(sel)';
X=X1(sel)';
Y=Y1(sel)';
U=[x;y;ones(1,3)];
V=[X;Y;ones(1,3)];
T=V/U;
cntrX=SZ/2;
cntrY=SZ/2;
im=zeros(SZ,SZ,3);
for i=1:size(pic2,1)
for j=1:size(pic2,2)
tmp=T*[j;i;1];
nx=round(tmp(1))+cntrX;
ny=round(tmp(2))+cntrY;
if nx>=1 && nx<=SZ && ny>=1 && ny<=SZ
im(ny,nx,:)=pic2(i,j,:);
```

```
end
end
end
im=imresize(im,1,'bicubic');
tpic1=zeros(SZ,SZ,3);
tpic1(1+cntrY:size(pic1,1)+cntrY,1+cntrX:size(pic1,2)+cntrX,:)=pic1;
re=rgb2gray(uint8(im))-rgb2gray(uint8(tpic1));
for k=1:3
ta=im(:,:,k);
tb=tpic1(:,:,k);
ta(re==0)=tb(re==0);
im(:,:,k)=ta;
end
clear ta tb re tpic1
im=getPicture(im,SZ);
im=uint8(im);
if length(size(pic1))==2
im=rgb2gray(im);
end
function im=getPicture(pic,SZ)
% Function: obtain the useful image region
% Input
% pic: the stitching image
% SZ: given size
% Output:
% im: useful image region
if length(size(pic))==2
pic=cat(3,pic,pic,pic);
end
k=1;
while k<SZ
if any(any(pic(k,:,:)))
break
end
k=k+1;
end
ceil=k; % Upper boundary
k=SZ;
while k>0
if any(any(pic(k,:,:)))
```

```
break
end
k=k-1;
end
bottom=k; % Lower boundary
k=1;
while k<SZ
if any(any(pic(:,k,:)))
break
end
k=k+1;
end
left=k; %left boundary
k=SZ;
while k>0
if any(any(pic(:,k,:)))
break
end
k=k-1;
end
right=k; %right boundary
%%obtain image
im=pic(ceil:bottom,left:right,:);
```

8.3.3   Auto-Sorting for Image Sequence

In order to stitch the images, the input image sequence [5] is ordered according to the actual scene content, that is to say, each adjacent two images must have overlapping parts, so that the correct panoramic image can be spliced. However, in the process of capturing images and storage or input, sequence of images could be confusing and couldn't stitch directly. In order to implement image sequence auto-sorting, there are 3 problems to solve firstly:

(1) Determine whether there is overlapping regions between two images, that is, whether two images are related;
(2) Determine the head and tail images of the sequence of images;
(3) Determine the relationship between the left and right position of two overlapping images.

In this section, we use phase correlation method to sort the image sequence.
The principle of phase correlation method is as follows:
Suppose there is an offset $(\Delta x, \Delta y)$ between image 1 $I_1(x, y)$ and image 2 $I_2(x, y)$:

$$I_1(x, y) = I_2(x - \Delta x, y - \Delta y) \tag{8.10}$$

According to shift characteristics of Fourier transformation, here is:

$$\hat{I}_1(u, v) = e^{-j2\pi(u\Delta x + v\Delta y)}\hat{I}_2(u, v) \tag{8.11}$$

Its normalized mutual power spectrum is represented as:

$$\tilde{Co}rr(u, v) = \frac{\hat{I}_1(u, v)\hat{I}_2^*(u, v)}{\left|\hat{I}_1(u, v)\hat{I}_2(u, v)\right|} = e^{-j2\pi(u\Delta x + v\Delta y)} \tag{8.12}$$

$\hat{I}_1$ and $\hat{I}_2$ are I_1 and I_2's Fourier transformation separately, $\hat{I}_2^*$ is the complex conjugate of I_2.

The phase of cross power spectrum density equals the phase difference of two images. Normalized cross power spectrum density is operated to get an impulse function through the inverse Fourier transformation:

$$Corr(x, y) = F^{-1}[e^{-j2\pi(u\Delta x + v\Delta y)})] = \delta(x - \Delta x, y - \Delta y) \tag{8.13}$$

This function takes max value at relative displacement $(\Delta x, \Delta y)$ (match point) of two images, anywhere else near 0. relative displacement $(\Delta x, \Delta y)$ was determined by finding out the position of the peek point in formula (4.4). In the case of only translation between images, the magnitude of the peak of the impulse function reflects the correlation between the two images and take a value in an interval [0, 1]. The larger the overlapping region between two images, the larger the value is. If two images have the same content, the value is 1, and the value is 0 when it is completely different. If there is still perspective, noise or moving target between two images, the energy of the impulse function will be distributed from a single peak to other small peaks, but its maximum peak position has some robustness.

According to the principle of phase correlation method, the automatic sorting algorithm is as follows:

(1) Determine the head and tail images (leftmost and rightmost images) and adjacent image.

For a given image sequence with n images, any image can be computed by the remaining $N - 1$ images to get the $N - 1$ correlation degree. Since the image will be adjacent to up to two images (an intermediate image), it will at least be adjacent to one of the images (head and tail images). Therefore, if the first two largest is selected from the $N - 1$ correlation calculated from the image, the image will overlap with the two or one of the two correlations. Operated on the left $N - 1$ images, $2N$ largest correlation degrees are obtained. For the head image and tail image, their corresponding two correlation degrees will have one not eligible. Obviously, the corresponding correlation degrees of the head image and the tail image are smaller than the other degrees. When finding out the smallest correlation degrees of the $2N$ degrees, the head and tail image are obtained correspondly. Finally, the head and tail image differ from the adjacent images.

(2) Determine the relationship between the left and right positions of two adjacent images.

In the method of phase correlation algorithm, the measure results show the δ pulse function with very sharp correlation peaks when two images are really relevant. The horizontal translation parameters of two images can be calculated by the corresponding pixel points of the peak. When horizontal translation parameter x is greater than half of the image width, you can subtract it from the image width and then take the negative. If the horizontal translation between image **A** and **B** is negative, Image **A** is on the left side of image **B**, and conversely, image **A** is on the right side of image **B**.

Therefore, the automatic sorting of image sequences is completed.

The MATLAB code is shown in PROGRAMME 8.6, where the poc_2pow function has described in PROGRAMME 8.3:

PROGRAMME 8.6: Image sequence automatic sorting

```
clear;clc;
%%%%%%%%%%%%%%%%%%input images, tectonic image pyramid, store in an array
tic
fprintf('image input, image pyramid building...');
level=2;
T0=uint8([]);T=uint8([]);
%%%%%%%%%%%%%%%%%%%%%%%
i1=imread('X1.jpg');T0(:,:,:,1)=i1;
i2=imread('X2.jpg');T0(:,:,:,2)=i2;
i3=imread('X3.jpg');T0(:,:,:,3)=i3;
[h,w,d]=size(T0(:,:,:,1));%%% same size of images is required
% [T,Terr]=multi_resolution(T0,level);
T=multi_resolution(T0,level);
toc
% %%%%%%%%%%%%%%%%%%%%% calculate offset through phase correlation
tic
fprintf(' calculate offset through phase correlation...');
M=3;%%number of images
% L=M*w;%%total length of images
% suml=0;%%total length of the overlapping region
for N=1:M
if N<M [i,j]=poc_2pow(T(:,:,:,N),T(:,:,:,N+1));
elseif N==M [i,j]=poc_2pow(T(:,:,:,N),T(:,:,:,1));
end
```

```
coor_shift(N,1)=i;
coor_shift(N,2)=j;
%       sum1=sum1+j;
end
coor_shift=coor_shift*2^level;%%% convert the offsets in the pyramid hierarchy to the offset of the
    original
toc
%%%%%%%%%%%%%%%%% Transform to cylindrical coordinate system
tic
fprintf(' Transform to cylindrical coordinate system...');
f=sqrt(h^2+w^2);
[T1,coor_shift02]=coortransf(T0,f,coor_shift);
toc
%%%%%%%%%%%%%%%merge overlapping parts
tic
fprintf(' merge overlapping parts, stitching image...');
panorama1=T1(:,:,:,1);
for N=1:M
if N<M panorama1=mosaic(panorama1,T1(:,:,:,N+1),coor_shift02(N,1),coor_shift02(N,2));
end
end
toc
%%%%%%%%%%%%%%%%%%%%%reconstruct image
tic
fprintf('save and show result...');
imwrite(panorama1,'pic1.jpg','jpg');
imshow(panorama1,[]);
toc
```

A complete picture is obtained after sorting the input images.

8.3.4 Harris Point Registration Based on RANSAC Algorithm

Harris point registration based on Random Sample Consensus (RANSAC) is a kind of matching method based on features. Harris points detection is firstly performed and then make a rough matching according to local characteristics of extracted points to find out correspondence between sets of points to be matched. After rough matching, most wrong matching points pairs are removed, but there remain many points missing the requirements. These point pairs with large errors in geometrical relationship remains mainly because gray scale information similarity. These points are called pseudo matching pairs. The RANSAC algorithm is used to remove the pseudo matching pairs.

RANSAC is a kind of iteration algorithm to estimate mathematical model parameters. The main idea is to calculate the parameters to make the majority of samples (feature points) can meet the mathematical model. At iteration, the minimum number of samples is used to sample the model and calculate the parameters and the number of samples confirming to the model is counted. And the maximum sample parameters are considered as the values of the final model. The sample point that conforms to the model is called the inliers, and the sample point that does not conform to the model is called the outer point or the wild point.

RANSAC's basic ideas are as follows:

Consider a model required a minimum sampling set with n samples (n for the minimum number of samples required to initializing the model parameters) and a sample set P, numbers of samples of set P #(P)>n. Subset S with n samples which are random extracted from P is used to initialize model M:

The samples in Complement set $SC = P/S$ whose error is less than a set threshold t, along with set S constitute S^*. S^* is considered as inliers set and construct S's Consensus Set.

If #(S^*) * N, right parameters are considered obtained and Least Squares method and so on are used to estimate new model M^* on inliers set S^*; or resample new S and repeat.

After a certain number of sampling, if no consistent set is found, the algorithm fails, otherwise the maximal consensus set obtained after sampling, and the algorithm ends. The code is shown in PROGRAMME 8.7.

PROGRAMME 8.7: Harris point registration based on RANSAC algorithm

```
function points = kp_harris(im)

    % Extract keypoints using Harris algorithm (with an improvement

    % version)

    % Author :: Vincent Garcia

    % Date     :: 05/12/2007

    % INPUT

    % im       : the graylevel image

    % OUTPUT

    % points : the interest points extracted
```

```
% REFERENCES
% C.G. Harris and M.J. Stephens. "A combined corner and edge detector",
% Proceedings Fourth Alvey Vision Conference, Manchester.
% pp 147-151, 1988.
% Alison Noble, "Descriptions of Image Surfaces", PhD thesis, Department
% of Engineering Science, Oxford University 1989, p45.
% C. Schmid, R. Mohrand and C. Bauckhage, "Evaluation of Interest Point Detectors",
% Int. Journal of Computer Vision, 37(2), 151-172, 2000.
% EXAMPLE
% points = kp_harris(im)
% only luminance value
im = double(im(:,:,1));
sigma = 1.5;
% derivative masks
s_D = 0.7*sigma;
x   = -round(3*s_D):round(3*s_D);
dx = x .* exp(-x.*x/(2*s_D*s_D)) ./ (s_D*s_D*s_D*sqrt(2*pi));
dy = dx';
% image derivatives
Ix = conv2(im, dx, 'same');
Iy = conv2(im, dy, 'same');
% sum of the Auto-correlation matrix
s_I = sigma;
g = fspecial('gaussian',max(1,fix(6*s_I+1)), s_I);
Ix2 = conv2(Ix.^2, g, 'same'); % Smoothed squared image derivatives
Iy2 = conv2(Iy.^2, g, 'same');
Ixy = conv2(Ix.*Iy, g, 'same');
% interest point response
cim = (Ix2.*Iy2 - Ixy.^2)./(Ix2 + Iy2 + eps);          % Alison Noble measure.
% k = 0.06; cim = (Ix2.*Iy2 - Ixy.^2) - k*(Ix2 + Iy2).^2;    % Original Harris measure.
% find local maxima on 3x3 neighborgood
[r,c,max_local] = findLocalMaximum(cim,3*s_I);
% set threshold 1% of the maximum value
t = 0.1*max(max_local(:));
% find local maxima greater than threshold
[r,c] = find(max_local>=t);
% build interest points
points = [r,c];
end
```

```
function [row,col,max_local] = findLocalMaximum(val,radius)
    % Determine the local maximum of a given value
    %
    % Author :: Vincent Garcia
    % Date     :: 09/02/2007
    %
    % INPUT
    % val      : the NxM matrix containing values
    % radius : the radius of the neighborhood
    % OUTPUT
    % row         : the row position of the local maxima
    % col         : the column position of the local maxima
    % max_local : the NxM matrix containing values of val on unique local maximum
    % EXAMPLE
    % [l,c,m] = findLocalMaximum(img,radius);
    % FIND LOCAL MAXIMA BY DILATION (FAST) /!\ NON UNIQUE /!\
    % mask = fspecial('disk',radius)>0;
    % val2 = imdilate(val,mask);
    % index = val==val2;
    % [row,col] = find(index==1);
    % max_local = zeros(size(val));
    % max_local(index) = val(index);
    % FIND UNIQUE LOCAL MAXIMA USING FILTERING (FAST)
    mask    = fspecial('disk',radius)>0;
    nb      = sum(mask(:));
    highest           = ordfilt2(val, nb, mask);
    second_highest    = ordfilt2(val, nb-1, mask);
    index             = highest==val & highest~=second_highest;
    max_local         = zeros(size(val));
    max_local(index) = val(index);
    [row,col]         = find(index==1);
    % FIND UNIQUE LOCAL MAXIMA (FAST)
    % val_height   = size(val,1);
    % val_width    = size(val,2);
    % max_local    = zeros(val_height,val_width);
    % val_enlarge = zeros(val_height+2*radius,val_width+2*radius);
    % val_mask    = zeros(val_height+2*radius,val_width+2*radius);
    % val_enlarge( (1:val_height)+radius , (1:val_width)+radius ) = val;
    % val_mask(    (1:val_height)+radius , (1:val_width)+radius ) = 1;
    % mask    = fspecial('disk',radius)>0;
```

```
% row = zeros(val_height*val_width,1);
% col = zeros(val_height*val_width,1);
% index = 0;
% for l = 1:val_height
%       for c = 1:val_width
%             val_ref = val(l,c);
%             neigh_val   = val_enlarge(l:l+2*radius,c:c+2*radius);
%             neigh_mask = val_mask(    l:l+2*radius,c:c+2*radius).*mask;
%             neigh_sort = sort(neigh_val(neigh_mask==1));
%             if val_ref==neigh_sort(end) && val_ref>neigh_sort(end-1)
%                   index            = index+1;
%                   row(index,1)     = l;
%                   col(index,1)     = c;
%                   max_local(l,c) = val_ref;
%             end
%       end
% end
      % row(index+1:end,:) = [];
      % col(index+1:end,:) = [];
end
```

RANSAC code:

```
function [final_inliers flag bestmodel] = AffinePairwiseRansac(frames_a1, frames_a2, all_matches)
% iterations = 0
% bestfit = nil
% besterr = something really large
% while iterations < k {
%       maybeinliers = n randomly selected values from data
%       maybemodel = model parameters fitted to maybeinliers
%       alsoinliers = empty set
%       for every point in data not in maybeinliers {
%             if point fits maybemodel with an error smaller than t
%                   add point to alsoinliers
%       }
%       if the number of elements in alsoinliers is > d {
%             this implies that we may have found a good model
%             now test how good it is
%             bettermodel = model parameters fitted to all points in maybeinliers and alsoinliers
%             thiserr = a measure of how well model fits these points
%             if thiserr < besterr {
%                   bestfit = bettermodel
%                   besterr = thiserr
```

```
%              }
%       }
%       increment iterations
% }
% return bestfit
%
%first decide how many matches we have
MIN_START_VALUES = 4;
num_matches = size(all_matches,2);
if (num_matches < MIN_START_VALUES)
     final_inliers = [];
     bestmodel = [];
     flag = -1;
     return
end
% Todo? These might have to be changed if the values are different.
Z_OFFSET = 640;
COND_THRESH = 45;
% RANSAC parameters
NUM_START_VALUES = 3;    % only 3 corrospondences needed for determining affine model
K = 50;
ERROR_THRESHOLD = 10; % fairly high threshold - this is in number of pixels
D = 1; % additional points must fit any given affine model
N = NUM_START_VALUES;
RADIUS = 30; %changed to 30 by vijay
MIN_NUM_OUTSIDE_RADIUS = 1;
%best error, best fit
iteration = 0;
besterror = inf;
bestmodel = [];
final_inliers = [];
max_inliers = 0;
while (iteration < K)
     %start with NUM_START_VALUES unique values
     uniqueValues = [];
     max_index = size(all_matches, 2);
     while (length(uniqueValues) < NUM_START_VALUES)
         value = ceil(max_index*rand(1,1));
         if (length(find(value == uniqueValues)) == 0)
             %unique non-zero value
```

```
            uniqueValues = [uniqueValues value];
        end
    end
    %uniqueValues are the indices in all_matches
    maybeinliers = all_matches(:, uniqueValues);      %start with NUM_START_VALUES unique
random values
    % make sure points are well distributed
    point_matrix = [frames_a1(:, maybeinliers(1, :)); Z_OFFSET*ones(1, NUM_START_VALUES)];
    if (cond(point_matrix) > COND_THRESH)
        iteration = iteration + 1;
        continue;
    end
    M_maybemodel = getModel(maybeinliers, frames_a1, frames_a2);
    if (prod(size(M_maybemodel)) == 0)
        iteration = iteration + 1;
        continue;
    end
    alsoinliers = [];
    %figure out other inliers
    for i = 1:size(all_matches, 2)
        temp = find(all_matches(1,i) == maybeinliers(1,:));
        if (length(temp) == 0)
            %this means, point not in maybeinlier
            a1 = frames_a1(1:2, all_matches(1,i));
            a2 = frames_a2(1:2, all_matches(2,i));
            if (getError(M_maybemodel, a1, a2) < ERROR_THRESHOLD )
                alsoinliers = [alsoinliers all_matches(:,i)];
            end
        end
    end
    if (size(alsoinliers,2) > 0)
        num = 0;
        dist = [];
        for i = 1:NUM_START_VALUES
            diff =        frames_a1(1:2,   alsoinliers(1, :)) - ...
                    repmat(frames_a1(1:2, maybeinliers(1, i)), [1, size(alsoinliers, 2)]);
            dist = [dist; sqrt(sum(diff.^2))];
        end
        num = sum(sum(dist > RADIUS) == NUM_START_VALUES);
        if (num < MIN_NUM_OUTSIDE_RADIUS)
            iteration = iteration+1;
```

```
            continue;
        end
    end
%see how good the model is
%fprintf('Number of elements in also inliers %d\n', size(alsoinliers,2));
if (size(alsoinliers,2) > D)
        %this implies that we have found a good model
        %now let's see how good it is
        %find new model
        all_inliers = [maybeinliers alsoinliers];
        M_bettermodel = getModel(all_inliers, frames_a1, frames_a2);
        %the new model could be bad
        if (prod(size(M_bettermodel)) == 0)
            iteration = iteration+1;
            continue;
            end
        %find error for the model
        thiserror = getModelError(M_bettermodel,all_inliers, frames_a1, frames_a2);
        if max_inliers < size(all_inliers, 2) | (thiserror < besterror & max_inliers == size(all_inliers, 2))
                bestmodel = M_bettermodel;
                besterror = thiserror;
                final_inliers = all_inliers;
                max_inliers = size(final_inliers, 2);
            end
    end
    %do it K times
    iteration = iteration + 1;
end
%bestmodel has the best Model
if (prod(size(bestmodel)) ~= 0)
    % a model was found
    fprintf('Error of best_model ~%f pixels\n', besterror);
    flag = 1;
else
    flag = -1;
    final_inliers = [];
    bestmodel = [];
    fprintf('No good model found !\n');
end
end
```

```
function error = getModelError(M ,matches, frames_a1, frames_a2)
        nummatches = size(matches,2);
        error = 0;
        for i = 1:nummatches
                a1 = frames_a1(1:2, matches(1,i));
                a2 = frames_a2(1:2, matches(2,i));
                error = error + getError(M, a1, a2);
        end
        error = error/nummatches;
   end
function M = getModel(matches, frames_a1, frames_a2)
        %let's go from 1 to 2 -- changed on Apr 28 to be consistent
        %with epipolar and perspective models
        singular_thresh = 1e-6;
        scaling_ratio_thresh = 5;

        scale_thresh = 0.005;
        % approximate M
        M = zeros(3,3);
        Y = []; X = [];
           for i = 1:size(matches,2)
               a1 = frames_a1(1:2, matches(1,i));
               a2 = frames_a2(1:2, matches(2,i));
               Y = [Y; a2];
               X = [X; a1(1) a1(2) 1 0 0 0; 0 0 0 a1(1) a1(2) 1];
        end
          %to check if matrix is singular
        if (1/cond(X) < singular_thresh)
               M = [];
               return
        end
         M = X\Y;
        %we need to return M - a 3X3 matrix, where the last row is (0 0 1)
        M = [reshape(M, 3,2)'; 0 0 1];
        %let's add some rules to remove any crazy map
        %we definitely cannot have reflection
        [u, s, v] = svd(M(1:2,1:2));
        if (det(u*v') < 0)
               %==> there is a reflection
               M=[];
```

```
%                  fprintf('Special case to avoid reflection\n');
                return
            end
            %we cannot have crazy ratios of scaling in the two dimensions.
            if (cond(M(1:2,1:2)) > scaling_ratio_thresh)
                %==> the matches are bad
                M=[];
%                  fprintf('Special case to avoid crazy scaling ratio\n');
                return
            end


            %check for crazy zoom
            if (s(1,1) < scale_thresh | s(2,2) < scale_thresh)
                M = [];
            end
end
function error = getError(M, a1, a2)
            %a2_model is the value of a2 that comes from the model
            %calculate mapping error
            a2_model   = M*([a1;1]);   %3x1 vector, only the first two values matter
            error = dist(a2, a2_model(1:2));
end
function d =   dist(one, two)
        d = sqrt(sum((one-two).^2));
end
function [final_inliers flag bestmodel] = PerspectivePairwiseRansac(frames_a1, frames_a2, all_matches)
        % iterations = 0
        % bestfit = nil
        % besterr = something really large
        % while iterations < k {
        %       maybeinliers = n randomly selected values from data
        %       maybemodel = model parameters fitted to maybeinliers
        %       alsoinliers = empty set
        %       for every point in data not in maybeinliers {
        %             if point fits maybemodel with an error smaller than t
        %                   add point to alsoinliers
        %       }
        %       if the number of elements in alsoinliers is > d {
        %             this implies that we may have found a good model
        %             now test how good it is
```

```
%                bettermodel = model parameters fitted to all points in maybeinliers and alsoinliers
%                thiserr = a measure of how well model fits these points
%                if thiserr < besterr {
%                        bestfit = bettermodel
%                        besterr = thiserr
%                }
%        }
%        increment iterations
% }
% return bestfit
%
%finds the model from the first image to the second image
%
%        [XW] = [a b c][x]
%        [YW] = [d e f][y]
%        [W]   = [g h 1][1]
%(x,y,1) are points in the first image and they map to (XW, YW, W) in
%the second image
%first decide how many matches we have
MIN_START_VALUES = 20;
num_matches = size(all_matches,2);
if (num_matches < MIN_START_VALUES)
    final_inliers = [];
    bestmodel = [];
    flag = -1;
    return
end
%RANSAC parameters
K = 150;
NUM_START_VALUES = 4;    % using 4 point least squares solution
ERROR_THRESHOLD = 10; % this is in the number of pixels.
D = 8; %start with 4 points and fit atleast 8 more fit the model
%best error, best fit
iteration = 0;
besterror = inf;
bestmodel = [];
final_inliers = [];
max_inliers = 0;
while (iteration < K)
    %start with NUM_START_VALUES unique values
```

```
uniqueValues = [];
max_index = size(all_matches, 2);
while (length(uniqueValues) < NUM_START_VALUES)
    value = ceil(max_index*rand(1,1));
    if (length(find(value == uniqueValues)) == 0)
        %unique non-zero value
        uniqueValues = [uniqueValues value];
    end
end
%uniqueValues are the indices in all_matches
maybeinliers = all_matches(:, uniqueValues);    %start with NUM_START_VALUES unique
random values
M_maybemodel = getModel(maybeinliers, frames_a1, frames_a2);
if (prod(size(M_maybemodel)) == 0)
    iteration = iteration + 1;
    continue;
end
alsoinliers = [];
%figure out other inliers
for i = 1:size(all_matches, 2)
    temp = find(all_matches(1,i) == maybeinliers(1,:));
    if (length(temp) == 0)
        %this means, point not in maybeinlier
        a1 = frames_a1(1:2, all_matches(1,i));
        a2 = frames_a2(1:2, all_matches(2,i));
        if (getError(M_maybemodel, a1, a2) < ERROR_THRESHOLD )
            alsoinliers = [alsoinliers all_matches(:,i)];
        end
    end
end
%see how good the model is
%fprintf('Number of elements in also inliers %d\n', size(alsoinliers,2));
if (size(alsoinliers,2) > D)
    %this implies that we have found a good model
    %now let's see how good it is
    %find new model
    all_inliers = [maybeinliers alsoinliers];
    M_bettermodel = getModel(all_inliers, frames_a1, frames_a2);
    %the new model could be bad
    if (prod(size(M_bettermodel)) == 0)
```

```
                    iteration = iteration+1;
                    continue;
            end
            %find error for the model
            thiserror = getModelError(M_bettermodel,all_inliers, frames_a1, frames_a2);
            if  max_inliers  <  size(all_inliers,  2)  |  (thiserror  <  besterror  &  max_inliers  ==
size(all_inliers, 2))
                    bestmodel = M_bettermodel;
                    besterror = thiserror;
                    final_inliers = all_inliers;
                    max_inliers = size(final_inliers, 2);
            end
        end
        %do it K times
        iteration = iteration + 1;
    end

    if (prod(size(bestmodel)) ~= 0)
        % a model was found
        fprintf('Error of best_model ~%f pixels\n', besterror);
        flag = 1;
      else
        flag = -1;
        final_inliers = [];
        bestmodel = [];
        fprintf('No good model found !\n');
    end
end
function error = getModelError(M ,matches, frames_a1, frames_a2)
        nummatches = size(matches,2);
        error = 0;
        for i = 1:nummatches
                a1 = frames_a1(1:2, matches(1,i));
                a2 = frames_a2(1:2, matches(2,i));
                error = error + getError(M, a1, a2);
        end
        error = error/nummatches;
end
function P = getModel(matches, frames_a1, frames_a2)
    %goes from 1 to 2
    %let's use a least squares approach. Referenced from
```

```
%http://alumni.media.mit.edu/~cwren/interpolator
nummatches = size(matches,2);
LHS= [];
for i = 1:nummatches
    a1 = frames_a1(1:2, matches(1,i)); x = a1(1); y=a1(2);
    a2 = frames_a2(1:2, matches(2,i)); X = a2(1); Y=a2(2);
    LHS = [LHS; x y 1 0 0 0 -X*x -X*y; 0 0 0 x y 1 -Y*x -Y*y];
end
RHS = reshape(frames_a2(1:2, matches(2, :)), nummatches*2, 1);
P = reshape([(LHS\RHS);1], 3,3)';
%to get P in the form
%       [a b c; d e f; g h 1];
end
function error = getError(P, a1, a2)
    %the model F goes from image 1 to image 2
    %the corresponding point for a1
    temp = P*[a1;1];
    a2_model(1) = temp(1)/temp(3);
    a2_model(2) = temp(2)/temp(3);
    error = dist(a2_model', a2);
end
  function d =    dist(one, two)
    d = sqrt(sum((one-two).^2));
end
```

8.4 Panoramic Image Stitching

Panoramic image stitching aims to do seamless stitching on image sequence taken from the same scene, different perspective, different focal lengths, from the same optical center with partially overlapping. This means that image registration algorithm is used to calculate the motion parameters between each frame and then synthetic a large static wide-angle image. Moreover, the stitching image requires to be as close as the real scene without obvious seam. According to different viewpoints, image stitching can be divided into algorithm based on single viewpoint and algorithm based on multiple viewpoints. To obtain single viewpoint image sequences, a camera is fixed in a position and rotate it around; or, set cameras in a circle, the optical axis of the camera is on the same plane and intersects with one point, and the video is collected in real time. To obtain multiple viewpoints image sequences, usually use a camera capture a set of image sequences on a horizontal level or set multiple cameras in different positions and capture at the same time. Image stitching algorithm based on multiple viewpoints is commonly used in stitching banded panoramic images.

This section introduces an image stitching algorithm based on image projection transformation without active objects.

Discrete image information can only express information on a part of the visual environment. The panoramic view based on image rendering is to show the discrete image information in an image completely. Build a complete graphics environment for better 3D visual effects.

(1) Image positioning

Automatically find overlapping locations for images. Proposing there are two rectangle regions A and B. B contains a region A_2. A_2 and A are same module, the position of A_2 in B is to solve. The typical algorithm is to search from the lower left corner of B, where each piece is compared to a with the same area C and A, and the value of the evaluation function, the smallest area is A_2.

(2) Image stitching

After image positioning, if splice the two images simply, there will be a clear seam due to the difference of brightness. The color fitting method can be used to reconcile the brightness of adjacent images and produce seamless synthetic images.

(3) Implementation of cylindrical projection

Shot at the same point as the rotating camera, the cylindrical panoramic images are not in the same coordinate system. There is a certain angle on the projection surface. In order to generate a panoramic image, we must transform these images into a unified cylindrical coordinate system and use image stitching technology to remove the overlap of every two images. In this way, a complete cylindrical panoramic picture is obtained.

As shown in Fig. 8.13, it's the positive projection diagram of cylindrical surface. I is a frame extracted from the video, P is any point on the captured image, Q is the point where P maps to the cylinder coordinate.

Fig. 8.13 The positive projection diagram of cylindrical surface

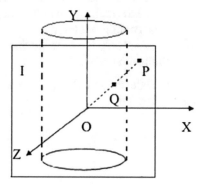

Assuming that W and H are the width and height of the Image I respectively, f is the radius of the cylinder, so that P's coordinate in 3D coordinate system is represented as $(x - W/2, y - H/2, f)$. Using the combination of the parametric equation and the cylindrical equation, assuming Q's coordinate is (x', y', z') and P as well as Q are in the same line, which satisfies the parametric equation:

$$\begin{cases} x' = t(x - W/2) \\ y' = t(y - H/2) \\ z' = tf \end{cases} \tag{8.14}$$

where t is the parameter, coupled with the cylindrical surface equation: $x'^2 + z'^2 = f^2$.

Thus, we can get the coordinates of Q point because the coordinates of Q point is three-dimensional, we convert it into two dimensions to get:

$$\begin{cases} x' = f * \arctan\left[(x - W/2)/f\right] + f * \arctan(W/2f) \\ y' = f(y - H/2)/\sqrt{(x - W/2)^2 + f^2} + H/2 \end{cases} \tag{8.15}$$

After the image is projected to the cylindrical plane, the images of the same coordinate system are obtained. Then by looking for the transformation between adjacent images, the sequence images are spliced together to form a cylindrical panoramic image under the same scene.

The steps of using IBR method to splice video panoramic image are as follows:

(1) Extract key frames of the video and use images to represent information in videos.
(2) Find out the overlapping region of images, this means that extract feature position.
(3) Image registration, match feature points. Use fine match algorithm to remove wrong point pairs and moving corner point pairs. The coordinate transformation function is obtained by calculating the transformation matrix between the datum image and the image to be matched. Finally, the coordinate transformation function is used to transform the image to the datum coordinate system and realize the registration of the image to be matched with the datum image in the same coordinate system.
(4) The final step is image stitching, involving the fusion of two images and the elimination of seams.

The code for reading a video and extracting key frames is shown in PRO-GRAMME 8.8.

PROGRAMME 8.8: Read video and extract the key frames

```
xyloObj = VideoReader('test.avi');%read video
nFrames = xyloObj.NumberOfFrames;      %number of frames
count=1;        %count the extracted frames
letter='a'; %tag，to make subsequent read sequences normal, precede the Arabic numerals with letters
for k = 1 : 5: nFrames
    mov(k).cdata =   read(xyloObj,k);        %image color data
    strtemp = strcat('images/',letter+count/10);
    strtemp = strcat(strtemp, int2str(count));
    strtemp = strcat(strtemp,'.jpg');
    count=count+1;
    imwrite(mov(k).cdata,strtemp);%save as strtemp.jpg
end
%feature transformation method based on color——using color histogram to measure color feature
%pop up a few key frames
filenames=dir('images/*.jpg');        %image source
num=size(filenames,1);                    %number of images
key=zeros(1,num);                       % (0,1) key frame array
count=0;                                  %save a few key frames
threshold=0.75;                       %set threshold
if num==0
    error('Sorry, there is no pictures in images folder!');
else                                     %set the first frame as key frame
    img=imread(strcat('images/',filenames(1).name));
    key(1)=1;
    count=count+1;
    %obtain RGB histogram
    [preCountR,x]=imhist(img(:,:,1));     %red histogram
    [preCountG,x]=imhist(img(:,:,2));     %green histogram
    [preCountB,x]=imhist(img(:,:,3));     %blue histogram
    %show first key frame
    figure(count);
    imshow('images/a1.jpg');
    for k=2:num
        img=imread(strcat('images/',filenames(k).name));
        [newCountR,x]=imhist(img(:,:,1));     %red histogram
        [newCountG,x]=imhist(img(:,:,2));     %green histogram
        [newCountB,x]=imhist(img(:,:,3));     %blue histogram
        sR=0;
        sG=0;
        sB=0;
```

```
%use method of color histograms
for j=1:256
   sR=min(preCountR(j),newCountR(j))+sR;
   sG=min(preCountG(j),newCountG(j))+sG;
   sB=min(preCountB(j),newCountB(j))+sB;
end
dR=sR/sum(newCountR);
dG=sG/sum(newCountG);
dB=sB/sum(newCountB);
            %YUV,persons are sensitive to Y
            d=0.30*dR+0.59*dG+0.11*dB;
            if d<threshold              %small similarity, new keyframes found
                key(k)=1;               %set as keyframes
                count=count+1;
                figure(count);
                imshow(strcat('images/',filenames(k).name));
                %nearest update color histogram
                preCountR=newCountR;
                preCountG=newCountG;
                preCountB=newCountB;
            end
       end
end
keyFrameIndexes=find(key)
```

Panoramic image stitching based on IBR method using key frames, whose code is programme 8.9, and the functions involved are described in programme 8.3.

PROGRAMME 8.9: Panoramic image stitching based on IBR method

```
clear;clc;
%%%%%%%%%%%%%%%%%input images, tectonic image pyramid, save an array
tic
fprintf(' input images, tectonic image pyramid,...');
level=3;
T0=uint8([]);T=uint8([]);
%%%%%%%%%%%%%%%%%%%%%
i1=imread('1.jpg');T0(:,:,:,1)=i1;
i2=imread('2.jpg');T0(:,:,:,2)=i2;
i3=imread('3.jpg');T0(:,:,:,3)=i3;
```

```
i4=imread('4.jpg');T0(:,:,:,4)=i4;
i5=imread('5.jpg');T0(:,:,:,5)=i5;
i6=imread('6.jpg');T0(:,:,:,6)=i6;
i7=imread('7.jpg');T0(:,:,:,7)=i7;
i8=imread('8.jpg');T0(:,:,:,8)=i8;
[h,w,d]=size(T0(:,:,:,1));%%%same image size is required
% [T,Terr]=multi_resolution(T0,level);
T=multi_resolution(T0,level);
toc
% %%%%%%%%%%%%%%%%%%% calculate offset through phase correlation
tic
fprintf(' calculate offset through phase correlation..');
M=8;%% number of images
% L=M*w;%% total length of images
for N=1:M
    if N<M [i,j]=poc_2pow(T(:,:,:,N),T(:,:,:,N+1));
    elseif N==M [i,j]=poc_2pow(T(:,:,:,N),T(:,:,:,1));
    end
    coor_shift(N,1)=i;
    coor_shift(N,2)=j;
end
coor_shift=coor_shift*2^level;%%% convert the offsets in the pyramid hierarchy to the offset of the
    original
toc
%%%%%%%%%%%%%%%% Transform to cylindrical coordinate system
tic
fprintf(' Transform to cylindrical coordinate system..');
f=sqrt(h^2+w^2);
[T1,coor_shift02]=coortransf(T0,f,coor_shift);
toc
%%%%%%%%%%%% merge overlapping parts
tic
fprintf(' merge overlapping parts, stitching image....');
panorama1=T1(:,:,:,1);
for N=1:M
    if N<M
panorama1=mosaic(panorama1,T1(:,:,:,N+1),coor_shift02(N,1),coor_shift02(N,2));
    end
```

end

toc

%%%%%%%%%%%% reconstruct image

tic

fprintf(' save and show result..');

toc

image1=rgb2gray(panorama1);

index=find(image1(:,1)>=255);

aa=max(index);

[r,c]=size(image1)

image1=imcrop(panorama1,[1,aa,c,r]);

imshow(image1);

imwrite(image1,'pic1.jpg','jpg');

Key frames extracted is shown in Fig. 8.14.
Panoramic stitching image is shown in Fig. 8.15.

Fig. 8.14 Key frames extracted

Fig. 8.15 The panoramic stitching image

References

1. Yanyan L, Xu S (2008) Study of image stitching algorithm based on ratio matching. Electron Meas Technol
2. Le-Fu WU, Ding GT (2010) Region-based images stitching algorithm. Comput Eng Design 31(18):4043–4044
3. Zhou DF, Ming-Yi HE, Yang Q (2009) A robust seamless image stitching algorithm based on feature points. Meas Control Technol 28(6):32–36
4. Chandratre R, Chakkarwar VA (2014) Image stitching using Harris and RANSAC. Int J Comput Appl 89(15):14–19
5. Zhao WJ, Gong SR, Liu Q et al (2007) An auto-sorting arithmetic for image sequence used in image Mosaics. J Image Graph 12(10):1861–1864

Chapter 9
Image Watermarking

Abstract In this chapter we firstly introduce the application background of digital watermarking, then represent fragile watermarking, robust watermarking, and semi-fragile watermarking embedding methods respectively.

9.1 Introduction

Digital Watermarking is a technology which embedding [1] the symbolic information into the multimedia works directly through a certain algorithm [2], but it will not affect the value of the original content and using. It cannot be noticed by human perception system unless through a dedicated detector or reader. The watermark may be the serial number of the author, the logo of the company, the special text, and so on. It can be used to identify the sources and versions of documents, images or music products, the author, the owner, the issuer and the ownership of the digital products. Figure 9.1a is the original image, also known as the host image, Fig. 9.1b is the watermark image, Fig. 9.1c is the image after watermarking. And Fig. 9.1a, c shall not cause any visual differences in human eyes.

Digital watermarking is an information security technology which developed in the 1990s. It provides a new solution for protecting the copyright of multimedia information and ensuring the safe use of multimedia information, and has become one of the fastest growing hot spot in the field of multimedia information security. It has received great attention from both the international academic and the business. Digital watermarking is used to solve the problem of intellectual property protection and it is one of the most potential multi-disciplinary cross technology.

In fact, digital watermarking is one who embeds the label with particular significance into digital image, audio, document, book, video and other digital product by using digital insertion method for copyright protection, information hiding, tamper proof, data file authenticity identification and so on. At the same time, the integrality of digital information is ensured by the detection and analysis of the watermark. In practice, the following several issues constitute the background of digital watermarking [3–5]:

© Springer International Publishing AG, part of Springer Nature 2019
329
S. Gong et al., *Advanced Image and Video Processing Using MATLAB*,
Modeling and Optimization in Science and Technologies 12,
https://doi.org/10.1007/978-3-319-77223-3_9

(a) The original image   (b) The watermark image   (c) The embedded image

Fig. 9.1 Digital watermark embedded in the image

(1) Intellectual Property Protection of Digital Works

At present, copyright protection of digital works (such as computer art, scanned images, digital music, video, 3D animation) is a hot issue, which is the most important application of watermarking. As the copy and modification of digital works is very easy, and it can be exactly the same as the original one. So, the originator has to adopt some measures that may damage the quality of original works seriously to add some copyright logos in works for protection, but these visible signs can be tampered easily. The digital watermarking utilizes data hiding principles to make copyright logos invisible or inaudible, which does not damage the quality of the original work, but also achieves the purpose of copyright protection. This application requires very high robustness. At present, the digital watermarking technology for copyright protection has entered the initial stage of practical application. The "digital library" software of IBM has provided a digital watermarking function, and Adobe also integrated the Digimarc company's digital watermark plugin in its famous Photoshop. In general, the digital watermark products on the market are not yet mature in technology, and are easy to be destroyed or cracked, so there is still a long way to go from the real utility.

(2) Anti-counterfeiting of bills in business transaction

With the development of high-quality image input or output devices, especially the appearance of color inkjet with precision over 1200 dpi, laser printers and high-precision color copiers, which makes the counterfeiting of money, checks and other notes easier. On the other hand, there will be many transitional electronic documents, such as scanning images of various paper notes during the transition from traditional business to e-commerce. Even after the network security technology is ripe, all kinds of electronic bills also need some non-password authentication methods. Digital watermarking technology can provide invisible certification marks for various bills, which greatly increases the difficulty of forgery.

(3) The hidden identification and tampering tips of the audiovisual data

The identification information of the data is often more valuable than the data itself, such as the date, longitude and latitude of the remote-sensing images and so on. Data that without any identification information is sometimes even unusable, but it is dangerous to mark the important information directly on the original file. Digital watermarking provides a way to hide the identities and the identification information is not visible on the original document, it only can be read through a special reading program.

According to the function of digital watermarking, it can be divided into robust watermarking, fragile watermarking and semi-fragile watermarking. The main purpose of robust watermarking [6] is to protect the copyright of digital works, which requires that embedded watermarks should sustain a variety of common signal processing operations, including unintentional or malicious processing, such as lossy compression, filtering, smoothing, signal reduction, image enhancement, resampling, geometric deformation, and so on. After all kinds of processing, the robust watermark should be able to detect after as long as the host information is not destroyed greatly. Therefore, it needs a higher demand of robustness. Fragile watermarking is also known as the fully fragile watermarking, which can detect any changes of image pixel values. The purpose of fragile watermarking is to protect the integrity of digital works and to identify the authenticity of digital works. Semi-fragile watermarking needs to resist a certain degree of beneficial digital signal processing operations such as JPEG compression, etc. This type of watermarking is more robust than the fully fragile watermarking slightly, which allow the image to have a certain change, and it may be a check of integrity to some extent.

According to the implementation method of watermarking, it can be divided into spatial domain digital watermarking and frequency domain digital watermarking. Spatial domain digital watermarking superimposes watermark signal on the signal space directly, while frequency domain digital watermarking often uses the technique which likes spread spectrum image technology to hide the watermark information. Such techniques are generally based on common image transformations, including discrete cosine transform (DCT), discrete wavelet transform (DWT), Fourier transform (DFT or FFT) and so on.

A digital watermarking system generally includes three basic aspects: the generation of watermarks, the embeddedness of watermarks, and the extraction or detection of watermarks. Digital watermarking is a quasi-optimal problem that seeks to satisfy the demands of imperceptibility, reliability and robustness through the analysis of image host, the pretreatment of embedded information, the selection of position of insertion, the design of embedded model, the control of embedded modulation and so on. As an important part of watermark information, the key is often embedded in different steps such as information preprocessing, embedded point selection and modulation control, etc.

The basic frameworks diagram of the general process of digital watermark embedding and detection are shown in Figs. 9.2 and 9.3.

Figure 9.2 shows the embedding process of the watermark. Set the watermark information W as input, the multimedia products such as images, documents, audios, videos as original carrier data I and K as the optional private key (or public key). The watermark information W may be data of any forms, such as characters, binary images, grayscale images or color images, 3D images, and so on. The watermark generation algorithm G should ensure the uniqueness, validity and irreversibility of

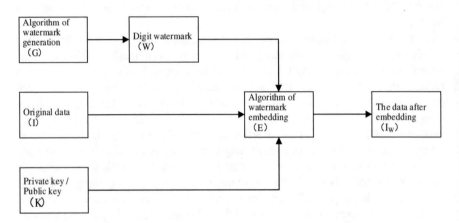

Fig. 9.2   The basic framework of the general process of watermark embedding

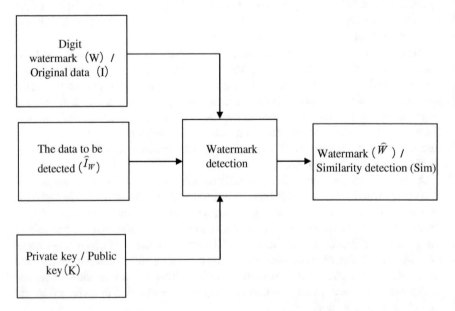

Fig. 9.3   Basic framework of the general process of watermarking detection

the digital watermark. The key K can be used to enhance the security to avoid the unauthorized restoration and repair of watermarks. All the utility systems must use a key, and some even use a combination of several keys.

There are many algorithms for watermark embedding, and Eq. (9.1) gives a general formula for the embedding process of the watermark:

$$I_W = E(I, W, K) \tag{9.1}$$

where I_W denotes the data after embedding the watermark (i.e. the watermark carrier data), I denotes the original carrier data, W denotes the watermark sets, and K is a key set. Where K is an optional term, which is generally used for the generation of watermark signals.

Figure 9.3 indicates the process of watermark detection. It can be divided into the following three types according to whether the original information is needed

(1) Require the original carrier data I:

$$\widehat{W} = D\left(\widehat{I}_W, I, K\right) \tag{9.2}$$

(2) Requires the original watermark W:

$$\widehat{W} = D\left(\widehat{I}_W, W, K\right) \tag{9.3}$$

(3) Without the original information:

$$\widehat{W} = D\left(\widehat{I}_W, K\right) \tag{9.4}$$

where $\widehat{W}$ is the extracted watermark while D is the watermark detection algorithm, and $\widehat{I}_W$ is the watermark carrier data that has been attacked during the transmission. There are 2 means of detection: one is the extraction of the embedded signal or correlation verification based on the given original information, and the second is whole search or distribution hypothesis testing for embedded information without the original information. If the signal is a random signal or a pseudo-random signal, it is proved that the general method of proving that the detection signal is the watermark signal is to do the similarity test. The general formula for watermark similarity test are as follows:

$$Sim = \frac{W * \widehat{W}}{\sqrt{W * W}} \text{ or } Sim = \frac{W * \widehat{W}}{\sqrt{W * W}\sqrt{\widehat{W} * \widehat{W}}} \tag{9.5}$$

where $\widehat{W}$ is the extracted watermark, W is the original watermark, Sim represents the similarity of different signals.

9.2 Fragile Watermarking Based on Spatial Domain

The fragile watermarking algorithm based on spatial domain usually loads the watermark information on original data directly by modifying the pixel value of the image. The most representative one is Least Significant Bit (LSB) method, which modify the minimum valid bit of image pixel value to achieve the purpose of embedding watermark information into the host image. Once the image has been tampered with, the information of the minimum valid bit is also changed, so that we can locate the tampered area through the corresponding detection program.

The LSB method is one of the earliest and most basic way of image information hiding method based on spatial domain, and many other methods are developed based on LSB. Nowadays, some simple information hiding softwares, such as Hide and Seek, Stego-Dos, White Noise S-tools and so on, often use LSB algorithm and palette adjustment to hide the information into 24-bit images or 256-color images.

The LSB refers to the zeroth bit (or the lowest bit) of a binary number, with a weight of 2^0, which can be used to detect the parity of numbers. The LSB algorithm make the use of the principle of bit-plane in digital image processing, i.e. change the information of the lowest bit of the image. So that the influence on image information is very small, and even the visual perception system of human eyes cannot perceive it. Taking a 256-grayscale image as an example, it requires 8 bits to represent the 256-level grayscale, but the effect of each bit is different, the higher bit has more effect on the image, whereas the lower bit effect weakly, even cannot be perceived.

The implementation of the LSB algorithm is relatively easy. Firstly, we need to consider the number of watermark information. If only the least one bit is available, the amount of the watermark information that can be embedded is the $\frac{1}{8}$ of the original image. If the lowest two bits are available, the amount of the watermark information is $\frac{1}{4}$ of the original image, and so on. The more the lowest bit is available, the more information can be embedded in the original image, and will also has a greater impact on the visual perception of the image. Then, adjust the size and bit of the digital watermark appropriately to meet the demand of digital watermarking data size of the image. Finally, set the lowest position of the original image to 0, and put the digital watermark data at the lowest bit of the original image.

The code based on LSB algorithm is shown in PROGRAMME 9.1 and PRO-GRAMME 9.2:

PROGRAMME 9.1: Watermark Embedding

```
[C,map]=imread('color.bmp');              % read the original image
[m,map1]=imread('word.bmp');              % read the watermarking image
Mc=size(C,1);                             % the rows of the original image
Nc=size(C,2);                             % the columns of the original image
Mm=size(m,1);                             % the rows of the watermarking
Nm=size(m,2);                             % the columns of the watermarking
w_i=C;                                    % assign original image to w_i
for ii=1:Mc                               % change the LSB value of the original
 for jj=1:Nc
  w_i(ii,jj)=bitset(w_i(ii,jj),1,m(ii,jj));   % call the bitset () function
 end
end
imwrite(w_i,'lsb_watermark.bmp','bmp');   % write the embedded watermark image into lsb_watermark.bmp
figure(1);
imshow(w_i,[]);title('After Embedding')   % show the picture after embedding
figure(2);
imshow(C,[]);title('Before Embedding')    % show original image
figure(3);
imshow(m,[]);title('Watermarking Image')  % show watermarking image
```

Given a 200×200 image, and the digital watermark is a pure text binary image. We use bitset () function in MATLAB to set bit plane to 0 and embed digital watermark data, we call for function bitset (A, bit) to set bit plane to 0, where A indicates the image to be set to 0, bit indicates which position to be on 0. If we want to set the least bit to 0, it can be indicated as bitset (A, 1). The way of embedding is: w_i (ii, jj) = bitset (w_i (ii, jj), 1, w (ii, jj)), where w_i indicates the image to be embedded. 1 indicates the least bit to be embedded and 2 represents to embed in the second bit plane, and the rest can be done in the same manner, where w represents the watermark image.

PROGRAMME 9.2: **Extraction of Digital Watermarking**

```
file_name='lsb_watermark.bmp';
watermark_image=imread(file_name);                    % read the embedded image
Mw=size(watermark_image,1);                           % the raws of embedded image
Nw=size(watermark_image,2);                           % the columns of embedded image
file_name='word.bmp';
orig_watermark=imread(file_name);                     % read the original watermark image
Mm=size(orig_watermark,1);                            % the raws of the watermark image
Nm=size(orig_watermark,2);                            % the columns of watermark image
for ii=1:Mw                  % reconstruct watermark by using LSB of the embedded image
 for jj=1:Nw
  watermark(ii,jj)=bitget(watermark_image(ii,jj),1); % reconstruct watermark by using bitget()
 end
end
watermark=2*double(watermark);   % turn the extracted watermark into the original watermark size
for ii=1:Mm-1
 for jj=1:Nm-1
  watermark1(ii+1,jj+1)=watermark(ii,jj);
 end
end
watermark1(1,1)=watermark(Mm,Nm);
figure(1)
subplot(1,2,1)
imshow(watermark_image,[]);title('After Embedding')      % show the picture after embedding
figure(2)
subplot(1,2,2)
imshow(orig_watermark,[]);title('Original Watermark')    % show original image
```

9.3   Robust Watermarking Based on DCT

In recent years, many different types of digital watermarking technologies have been proposed. According to differences of embedding domain, they can be divided into two categories: spatial domain and transform domain. The former embeds the information into host images in spatial domain, and the latter embeds the information into the transform domain by changing the coefficients of the transform domain. Next, we will introduce the robust digital watermarking technology and fragile digital watermarking technology based on the transform domain respectively.

The DCT transformation is the abbreviation of Discrete Cosine Transform. The main idea is to select the medium frequency and low frequency coefficients on the DCT transform domain to superimpose the watermark information, because the human vision perception is mainly concentrated on these frequency bands. When

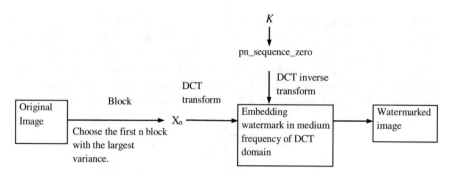

Fig. 9.4 The flow chart of robust watermark embedding based on DCT

attackers damage the watermark, it is inevitably causes a serious decline in the quality of the image, and the general processing will not change the data of this part. Moreover, since JPEG, MPEG and other compression algorithms are quantification in the DCT transform domain, therefore, it can resist a certain lossy compression through the clever fusion of watermark and quantification. In addition, the statistical distribution of DCT transform domain coefficients owns a good mathematical model, which can estimate the information content of watermark theoretically. Digital watermarking based on DCT transform will distributed in the whole image space during the inverse transformation, so unlike spatial domain based technology which is easy affected by the attacks such as cutting, low pass filter, etc. Because of its good robustness and concealing, the image digital watermarking algorithm based on DCT transform is the hot topic of research at home and abroad.

The flow chart of the robust watermark embedding based on DCT is shown in Fig. 9.4.

The original image is divided into 8×8 blocks. Firstly, calculate the variances of all the sub-blocks, and select the front n blocks with the maximum variance. Then, embed the random sequence pn_sequence_zero in the medium frequency of DCT domain according to the system key K. Finally, the result image is generated by the inverse DCT transform of the sub-blocks. K and pn_sequence_zero are used in combination to select the embedding position.

Specific steps are as follows:

(1) Perform DCT transform on blocks of the original image

To be compatible with the international compression standard, so that the algorithm can be implemented in the compressed domain, we divide the original image into non-overlapping 8×8 sub-blocks and then perform the DCT transformation on each sub-block.

(2) Block classification based on texture masking feature

According to the illumination masking characteristics and texture masking properties of human vision system (HVS), we know that the higher the brightness of background, the more complex the texture, the less sensitive human vision is to its slight transformation. Therefore, to achieve the perceived similarity between the original image and the processed image, the watermark signal should be embedded as much as possible to the more complex sub-blocks in the image. Here we take the variance σ^2 of the sub-block to measure the complexity of the texture. Calculate the mean gray value m and variance of sub-block. The equations are as follows:

$$m = \frac{1}{n^2} \sum_{i=0}^{n-1} \sum_{j=0}^{n-1} x(i, j) \tag{9.5}$$

$$\sigma^2 = \frac{1}{n^2} \sum_{i=0}^{n-1} \sum_{j=0}^{n-1} [x(i, j) - m]^2 \tag{9.6}$$

The variance σ^2 reflects the smoothness of blocks. When σ^2 is small, blocks are relatively uniform, on the contrary, blocks contain more complex textures or edges. When too much information is embedded into the smooth area, it will cause the phenomenon of block effect, which will result in a decline in image quality. According to the analysis of human visual model, embedding the watermark into the complex area of the texture conforms to the watermark algorithm. Specifically, the SORT function of MATLAB can be used to sort the variance values from small to large to embed watermarks into complex texture sub-block.

(3)   The generation and embeddedness of watermark

The binary watermark image (Fig. 9.1b) is connected to be a one-dimensional row vector as the watermark information. When using the digital watermarking algorithm based on DCT, because the human eye is relatively sensitive to low-frequency noise, we should embed the watermark in the higher frequency part for the watermark is not easy to detect. But it is easy to lose information because of quantization and low-pass filtering, which affects the robustness of watermarking. To solve the contradiction between low frequency and high frequency, the watermark information is embedded in the middle frequency part of the host image by using a compromise method. Figure 9.5 is the medium frequency position of the sub-block. The specific embedding location is determined by parameters K and sequence.

(4)   Block DCT inverse transform

According to above steps, the embeddedness programme of digital watermark is shown in PROGRAMME 9.3:

Fig. 9.5 The position 8 medium frequency block with DCT coefficient embedded

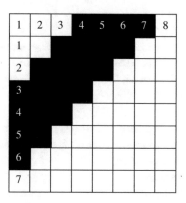

PROGRAMME 9.3: Digital Watermark Embedding Programme

```
k=20;                          % set the watermark strength
blocksize=8;                   % set the block size of the image to be 8
midband=[0,0,0,1,1,1,1,0;      % define the frequency coefficients of DCT
         0,0,1,1,1,1,0,0;
         0,1,1,1,1,0,0,0;
         1,1,1,1,0,0,0,0;
         1,1,1,0,0,0,0,0;
         1,1,0,0,0,0,0,0;
         1,0,0,0,0,0,0,0;
         0,0,0,0,0,0,0,0];
message=double(imread('copyright.bmp'));% read the watermark image and convert it to double-precision array
Mm=size(message,1);            % calculate the height of image
Nm=size(message,2);            % calculate the width of image
Qm=size(message,3);            % the number of image channels
n=Mm*Nm;
%transform the watermarked image into one-dimensional row vector
message=round(reshape(message,1,n*Qm)./256);
%read the original host image and convert it into a double-precision array
cover_object=double(imread('lena.bmp' ));
Mc=size(cover_object,1);Nc=size(cover_object,2);% calculate the height and width of original image
c=Mc/8;  d=Nc/8;  m=c*d;                % compute blocks for image segmentation
% calculate the variance of each piece of host image
xx=1;
```

```
for j=1:c
   for i=1:d
      pjhd(xx)=1/64*sum(sum(cover_object((1+(j-1)*8):j*8,(1+(i-1)*8):i*8)));
      fc(xx)=1/64*sum(sum((cover_object((1+(j-1)*8):j*8,(1+(i-1)*8):i*8)-pjhd(xx)).^2));
      xx=xx+1;
   end
end
A=sort(fc);B=A((c*d-n+1):c*d);        % selete the top n of the variance
% embed the watermark information into the former n block with the largest variance
fc_o=ones(1,c*d);
for g=1:n
   for h=1:c*d
      if B(g)==fc(h)
         fc_o(h)=message(g);
         h=c*d;
      end
   end
end
message_vector=fc_o;
watermarked_image=cover_object;
% set the MATlAB random number generator state J as the system key K
rand('state',7);
% based on current random number generator state J,a pseudo-random sequence of 0,1 is generated
pn_sequence_zero=round(rand(1,sum(sum(midband))));
% embed the watermark
x=1;y=1;
for (kk = 1:m)
% block DCT transform
dct_block=dct2(cover_object(y:y+blocksize-1,x:x+blocksize-1));
   ll=1;
   if (message_vector(kk)==0)
      for ii=1:blocksize
         for jj=1:blocksize
            if (midband(jj,ii)==1)
               dct_block(jj,ii)=dct_block(jj,ii)+k*pn_sequence_zero(ll);
               ll=ll+1;
            end
         end
      end
   end
```

```
watermarked_image(y:y+blocksize-1,x:x+blocksize-1)=idct2(dct_block);
if (x+blocksize) >= Nc
    x=1;   y=y+blocksize;
else
    x=x+blocksize;
end
end
watermarked_image_int=uint8(watermarked_image);
% generate and output the image embeded with watermark
imwrite(watermarked_image_int,'dct2_watermarked.bmp','bmp');
% show the PSNR
xsz=255*255*Mc*Nc/sum(sum((cover_object-watermarked_image).^2));
psnr=10*log10(xsz)
% show the image after embedding the watermark
figure(1)
imshow(watermarked_image_int,[])
title('Watermarked Image')
```

Several one-dimensional arrays are involved in the embedding process: message and B are one-dimensional array of 1 row and n columns; fc, fc_o are one-dimensional arrays of 1 row and m columns, while pn_sequence_zero is a one-dimensional array of 1 row 22 columns. The message is determined by the watermark image, and pn_sequence_zero is uniquely determined by the current pseudo-random number generator state J of the system, both message and pn_sequence_zero are composed of 0 and 1.

Specifically, we first set all the elements of the one-dimensional array fc_o to 1, the variance array fc is sorted in descending order to obtain the top n value with variance to form the array B; Then, modify the value of fc_o(i) which refers to the largest variance image block and make fc_o(i) = message(1); Modify the value of fc_o(i) which refers to image block with the second largest variance and makes fc_o(i) = message(2); And so on, modify the m values to get the one-dimensional value message vector; At last, the image block message_vector(i) whose value is 0 is selected as the actual image block that actually embedded in the watermark. When the 22 coefficients of the selected image block in the DCT medium frequency are embedded in the K times of the pseudo random sequence pn_sequence_zero, all image blocks are transformed by inverse DCT to generate a watermarked image.

Figure 9.6 shows a case of watermark embedding. Figure 9.6a is a 480 × 480 8-bit grayscale image 'Lena'. Figure 9.6b is a binary watermark image with a size of 50 × 20 (only 0, 1). Figure 9.6c is an image after embedding a watermark in Lena.

It can be seen from the results that the original host image has no visible distortion after embedding the watermark, and its PSNR is 45.6286 dB. The larger the PSNR value is, the better the invisibility is, so the method has better invisibility.

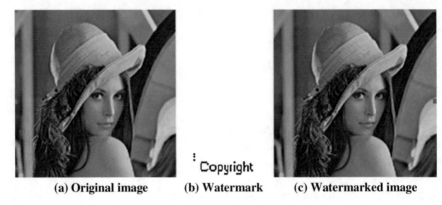

(a) Original image   **(b) Watermark**   **(c) Watermarked image**

Fig. 9.6 Embedding of digital watermarking

The extraction process of digital watermarking based on DCT is as follows:

(1) Original image and the image to be measured are evaluated in the DCT domain, then, compare the correlation and determine the sequence message_vector;

(2) The texture block is determined by the variance of the image block, then we can determine the embedding position of the watermark;

(3) Similar to the steps at the time of embedding, a one-dimensional watermarking sequence is formed according to the sequence message vector and the order of the texture block complexity;

(4) Reconstruct watermarking sequence into two-dimensional watermark recovery image, and the copyright authentication of the image is carried out accordingly.

According to the above steps, the digital watermark extraction programme is shown in PROGRAMME 9.4:

PROGRAMME 9.4: The Digital Watermark Extraction Programme with MATLAB

```
blocksize=8;
  midband=[  0,0,0,1,1,1,1,0;
             0,0,1,1,1,1,0,0;
             0,1,1,1,1,0,0,0;
             1,1,1,1,0,0,0,0;
             1,1,1,0,0,0,0,0;
             1,1,0,0,0,0,0,0;
             1,0,0,0,0,0,0,0;
             0,0,0,0,0,0,0,0   ];
```

```matlab
cover_object=double(imread(lena.bmp));                                   %read the original host image
watermarked_image=double(imread(dct2_watermarked.bmp));                 %read the image to be detected
Mw=size(watermarked_image,1); Nw=size(watermarked_image,2);
c=Mw/8;  d=Nw/8;  m=c*d;
orig_watermark=double(imread(copyright.bmp));                           %read the watermarking image
Mo=size(orig_watermark,1);          No=size(orig_watermark,2); n=Mo*No;
rand('state',7);       %set the same random number generator state J as the system key K when in detection
pn_sequence_zero=round(rand(1,sum(sum(midband))));        %generate    the    same    pseudo-random sequence
%extract the watermark
x=1;   y=1;
for (kk = 1:m)
   dct_block1=dct2(watermarked_image(y:y+blocksize-1,x:x+blocksize-1));
   dct_block2=dct2(cover_object(y:y+blocksize-1,x:x+blocksize-1));
   ll=1;
   for ii=1:blocksize
      for jj=1:blocksize
         if (midband(jj,ii)==1)
            sequence(ll)=dct_block1(jj,ii)-dct_block2(jj,ii);
            ll=ll+1;
         end
      end
   end
   %calculate the correlation of the two sequences
   if (sequence==0)
      correlation(kk)=0;
   else
   correlation(kk)=corr2(pn_sequence_zero,sequence);
   end
   %line feed
   if (x+blocksize) >= Nw
      x=1;   y=y+blocksize;
   else
      x=x+blocksize;
   end
end
for (kk=1:m)
    if (correlation(kk) >0.5)
       message_vector(kk)=0;
    else
       message_vector(kk)=1;
    end
   end
```

```
%calculate the variance of the original image
xx=1;
for j=1:c
   for i=1:d
      pjhd(xx)=1/64*sum(sum(cover_object((1+(j-1)*8):j*8,(1+(i-1)*8):i*8)));
      fc(xx)=1/64*sum(sum((cover_object((1+(j-1)*8):j*8,(1+(i-1)*8):i*8)-pjhd(xx)).^2));
      xx=xx+1;
   end
end
%take out the top n with the largest value of variance
A=sort(fc);   B=A((c*d-n+1):c*d);
%extract the watermark
fc_o=ones(1,n);
for g=1:n
   for h=1:c*d
      if B(g)==fc(h)
         fc_o(g)=message_vector(h);
         h=c*d;
      end
   end
end
message_vector=fc_o;
%reorganize the embedded image information
message=reshape(message_vector(1:Mo*No),Mo,No);
%the similarity between original watermark and extracted watermark image
sim=corr2(orig_watermark,message)
imwrite(message,'message.bmp','bmp');
```

The extracted watermark image is shown in Fig. 9.7. Figure 9.7a shows the watermark image to be extracted, and Fig. 9.7b is the watermark image extracted from the above process.

9.4   Semi-fragile Watermarking Based on DWT

In practice, there is no need for a fragile watermarking to be very sensitive to all modifications. While the semi-fragile watermarking requires the watermark to resist a certain degree of beneficial digital signal process, such as JPEG compression, etc. This type of watermarking is slightly more robust than the fully fragile watermark, which allows some changes in the image, and it is a certain degree of integrity test of the image.

Semi-fragile watermark combines the characteristics of robust watermarking and fragile watermarking, which is mainly used in the image content certification, and requires that it must have two basic characteristics:

(a) Watermark image to be extracted   **(b) Watermark**

Fig. 9.7 The result of the digital watermark extraction

(1) Transparency: the process of embedding is imperceptible, and the image quality after embedding cannot cause qualitative changes;
(2) Blind detection: the original image is not necessary at the time of authentication.

In recent years, many semi-fragile watermarking methods have been proposed. It can be divided into spatial domain algorithm and transform domain algorithm. The watermark is embedded in the spatial domain when refers to the spatial algorithm. The frequency domain algorithm is based on image transformation, namely local or all transformations, these transformations include discrete cosine transform (DCT), discrete wavelet transform (DWT) [7], Fourier transform (FT or FFT) [8–10], and Hadamard transform. Many researchers believe that the watermarking algorithm of transform domain has many advantages, including the ability to embedding more data without affecting the visual effects of the carrier, and it can be combined with some compression coding processes (such as DCT domain and JPEG, DWT domain and JPEG2000), and the embedded watermark has a stronger robustness (often for compression). But compared with the frequency domain algorithm, the spatial domain algorithm also has the advantages of small amount of calculation and convenient implementation. So, a method should be evaluated depend on the application and its performance, rather than the spatial domain or frequency domain algorithm, especially for semi-fragile watermarking.

DWT is the abbreviation of Discrete Wavelet Transform, its basic idea is to decompose the image into multi-resolution, which decompose the image into different spatial and frequency sub images, thus more conforms to the visual mechanism of the human eye. DWT not only has the good local spatial frequency analysis characteristics and multi-resolution analysis characteristics, but also has more outstanding ability of anti-filtering and anti-compression attack. In the static image compression standard JPEG2000, DWT replaced the DCT which used in JPEG. So, the

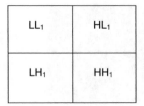

Fig. 9.8   The schematic diagram of each component of a primary decomposition

DWT-based digital watermarking technology is currently the hotspot in watermarking technology.

Generally speaking, DWT uses a multi-resolution decomposition method to decompose the image, and adds the watermark in the corresponding sub band coefficient image. The wavelet coefficients image consists of several sub bands coefficients images, and the wavelet coefficients of different sub bands reflect the characteristics of different spatial resolution of the image. Through the multi-level wavelet decomposition, the wavelet coefficients can not only represent the high frequency information of the local areas in the image, but also express the low frequency information of the image slices. Thus, by decoding the different series of coefficient images, images with different spatial resolutions can be obtained. The DWT transform can locate the local features of the image well, and the coefficients of the sub bands after wavelet decomposition can reflect this characteristic.

As a digital watermark embedding method, the DWT has getting more and more attention by researchers. The advantage of DWT method is that it can decompose the image into the frequency domain, and preserve the spatial distribution of image, which is very effective for strengthening the robustness of digital watermarking, lossy compression and local clipping. On the other hand, the multi-resolution analysis of the wavelet transform and the human visual characteristics can match well. Therefore, from the perspective of watermark visibility, DWT is also closer to the human visual perception system (HVS) requirements.

The watermark embedding process is as follows:

(1)   Perform wavelet transform on images. The basic idea of wavelet transform in image processing is to decompose the image into sub-images of different spatial and independent bands, and then process the coefficients of sub-images. The schematic diagram of the primary decomposition of the image is shown in Fig. 9.8.

It can be seen that an image is decomposed into 4 sub-images of 1/4 sizes after a primary wavelet decomposition; LL_1 in the upper left corner is a smooth approximation, that is, the low-frequency approximation sub-image; HL_1 in the upper right corner is a horizontal component, LH_1 in the lower left corner is a vertical component, while HH_1 in the lower right corner is diagonal components, which represents the medium and high frequency detail subgraph of horizontal, vertical and diagonal direction respectively. The low frequency part continues to be decomposed and get a n-level decomposition, resulting in LH, HH_j (j $= 1, 2, \ldots$, n) three high frequency band sub-images and a LL_n low frequency band sub-image. Where the low frequency band represents the best approximation to the original image. Its statistical characteristic is similar to the original image, and most of the energy is concentrated there. The high frequency band represents the edges and textures of the image. Through wavelet transform, it can effectively extract the high and low frequency components of the image. Because the sensitivity of the human eye to high frequency information is lower than that of the low frequency information, the watermark embedded in the higher frequency region has less influence on the original image, that is, the transparency of the watermark is better. The embedding process of the fragile watermark is shown in Fig. 9.9.

(2) Embed the watermark: Quantify the DWT coefficients $C(i)$ to embed the watermarks.

The wavelet coefficients are divided into two categories, $[C(i)/\nabla]$ is even for the first class, and $[C(i)/\nabla]$ is odd for another class, that is:

$$Q(C(i)) = [C(i)/\nabla] \bmod 2 \qquad (9.7)$$

And $Q(C(i)) \in \{0, 1\}$, where ∇ is a positive real number called quantization coefficient. The specific quantification process is:

a. if $Q(i) = W(i)$, then the coefficient will not change;
b. else, change $C(i)$, and make $Q(C(i)) = W(i)$, that is

$$C(i) = \begin{cases} C(i) - \nabla & C(i) \geq 0 \\ C(i) + \nabla & C(i) < 0 \end{cases} \qquad (9.8)$$

(3) Reconstruct the watermarked image by discrete wavelet transform.

For a given image, perform the discrete wavelet transform. According to the coefficient $C(i)$ of the wavelet transform domain, calculates $W(i)$ by the Eq. (9.1), that is, $W(i) = Q(C(i))$. The process of extraction is shown in Fig. 9.10.

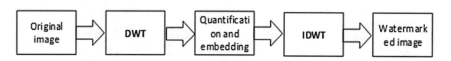

Fig. 9.9 Embedding process of the fragile watermark

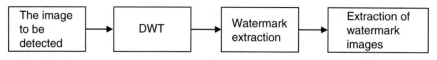

Fig. 9.10 The progress of watermark extraction

From the above algorithm, we can see that C records the high frequency coefficients after the wavelet transform, and Q is the high frequency coefficient after classification, while *step* is the quantization coefficient. The specific code is shown in PROGRAMME 9.5.

PROGRAMME 9.5: The Digital Watermarking Based on DWT

```
I= imread('lena.bmp');
M=imread('cameraman256.tif');
W=im2bw(M);              % Convert grayscale image to binary image
[CA,CH,CV,CD]=dwt2(I,'db1');
[length,width]=size(CA);
C=CD;
[M,N]=size(C);
Q=C;
step=5;
% the part of Watermark embedding
for i=1:M
   for j=1:N
      Q(i,j)=mod(round(C(i,j)/step),2);
      if Q(i,j)==W(i,j)
        C1(i,j)=C(i,j);
      else
        if C(i,j)>=0
          C1(i,j)=C(i,j)-step;
        else
          C1(i,j)=C(i,j)+step;
        end
      end
   end
end
WaterCD=C1(1:length,1:width);
IW=double(idwt2(CA,CH,CV,WaterCD,'db1'));
% the part of watermark extraction
[CA,CH,CV,CD]=dwt2(IW,'db1');
W1=zeros(64,64);
for i=1:M
   for j=1:N
      Q1(i,j)=mod(round(CD(i,j)/step),2);
      W1(i,j)=Q1(i,j);
   end
end
```

(a) Original image (b) Image after embedding (c) Watermark image (d) Watermark to extract

Fig. 9.11 DWT-based watermark results

subplot(3,4,1),imshow(I,[]),title(' The original image ');

subplot(3,4,2),imshow(IW,[]),title(' Image embeded in watermark ');

subplot(3,4,3),imshow(W,[]),title(' Embedded watermark ');

subplot(3,4,4),imshow(W1,[]),title(' The extracted watermark ');

[PSNR_OUT,Z] = psnr(I,IW);

str = sprintf('PSNR = %f',PSNR_OUT);

disp(str);

In the experiment, we select the 256×256 Lena grayscale image as original image, and takes the Cameraman 256 grayscale image whose size is 256×256 as watermark image. The results of the operation are shown in Fig. 9.11.

References

1. Rana R, Thangjam S, Singh S (2018) Performance analysis of video watermarking in transform domain using differential embedding. Inf Commun Technol Intell Syst (ICTIS 2017) 1
2. Joshi AM, Gupta S, Girdhar M et al (2017) Combined DWT–DCT-based video watermarking algorithm using Arnold transform technique
3. Alattar AM (2004) Reversible watermark using the difference expansion of a generalized integer transform. IEEE Trans Image Process 13(8):1147–1156
4. Podilchuk CI, Delp EJ (2001) Digital watermarking: algorithm and application 18(4):33–46
5. Li C, Ye B, Lai J et al (2015) A digital watermarking algorithm for trademarks based on U system. In: Image and graphics. Springer International Publishing, pp 43–52
6. Meenakshi K, Rao CS, Prasad KS (2014) A robust watermarking scheme based Walsh-Hadamard transform and SVD using ZIG ZAG scanning. In: International conference on information technology. IEEE Computer Society, pp 167–172
7. Wang J (2014) DWT-DFRFT combining image watermarking algorithm. In: International conference on information science, electronics and electrical engineering. IEEE, pp 750–753
8. Chen Z, Chen Y, Hu W et al (2015) Wavelet domain digital watermarking algorithm based on threshold classification. In: Advances in swarm and computational intelligence. Springer International Publishing, pp 129–136
9. Tsai FM, Hsue WL (2014) Image watermarking based on various discrete fractional fourier transforms. In: International Workshop on Digital Watermarking. Springer, Cham, pp 135–144
10. Othman MTB (2014) Digital image watermarking based on clustering. In: International conference on circuits, systems, communications, computers and applications

Chapter 10
Visual Object Recognition

Object recognition, one of the important tasks of image recognition, mainly aimed at the recognition of visible images. It can accurately define and describe objects by attributes and features of geometric appearance, texture and material of images. In a broad sense, the recognition process can distinguish objects from backgrounds and other suspicious objects, such as cars and roads, that is the object detection. In a narrow sense, the recognition process is to classify similar objects more specifically, such as different types of cars. This chapter employs three cases in face recognition, expression recognition and document image analysis, to briefly introduce the basic implementation steps of image recognition.

10.1 Face Recognition Based on Locality Preserving Projections

Face recognition [1–3] is one of the key technologies of biometric identification. Because of its natural, intuitive, non-contact, safe, fast and other characteristics, it has attracted much attention and has become the most promising technology. It has been used widely in such fields as electronic passports and identity cards, security, judicial and criminal investigation, self-help services, information security, and so on.

According to different classification criterias, face recognition can be divided into different methods. For example, on the basis of the linear nature of the algorithm, it can be divided into linear and nonlinear algorithms; According to whether or not the category information of the face is considered, it can be divided into supervised, semi-supervised and unsupervised algorithms; According to the characteristics of the architecture of the original data which is retained by the algorithm, it can be divided into local algorithm and global algorithm.

© Springer International Publishing AG, part of Springer Nature 2019   351
S. Gong et al., *Advanced Image and Video Processing Using MATLAB*,
Modeling and Optimization in Science and Technologies 12,
https://doi.org/10.1007/978-3-319-77223-3_10

The linear algorithm is to compute an explicit linear projection function, and project the original data from the high-dimensional space to the low-dimensional space. The nonlinear algorithm does not make assumptions about the projection, and the projection of the original data is implicit. It can only calculate the projection of the training data in the low-dimensional space, but cannot do anything about the new data. Linear algorithms include Principal Component Analysis (PCA), Linear Discriminant Analysis (LDA), Multidimensional Scaling Scaling (MDS), Neighborhood Preserving Embedding (NPE), etc. Among them, PCA and LDA are the global algorithms, while MDS and NPE retain the local architecture as the local algorithm. PCA, NPE and MDS are unsupervised algorithms, and LDA is a supervised algorithm. Nonlinear algorithms mainly include Locally Linear Embedding (LLE), Laplacian Eigenmaps (LE) and so on. LLE is an unsupervised global algorithm.

The following focuses on face recognition algorithms based on locality preserving projections and its various variants.

Let us consider a dataset $X = [x_1, x_2, \ldots, x_n]$, which is divided into c classes. Each class contains vectors n_i, $i = 1, 2, \ldots, c$, where $x_i \in R^m$ is a column vector of m dimension. The data projected into the low-dimensional space can be defined as $Y = [y_1, y_2, \ldots, y_n]$, in which $y_i \in R^d$, where d is the dimension of the low-dimensional space. The LPP [4–8] algorithm is based on the assumption that if the two vectors are very close in the high-dimensional space, and it is reasonable to believe that the projections of the two vectors are very close in the low-dimensional space. To ensure this hypothesis, the following objective function is defined:

$$\min_{W} \sum \|y_i - y_j\|^2 a_{ij} \tag{10.1}$$

$$a_{ij} = \begin{cases} \exp\left(-\frac{\|x_i - x_j\|^2}{t}\right) & x_i \ is \ a \ neighbor \ of \ x_j \ or \ x_j \ is \ a \ neighbor \ of \ x \\ 0 & other \end{cases} \tag{10.2}$$

$A = (a_{ij})_{n \times n}$ is a symmetric affinity matrix while t is a threshold.

Supposing that w is a projection column vector, and $y_i = w^T x_i$. Through simple derivation, we can convert the objective function Formula (10.1) into:

$$\frac{1}{2} \sum_{i,j} (y_i - y_j)^2 a_{ij} = \frac{1}{2} \sum_{i,j} (w^T x_i - w^T x_j)^2 a_{ij}$$

$$= \sum_i w^T x_i (\sum_j a_{ij}) x_i^T w - \sum_{i,j} w^T x_i a_{ij} x_j^T w$$

$$= w^T X(D - A)X^T w = w^T X L X^T w \tag{10.3}$$

where D is a diagonal matrix and $D_{ii} = \sum_j a_{ij}$. $L = D - A$ is a Laplace matrix. As well, the restriction conditions are added:

$$y^T D y = 1 \Rightarrow w^T X D X^T w = 1 \tag{10.4}$$

Thus, the LPP algorithm is converted to solve the problem of the optimal solution:

$$
\begin{cases}
\min\limits_{W \in R^{n \times d}} & Tr(W^T X L X^T W) \\
W^T(X D X^T)W = 1
\end{cases}
\tag{10.5}
$$

where W is a projection matrix:

The Lagrange multiplier algorithm is used to solve the upper formula. By simple calculation, the problem of Formula (10.4) can be converted into the problem to solve eigenvalues:

$$
X L X^T w = \lambda (X D X^T)w
\tag{10.6}
$$

Thus, the problem to solve the minimum value of the criterion function is converted into the process of obtaining eigenvalues and eigenvectors of the generalized characteristic equation in Formula (10.5). It can be proved that the eigenvectors corresponding to the minimum l nonzero eigenvalues constitute w.

LPP algorithm has many variations, such as Orthogonal Discriminant Locality Preserving Projection (ODLPP) [5, 9]. Considering that there are $C = \{n_1, n_2, \ldots, n_c\}$ classes in a high-dimensional Euclidean space R^n, and each class has N_i samples. $X = \{x_1, x_2, \ldots, x_N\}$ is a sample set with N samples, each sample belongs to one of the classes in C.

The algorithm seeks a projection matrix which makes the projected face image be in a lower dimension and has better separability. The LDA algorithm uses the inter-class scatter matrixes of the samples to represent the discreteness of samples which belong to different classes. The ODLPP algorithm draws on the ideas of LDA algorithm and introduces the inter-class scatter matrix into the criterion function of ODLPP. The criterion function defined by ODLPP is:

$$
\begin{cases}
\min\limits_{W \in R^{n \times d}} & \dfrac{Tr(W^T X L X^T W)}{Tr(W^T S_b W)} \\
W^T W = I
\end{cases}
\tag{10.7}
$$

where S_b is the inter-class discreteness matrix. Class information is added to the target function to ensure that the projection vector is orthogonal.

We can solve this problem by solving the eigenvalue problem:

$$
X L X^T w = \lambda S_b w
\tag{10.8}
$$

For the modification of the restrictive conditions, the overall dispersion matrix S_T is used to replace $X D X^T$, so that the projected data is uncorrelated. Thus, the Enhance Locality Preserving Projection (ELPP) algorithm is obtained:

$$\begin{cases} \min_{W \in R^{n \times d}} \quad Tr(W^T X L X^T W) \\ W^T S_T W = I \end{cases} \tag{10.9}$$

where S_T is the overall dispersion matrix:

$$S_T = \frac{1}{n} \sum_i (x_i - \overline{x})(x_i - \overline{x})^T \tag{10.10}$$

where $\overline{x}$ is the mean value of X which is represented as $\overline{x} = \frac{1}{n} \sum_i x_i$.

In order to solve this problem, the solver can be obtained by solving the eigenvalue problem:

$$X L X^T w = \lambda S_T w \tag{10.11}$$

Modifying the symmetric association matrix A by the following steps. Replacing the original matrix by the Pearson correlation coefficient matrix and use the adaptive method to select the nearest neighbor so that the LPP algorithm no longer contains parameters. Thus, the Parameterless Locality Preserving Projection (PLPP) [7] algorithm is implemented.

First, the Pearson correlation coefficient matrix p is defined. p_{ij} is the Pearson correlation coefficient of the vectors x_i, y_i

$$p_{ij} = \frac{m \sum x_i y_i - \sum x_i \sum y_i}{\sqrt{m \sum x_i^2 - (\sum x_i)^2} \sqrt{m \sum y_i^2 - (\sum y_i)^2}} \tag{10.12}$$

Then, since p_{ij} is between $[-1, 1]$, so the p_{ij} is normalized to $[0, 1]$.

$$\overline{p_{ij}} = \frac{p_{ij} - \min(P)}{1 - \min(P)} \tag{10.13}$$

Finally, the matrix A is defined as

$$a_{ij} = \begin{cases} \overline{p_{ij}} & \overline{p_{ij}} > m_i \text{ or } \overline{p_{ij}} > m_j \\ 0 \end{cases} \tag{10.14}$$

where $m_i = \frac{1}{n} \sum_i \overline{p_{ij}}$ is the average correlation coefficient.

Since the Parameterless Locality Preserving Projection (PLPP) algorithm does not consider class information, nor does it guarantee the orthogonality of projection vectors, we can add the class information on the basis of the original algorithm to obtain the Orthogonal Disciminant Parameterless Locality Preserving Projection (ODPLPP).

Define the matrix A as

$$a_{ij} = \begin{cases} \overline{P_{ij}}(\overline{P_{ij}} > m_i \text{ or } \overline{P_{ij}} > m_j) \text{ and } (i, j \text{ same type}) \\ 0 \end{cases} \quad (10.15)$$

Modify the object function:

$$\begin{cases} \min_{W \in R^{n \times d}} \dfrac{Tr(W^T X L X^T W)}{Tr(W^T S_b W)} \\ W^T W = I \end{cases} \quad (10.16)$$

In the above formula,

$$S_b = \frac{1}{n} \sum_{i=1}^{c} n_i (u_i - u)(u_i - u)^T$$

$$u_i = \frac{1}{n_i} \sum_{x_i} x_i \quad u = \frac{1}{n} \sum_{i=1}^{n} x_i \quad (10.17)$$

The solution of LPP algorithm is to convert the optimization problem into the eigenvalue problem, which selects the eigenvectors corresponding to the smallest d eigenvalues as the projection matrix, that is, a base of the low-dimensional linear space. LPP is a problem of small samples, so the dimension of data is first reduced through PCA in order to avoid singularity effectively. The steps are as follows:

(1) The original data is divided into the training set X and the test set Z. Assume that the training set has N samples, and each sample is a $m \times n$ matrix composed of a grayscale image. Connect each column of each grayscale image from left to right to form a column vector of $M = m \times n$ dimensional. Then the training data set is a $M \times N$ matrix which represents as $X = [x_1, x_2, \ldots, x_N]$, where $x_i \in R^M$ is a face image. The test data set is a $M \times N$ matrix, and it is represented as $Z = [z_1, z_2, \ldots, z_N]$, where $z_i \in R^M$ is a face image.
(2) Reduce the dimension of the original data X by PCA.
(3) Compute the symmetric affinity matrix A by k-nearest neighbor, where $k = 5$, and t takes the mean of the square of the distance between data points of the training set.
(4) Find the eigenvalues of $XLX^T w = \lambda(XDX^T)w$, select the smallest k eigenvalues, a projection matrix is made up of the eigenvectors corresponding to these eigenvalues. Among them, $W = [w_1, w_2, \ldots, w_N]$, where $w_i \in R^d$ is the eigenvector.
(5) Conduct the projection $Y = W^T X$, $Y = [y_1, y_2, \ldots, y_N]$, where $y_i \in R^d$ is the projection of the original data in low-dimensional space.
(6) The recognition phases. Project the test set into the low-dimensional space by the obtained projection matrix, $Z' = W^T Z$, and then NN (Nearest Neighbor) is used to classify the test set.

PROGRAMME 10.1 shows the MATLAB implementation of face recognition based on locality preserving projection. It includes the main function, the function D = Distance(fea_a,fea_b) which calculates the facial distance, and the LPP (Locality Preserving Projection) function which calculates the locality preserving projection.

PROGRAMME 10.1: Face recognition based on locality preserving projections

```
%Main function main()
load('ORL_32x32.mat') % Importing face data
%% split data
data=alls;
Label=unique(gnd);
TrainNum=5;
Tr_ind=[];
Te_ind=[];
for i=1:length(Label)% Take training set and test set for each type of face
     tempind=find(gnd==Label(i));
     Tr_ind=[Tr_ind, tempind(1:TrainNum) ];
     Te_ind=[Te_ind, tempind(TrainNum+1:end)];
end
%% label and data
Train_label=gnd(Tr_ind);
Test_label=gnd(Te_ind);
Trains=data(:,Tr_ind);
Tests=data(:,Te_ind);
%% performing PCA for speeding up
PCAoptions = [];
PCAoptions.PCARatio = 1;
[eigvector_PCA, eigvalue_PCA, meanData, new_X] = PCA(Trains',PCAoptions);
Trains=eigvector_PCA'*Trains;
Tests=eigvector_PCA'*Tests;
%% LPP, projection recognition
options = [];
options.Metric = 'Cosine';
options.NeighborMode = 'Supervised';
options.WeightMode = 'Cosine';
options.gnd = Train_label;
fea_Train=double(Trains');
W = constructW(fea_Train,options); % Weight matrix
options.bLDA = 1;
options.PCARatio = 1;
[eigvector, eigvalue] = LPP(fea_Train,W, options);
%%   recognition procedure, outputs the correct rate for a given data set
```

```
rates=[];
step=3;
count=0;
for i=1:step:length(eigvalue)
     train_proj=eigvector(:,1:i)'*Trains;
     test_proj=eigvector(:,1:i)'*Tests;
     count=count+1;
     rates(count) = NNC_Speed(train_proj,test_proj,Train_label,Test_label);
end
LPP_Accuracy=max(rates)
%    PCA:Principal Component Analysis. Solve eigenvalues and eigenvectors of covariance matrices
function [eigvector, eigvalue, meanData, new_data] = PCA(data, options)
if (~exist('options','var'))
    options = [];
else
    if ~strcmpi(class(options),'struct')
        error('parameter error!');
    end
    end
    %%%%% Go ahead and take the PCA dimension reduction
    bRatio = 0;
    if isfield(options,'PCARatio')
        bRatio = 1;
        eigvector_n = min(size(data));
    elseif isfield(options,'ReducedDim')
        eigvector_n = options.ReducedDim;
    else
        eigvector_n = min(size(data));
    end
      [nSmp, nFea] = size(data);
    eanData = mean(data);%%%mean: Row average
    data = data - repmat(meanData,nSmp,1);
    if nSmp >= nFea
        ddata = data'*data;
        ddata = max(ddata, ddata');
        if issparse(ddata)
            ddata = full(ddata);
        end
        %%%%%%%%%%Take The corresponding feature vectors according to the requirements
        if size(ddata, 1) > 1000 & eigvector_n < size(ddata, 1)/10    % using eigs to speed up!
```

```
        option = struct('disp',0);
        [eigvector, d] = eigs(ddata,eigvector_n,'la',option);
        eigvalue = diag(d);
    else
        [eigvector, d] = eig(ddata);
        eigvalue = diag(d);
        % ====== Sort based on descending order
        [junk, index] = sort(-eigvalue);% Take the minimum of eigenvalues which is in ascending order.
        eigvalue = eigvalue(index);
        eigvector = eigvector(:, index);
    end
    clear ddata;
    maxEigValue = max(abs(eigvalue));
    eigIdx = find(abs(eigvalue)/maxEigValue < 1e-12);%%% Find a subscript whose eigenvalue is zero
    eigvalue (eigIdx) = [];
    eigvector (:,eigIdx) = [];%%Empty the part that will be zero
else  % This is an efficient method which computes the eigvectors of
    % of A*A^T (instead of A^T*A) first, and then convert them back to
    % the eigenvectors of A^T*A.
    if nSmp > 700
        ddata = zeros(nSmp,nSmp);%%% Distance matrix, （exp type）
        for i = 1:ceil(nSmp/100)
            if i == ceil(nSmp/100)
                ddata((i-1)*100+1:end,:) = data((i-1)*100+1:end,:)*data';
            else
                ddata((i-1)*100+1:i*100,:) = data((i-1)*100+1:i*100,:)*data';
            end
        end
    elseif nSmp > 400
        ddata = zeros(nSmp,nSmp);
        for i = 1:ceil(nSmp/200)
            if i == ceil(nSmp/200)
                ddata((i-1)*200+1:end,:) = data((i-1)*200+1:end,:)*data';
            else
                ddata((i-1)*200+1:i*200,:) = data((i-1)*200+1:i*200,:)*data';
            end
        end
    else
        ddata = data*data';
    end
```

```
    ddata = max(ddata, ddata');%%% The symmetric affinity matrix is guaranteed to be symmetric,
    which
%should be paid more attention to!
    if issparse(ddata)
        ddata = full(ddata);
    end
    if size(ddata, 1) > 1000 & eigvector_n < size(ddata, 1)/10    % using eigs to speed up!
        option = struct('disp',0);
        [eigvector1, d] = eigs(ddata,eigvector_n,'la',option);
        eigvalue = diag(d);
    else
        [eigvector1, d] = eig(ddata);
        eigvalue = diag(d);
        % ====== Sort based on descending order
        [junk, index] = sort(-eigvalue);
        eigvalue = eigvalue(index);
        eigvector1 = eigvector1(:, index);
    end
    clear ddata;
    maxEigValue = max(abs(eigvalue));
    eigIdx = find(abs(eigvalue)/maxEigValue < 1e-12);
    eigvalue (eigIdx) = [];
    eigvector1 (:,eigIdx) = [];
    eigvector = data'*eigvector1;          % Eigenvectors of A^T*A
    clear eigvector1;
    eigvector = eigvector*diag(1./(sum(eigvector.^2).^0.5)); % Normalized eigenvector
end
if bRatio
    if options.PCARatio >= 1 | options.PCARatio <= 0%%%remain>1: represents the dimension to be
    retained; 0<remain<=1: represents the required contribution rate
        idx = length(eigvalue);
    else
        sumEig = sum(eigvalue);
        sumEig = sumEig*options.PCARatio;
        sumNow = 0;
        for idx = 1:length(eigvalue)
            sumNow = sumNow + eigvalue(idx);
            if sumNow >= sumEig
                break;
            end
```

```
                    end
                end
                eigvalue = eigvalue(1:idx);
                eigvector = eigvector(:,1:idx);
            else
                if eigvector_n < length(eigvalue)
                    eigvalue = eigvalue(1:eigvector_n);
                    eigvector = eigvector(:, 1:eigvector_n);
                end
            end
            if nargout == 4
                new_data = data*eigvector;
            end
            %% constructW
            function W = constructW(fea,options)
            if (~exist('options','var'))
                options = [];
            else
            if ~strcmpi(class(options),'struct')
                error('parameter error!');
            end
        end
        if ~isfield(options,'Metric')
            options.Metric = 'Euclidean';
        end
        switch lower(options.Metric)
            case {lower('Euclidean')}
            case {lower('Cosine')}
                if ~isfield(options,'bNormalized')
                    options.bNormalized = 0;
                end
            otherwise
                error('Metric does not exist!');
        end
        if ~isfield(options,'NeighborMode')
            options.NeighborMode = 'KNN';%%Nearest neighbor computing distance
        end
        switch lower(options.NeighborMode)
            case {lower('KNN')}    %For simplicity, we include the data point itself in the kNN
                if ~isfield(options,'k')
```

```matlab
            options.k = 5;
        end
        if options.k < 1
            options.k = 1;
        end
case {lower('epsilonNeighbor')}
    if ~isfield(options,'epsilon')
        options.epsilon = 0.5;
    end
case {lower('Supervised')}
    if ~isfield(options,'bLDA')
        options.bLDA = 0;
    end
    if options.bLDA
        options.bSelfConnected = 1;
    end
    if ~isfield(options,'gnd')
        error('Label(gnd) should be provided under "Supervised" NeighborMode!');
        end
    if length(options.gnd) ~= size(fea,1)
            error('gnd doesn"t match with fea!');
        end
    otherwise
        error('NeighborMode does not exist!');
end
if ~isfield(options,'WeightMode')
    options.WeightMode = 'Binary';
end
switch lower(options.WeightMode)
    case {lower('Binary')}
    case {lower('HeatKernel')}
        if ~strcmpi(options.Metric,'Euclidean')
            warning('"HeatKernel" WeightMode should be used under "Euclidean" Metric!');
            options.Metric = 'Euclidean';
        end
        if ~isfield(options,'t')
            options.t = 1;
        end
    case {lower('Cosine')}
        if ~strcmpi(options.Metric,'Cosine')
```

```
                          warning('"Cosine" WeightMode should be used under "Cosine" Metric!');
                          options.Metric = 'Cosine';
                    end
              if ~isfield(options,'bNormalized')
                          options.bNormalized = 0;
                    end
          otherwise
                    error('WeightMode does not exist!');
    end
    if ~isfield(options,'bSelfConnected')
          options.bSelfConnected = 1;
    end
    [nSmp, nFea] = size(fea);
    if          strcmpi(options.NeighborMode,'Supervised')           &           (options.bLDA        |
        strcmpi(options.WeightMode,'Binary'));
    else
          bDistance = 0;
          if strcmpi(options.Metric,'Euclidean')
          D = zeros(nSmp);
          for i=1:nSmp-1
                for j=i+1:nSmp
                          D(i,j) = norm(fea(i,:) - fea(j,:));
                    end
          end
          D = D+D';
          bDistance = 1;
    else
          if options.bNormalized
                D = fea * fea';
          else
                feaNorm = sum(fea.^2,2).^.5;
                fea = fea ./ repmat(max(1e-10,feaNorm),1,size(fea,2));
                D = fea * fea';
          end
    end
    end
    end
    switch lower(options.NeighborMode)
        case {lower('KNN')}
              if options.k >= nSmp
                    G = ones(nSmp,nSmp);
```

```
    else
        G = zeros(nSmp,nSmp);
        if bDistance
            [dump idx] = sort(D, 2); % sort each row
        else
            [dump idx] = sort(-D, 2); % sort each row
        end
        for i=1:nSmp
            G(i,idx(i,1:options.k+1)) = 1;
        end
    end
case {lower('epsilonNeighbor')}
    if bDistance
        [i,j] = find(D < options.epsilon);
    else
        [i,j] = find(D > options.epsilon);
    end
    G = sparse(i,j,1);
case {lower('Supervised')}
    G = zeros(nSmp,nSmp);
    Label = unique(options.gnd);
    nLabel = length(Label);
    if options.bLDA
        for idx=1:nLabel
            classIdx = find(options.gnd==Label(idx));
            G(classIdx,classIdx) = 1/length(classIdx);
        end
        W = sparse(G);
        return;
    else
        for idx=1:nLabel
            classIdx = find(options.gnd==Label(idx));
            G(classIdx,classIdx) = 1;
        end
    end
    if strcmpi(options.WeightMode,'Binary')
        if ~options.bSelfConnected
            G   = G - diag(diag(G));
        end
        W = sparse(G);
```

```
                    return;
            end
        otherwise
            error('NeighborMode does not exist!');
end
if ~options.bSelfConnected
    G   = G - diag(diag(G));
end
switch lower(options.WeightMode)
    case {lower('Binary')}
        W = max(G,G');
        W = sparse(W);
    case {lower('HeatKernel')}
        D = exp(-D.^2/options.t);
        W = D.*G;
        W = max(W,W');
        W = sparse(W);
    case {lower('Cosine')}
        W = D.*G;
        W = max(W,W');
        W = sparse(W);
    otherwise
        error('WeightMode does not exist!');
end
%%   LPP(Locality Preserving Projection) algorithm
function [eigvector, eigvalue, Y] = LPP(X, W, options)
if (~exist('options','var'))
    options = [];
else
    if ~strcmpi(class(options),'struct')
        error('parameter error!');
    end
end
%%%%% Go ahead and take the PCA dimension reduction
if ~isfield(options,'PCARatio')
    [eigvector_PCA, eigvalue_PCA, meanData, new_X] = PCA(X);
else
    PCAoptions = [];
    PCAoptions.PCARatio = options.PCARatio;
    [eigvector_PCA, eigvalue_PCA, meanData, new_X] = PCA(X,PCAoptions);
```

```
end
old_X = X;
X = new_X;
[nSmp,nFea] = size(X);%nSmp--row-- the number of faces
if nFea > nSmp
    error('X is not of full rank in column!!');
end
if ~isfield(options,'ReducedDim')
    ReducedDim = nFea;
else
    ReducedDim = options.ReducedDim;
end
if ReducedDim > nFea
    ReducedDim = nFea;
end
D = diag(sum(W));
%laplacian matrix L=D-W
L = W;
%% Take the corresponding feature vectors according to the requirements
DPrime = X'*D*X;
DPrime = max(DPrime,DPrime');
LPrime = X'*L*X;
LPrime = max(LPrime,LPrime');
dimMatrix = size(DPrime,2);
if dimMatrix > 1000 & ReducedDim < dimMatrix/10    % using eigs to speed up!
    option = struct('disp',0);
    [eigvector, eigvalue] = eigs(LPrime,DPrime,ReducedDim,'la',option);
    eigvalue = diag(eigvalue);
else
    [eigvector, eigvalue] = eig(LPrime,DPrime);%
    eigvalue = diag(eigvalue);
    [junk, index] = sort(-eigvalue);
    eigvalue = eigvalue(index);
    eigvector = eigvector(:,index); %%%% Reserve according to contribution rate
end
eigvalue = ones(length(eigvalue),1) - eigvalue;
if ReducedDim < size(eigvector,2)
    eigvector = eigvector(:, 1:ReducedDim);
    eigvalue = eigvalue(1:ReducedDim);
end
```

```
for i = 1:size(eigvector,2)
    eigvector(:,i) = eigvector(:,i)./norm(eigvector(:,i));
end
igvector = eigvector_PCA*eigvector;
if nargout == 3
    Y = old_X * eigvector;
end
%% NCC_Speed gives the accuracy
function [ all_rate ] = NNC_Speed(train_proj,test_proj,label_Train,label_Test)
%Nearest neighbor classifier gives the accuracy
train_num=length(label_Train);
test_num=length(label_Test);
Distance=zeros(test_num,train_num);
accu=0;
for i=1:test_num
    tempDist=sqrt(sum((train_proj-repmat(test_proj(:,i),1,train_num)).^2));% Seek European distance
    Distance(i,:)=tempDist;
    [dist,index]=sort(tempDist);
    if label_Train(index(1))==label_Test(i)
        % If the label of the minimum distance is the same as the original label, the recognition is
%considered correct
        accu=accu+1;
    end
end
all_rate=accu/(test_num)*100;
```

The following is a description of ODLPP (Oth_discriminant Locality Preserving Projection), PLPP (parameterless Locality Preserving Projection), ODPLPP (Othliscriminant parameterless Locality Preserving Projection), ELPP (Enhance Locality Preserving Projection), for reference.

```
%ODLPP : Oth_discriminant Locality Preserving Projection
function [eigvector, eigvalue, costTime] = ODLPP(data, options)
% orthogonal discriminant locality preserving projections—Neurocomputing 70 (2007) 1543 – 1546
% On the basis of preserving the original locality, the object function is modified, and the class information is
%added when the adjacency matrix is constructed.
% The between-class scatter matrix is added, and the feature vectors are orthogonal.
% Instead of using iterative methods in OLPP, the Schmidt orthogonalization is used
time_temp=cputime;
%%%% Go ahead and take the PCA dimension reduction
options_PCA.PCARatio = 0.98;
[eigvector_PCA, ~] = PCA(data,options_PCA);
```

```
data=data*eigvector_PCA;%%% Projecting data
%%%%Construct symmetric affinity matrix S%%%%
k=5;
[nSmp ,nFea]=size(data);% nSmp--row--the numbers of faces
temp_D = zeros(nSmp);% Temporary distance matrix
for i = 1:nSmp
    for j =1:nSmp
        temp_D(i,j) = norm(data(i,:) - data(j,:));% Distance between faces
    end
end
% Between-class scatter matrix Sb
classLable=unique(options.gnd);
classNo = length(classLable);        % Face classes
Sb=zeros(nFea,nFea);        % Between-class scatter matrix
classmean=zeros(classNo,nFea);    % Sample mean of each class
overallmean=mean(data); % Total sample mean
%%%%%%%%%%%%%%%%%%%%%%%%%%%%%%%%%%%%
for i=1:classNo
index=find(options.gnd==classLable(i));%% Find the same class in option.gnd as the class i
classmean(i,:)=mean(data(index,:));
Sb=Sb+length(index)*(classmean(i,:)-overallmean)'*(classmean(i,:)-overallmean);
end
%Construct adjacency matrixadj
adj=zeros(nSmp);% Adjacent to 1, not adjacent to 0
for i=1:classNo
    index = find(options.gnd == classLable(i));% Parts of the same class are adjacent
    adj(index,index)=1;
end
S = zeros(nSmp);% Distance matrix, （exp type）
t = (mean(mean(temp_D))).^2;
for i=1:nSmp
    for j=1:nSmp
        if adj(i,j)==1
            S(i,j) = exp(-temp_D(i,j).^2/t);
        end
    end
end
S=max(S,S');% symmetric affinity matrix
D = zeros(nSmp);        % Diagonal matrix
D = diag(sum(S));
```

```
L = D - S;%Laplacian matrix L=D-S
P = data' * L * data;
P = max(P , P');
Q = Sb;% Utilize between-class scatter matrix Sb
Q = max(Q , Q');
%%%%%%%%%%%%%%%%
[eigvector, eigvalue] = eig(P,Q);
eigvalue=diag(eigvalue);
[~ ,eindex]=sort(-eigvalue,'descend');% Take the minimum of eigenvalues which is in ascending order.
eigvector=eigvector(:,eindex);
eigvalue=eigvalue(eindex);
%%%%%%%%%%%%%%
eigvector = Schmidt(eigvector);% Eigenvectors are subjected to Schmidt orthogonalization
eigvector=eigvector*diag(1./(sum(eigvector.^2).^0.5));% Normalize eigenvectors
eigvector = eigvector_PCA*eigvector;
costTime=cputime-time_temp;
end
% PLPP (parameterless Locality Preserving Projection)
function [eigvector,eigvalue,costTime]=PLPP(data,options)
%PLPP : parameterless Locality Preserving Projection
time_temp=cputime;
%=======#======% Go ahead and take the PCA dimension reduction
options_PCA.PCARatio = 0.98;
[eigvector_PCA, ~] = PCA(data,options_PCA);
data=data*eigvector_PCA;%%% Projecting data,
%====#==========%Construct adjacency matrix W
%===#===%Construct correlation coefficient matrix P
[nSmp, nFea] = size(data);
% Correlation coefficient matrix
P = corrcoef(data');
% Normalize P to [0,1]
t = min(min(P));
temp = t*ones(nSmp);
P = (P-temp)/(1-t);
%===Construct W
W = zeros(nSmp);
for i=1:nSmp
    for j=1:nSmp
        r1 = mean(P(i,:));
        r2 = mean(P(j,:));
```

```
            if P(i,j)>r1 || P(i,j)>r2
                W(i,j)=P(i,j);
            end
        end
end
D = zeros(nSmp);
D = diag(sum(W));
L = D - W;% Laplacian matrixL=D-S
P = data' * L * data;
P = max(P , P');
Q = data' * D * data;
Q = max(Q , Q');
%===&
[eigvector, eigvalue] = eig(P,Q);
eigvalue=diag(eigvalue);
[~ ,eindex]=sort(-eigvalue,'descend');% Take the minimum of eigenvalues which is in ascending order.
eigvector=eigvector(:,eindex);
eigvalue=eigvalue(eindex);
%%%%%%%%%%%%%%
eigvector=eigvector*diag(1./(sum(eigvector.^2).^0.5));% Normalize eigenvectors
eigvector = eigvector_PCA*eigvector;
costTime=cputime-time_temp;
end
%ODPLPP : Oth_discriminant parameterless Locality Preserving Projection
function [eigvector,eigvalue,costTime]=ODPLPP(data,options)
time_temp=cputime;
%=======#======% Go ahead and take the PCA dimension reduction
options_PCA.PCARatio = 0.98;
[eigvector_PCA, ~] = PCA(data,options_PCA);
data=data*eigvector_PCA;%%% Projecting data,
%====#==========%Construct adjacency matrixW
%===#===%Construct correlation coefficient matrix P
[nSmp, nFea] = size(data);
% Between-class scatter matrix Sb
classLable=unique(options.gnd);
classNo = length(classLable);        % Face classes
Sb=zeros(nFea,nFea);        % Between-class scatter matrix
classmean=zeros(classNo,nFea);    % Sample mean of each class
overallmean=mean(data); % Total sample mean
%%%%%%%%%%%%%%%%%%%%%%%%%%%%%%%%%%
```

```
for i=1:classNo
    index=find(options.gnd==classLable(i));%% Find the same class in option.gnd as the class i.
        classmean(i,:)=mean(data(index,:));
        Sb=Sb+length(index)*(classmean(i,:)-overallmean)'*(classmean(i,:)-overallmean);
end
% Person correlation coefficient
P = corrcoef(data');
% Normalize P to [0,1]
t = min(min(P));
temp = t*ones(nSmp);
P = (P-temp)/(1-t);
%===Construct W, Add classLable at the same time.
% In addition to maintaining the original condition (P(i,j)>r1 || P(i,j)>r2), add another
%classLable(i)==classLable(j).
W = zeros(nSmp);
for i=1:nSmp
    for j=1:nSmp
        r1 = mean(P(i,:));
        r2 = mean(P(j,:));
        if (options.gnd(i)==options.gnd(j)) && (P(i,j)>r1 || P(i,j)>r2)
            W(i,j)=P(i,j);
        end
    end
end
D = zeros(nSmp);%%% Diagonal matrix
D = diag(sum(W));
L = D - W;%%% Laplacian matrixL=D-S
P = data' * L * data;
P = max(P , P');
Q = Sb;
Q = max(Q , Q');
%%%%%%%%%%%%%%%%%%
[eigvector, eigvalue] = eig(P,Q);
eigvalue=diag(eigvalue);
[~ ,eindex]=sort(-eigvalue,'descend');% Take the minimum of eigenvalues which is in ascending order.
eigvector=eigvector(:,eindex);
eigvalue=eigvalue(eindex);
%%%%%%%%%%%%%%%%
eigvector=Schmidt(eigvector);% Eigenvector rthogonalization
eigvector=eigvector*diag(1./(sum(eigvector.^2).^0.5));% Normalize eigenvectors
```

```
eigvector = eigvector_PCA*eigvector;
costTime=cputime-time_temp;
end
% Enhance Locality Preserving Projection
function [eigvector,eigvalue,costTime]=ELPP(data,options)
%ELPP : Enhance Locality Preserving Projection
time_temp=cputime;
%%%% Go ahead and take the PCA dimension reduction
options_PCA.PCARatio = 0.98;
[eigvector_PCA, ~] = PCA(data,options_PCA);
data=data*eigvector_PCA;%%% Projecting data
%%%% Construct adjacency matrix
k=5;
[nSmp ,nFea]=size(data);%nSmp—rows--the numbers of faces
temp_D = zeros(nSmp);% Temporary distance matrix
for i = 1:nSmp
    for j =1:nSmp
        temp_D(i,j) = norm(data(i,:) - data(j,:));% Distance between faces
        end
    end
    adj=zeros(nSmp);% Adjacency matrix, adjacent to 1, not adjacent to 0
     [~,index] = sort(temp_D, 2);
    for i=1:nSmp
        adj(i,index(i,2:k+1)) = 1;%k- The nearest neighbor cannot include itself, so 2:k+1, i.e. adj (I, I) =0;
    end
    S = zeros(nSmp);%%% Distance matrix, （exp types）
    t = (mean(mean(temp_D))).^2;
    for i=1:nSmp
        for j=1:nSmp
            if adj(i,j)==1
                S(i,j) = exp((-temp_D(i,j).^2)/(t));
            end
        end
    end
    S=max(S,S');%%% Symmetric affinity matrix
    D = zeros(nSmp);
    D = diag(sum(S));
    L = D - S;%%% Laplacian matrixL=D-S
    P = data' * L * data;
    P = max(P , P');
    %%%%%%%%%%%%%%%%%%%%%%%%%%%%%%%%%%%%%%%%%%%%%%%%%
```

```
ST=zeros(nFea,nFea);%% Overall dispersion matrix
overallmean=mean(data);%% Total sample mean
for i=1:nFea
       ST=ST+(data(i,:)-overallmean)'*(data(i,:)-overallmean);
end
Q = max(ST , ST');
%%%%%%%%%%%%%%%%%
[eigvector, eigvalue] = eig(P,Q);
eigvalue=diag(eigvalue);
[~ ,eindex]=sort(-eigvalue,'descend');% Take the minimum of eigenvalues which is in ascending order.
eigvector=eigvector(:,eindex);
eigvalue=eigvalue(eindex);
%%%%%%%%%%%%%%%%
eigvector=eigvector*diag(1./(sum(eigvector.^2).^0.5));% Normalize eigenvectors
eigvector = eigvector_PCA*eigvector;
costTime=cputime-time_temp;
end
```

We have simulated the above algorithms and their corresponding supervised algorithms on Yale, ORL and YaleB face databases respectively. There are 165 face images in Yale face database, including 15 people, each person has 11 images of different illumination and expressions [10], and the image resolution is 32×32; there are 400 face images in ORL face database, including 40 people, and each of whom has 10 images with 32×32 resolution and different illumination and expressions; there are 2432 face images in ORL face database, including 38 people, and each of whom has 64 images with 32×32 resolution and different illumination and expressions.

We randomly selected 6 images per person in Yale, ORL face database as the training set, and 40 images on YaleB face database were selected randomly for each person as the training set, and conducted 10 experiments. Figures 10.1, 10.2 and 10.3 show some samples of these face databases. Table 10.1 shows the maximum average recognition rate (%) and the corresponding standard deviation of LPP and its deformation algorithms on the three face databases. Figures 10.4, 10.5 and 10.6 shows the average of the 10 experimental recognition rates of the LPP and its deformation algorithms on the three face databases.

Fig. 10.1 Some samples of the ORL face database

Fig. 10.2 Some samples of the Yale face database

Fig. 10.3 Some samples of the YaleB face database

Table 10.1 The maximum average recognition rate and the corresponding standard deviation of various algorithms

Face database	Recognition accuracy (%)						
	Algorithms						
	LPP	SLPP	ELPP	SELPP	ODLPP	PLPP	ODPLPP
Yale	50.0 ± 6.23	76.0 ± 3.61	59.5 ± 4.36	75.9 ± 4.14	78.67 ± 4.12	50.9 ± 5.08	78.67 ± 3.97
ORL	84.6 ± 2.61	95.4 ± 1.65	90.6 ± 1.96	94.0 ± 2.07	97.63 ± 1.28	83.6 ± 3.03	97.5 ± 1.28
YaleB	81.8 ± 1.39	93.6 ± 0.89	86.5 ± 0.66	92.7 ± 0.67	93.42 ± 0.52	88.32 ± 0.90	93.17 ± 0.51

The best average recognition accuracy on three face databases

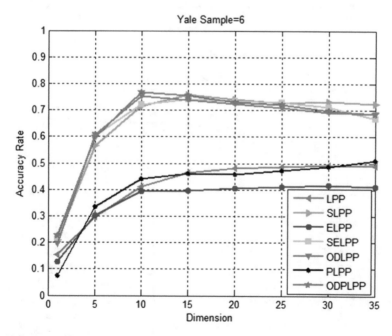

Fig. 10.4 Recognition rate on Yale face database

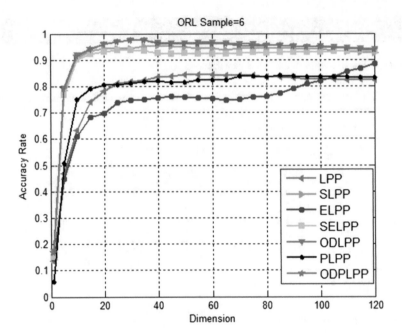

Fig. 10.5 Recognition rate on ORL face database

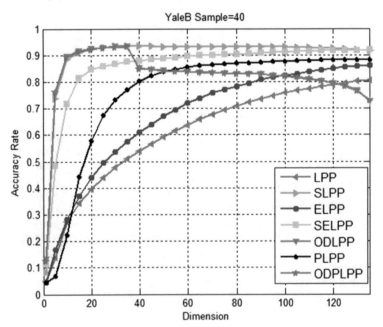

Fig. 10.6 Recognition rate on YaleB face database

10.2   Facial Expression Recognition Using PCA

Facial expression is an important way to express human emotions, it is also an effective means of human communication. Emotion, as an inner experience, is usually accompanied by corresponding nonverbal behaviors, such as facial expressions and body gestures, etc. People can express their thoughts and feelings accurately and subtly through expression. We can also understand the attitudes and feelings by identifying the expressions at the same time.

The process of facial expression recognition usually includes three nodes, which are face detection, facial feature extraction and emotion classification. As shown in Fig. 10.7, if we want to establish a facial expression recognition system, the first step is to detect and locate the human face; the second step is to extract features that can represent the essence of the input expression from the face image or image sequence, which can be divided into 3 modules: the generation of the original feature, the dimension reduction of features and the decomposition of the feature. The third step is to analyze the relationship between the features, and classify the emotional images of the input face to the corresponding categories.

The application of the PCA algorithm to emotion recognition assumes that facial expressions are in a low-dimensional linear space, and the expressions are separable. A new set of orthogonal bases is obtained after applying PCA algorithm into a space which is composed of several high-dimensional images. By preserving some orthogonal bases, the low-dimensional expression space can be generated, and an expression image can be represented as a linear combination of the set of bases.

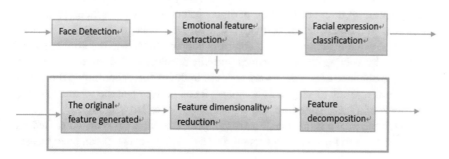

Fig. 10.7   Block diagram of facial expression recognition system

The images used for training are the face images which are normalized by the size and the gray level after face detection and preprocessing. The following describes the specific methods. Supposing that the size of the M training images is a $*$ b and each image is connected to a N dimensional vector by row or column, then the M vectors are put into a set S, as shown in the following formula:

$$S = \{\Gamma_1, \Gamma_2, \Gamma_3, \ldots, \Gamma_M\} \tag{10.18}$$

Compute the total average facial image $\Psi = \frac{1}{M} \sum_{n=1}^{M} \Gamma_n$ of the M training images. The difference Φ between each image and the average image can be calculated by subtracting the total average expression image from each training image.

$$\Phi_i = \Gamma_i - \Psi \tag{10.19}$$

Find the M orthogonal unit vectors u_n, and the $k(k = 1, 2, 3, \ldots, M)$ vectors in u_k are calculated by the following formula:

$$\lambda_k = \frac{1}{M} \sum_{n=1}^{M} (u_k^T \Phi_n)^2 \tag{10.20}$$

When the eigenvalue λ_k takes the minimum, u_k is basically determined. Actually, computing u_k is to calculate the eigenvectors of the covariance matrix.

$$C = \frac{1}{M} \sum_{n=1}^{M} \Phi_n \Phi_n^T = A A^T \tag{10.21}$$

In the above formula, $A = \{\Phi_1, \Phi_2, \Phi_3, \ldots, \Phi_n\}$. However, due to a large amount of computation in direct calculation of eigenvectors, it is quite difficult to find the eigenvalues and eigenvectors of matrix C with such a large dimension. Instead, the singular value decomposition theorem is adopted to reduce the computational complexity by solving the eigenvalues and eigenvectors of the alternative matrix $A^T A$. If the obtained eigenvectors are restored to the matrix according to the size of the sample image and displayed as an image, it can be seen that the feature vector is in the shape of a face. Therefore, the algorithm is also called 'Eigenface.'

Through the above steps, the dimension of the face image is reduced to find the appropriate vectors for facial expression. For a new face image, it can be expressed with Eigenface:

$$w_k = u_k^T(\Gamma - \Psi) \tag{10.22}$$

where $k = 1, 2, \ldots, M$. For the k th eigenface u_k, the corresponding weights can be calculated with the upper formula, and M weights can form a vector.

$$\Omega^T = [w_1, w_2, \ldots, w_M]$$
$$\varepsilon_k = ||\Omega - \Omega_k||^2 \tag{10.23}$$

After obtaining the representation of Eigenface to the face, the recognition of the face is as follows:

$$\varepsilon_k = ||\Omega - \Omega_k||^2 \tag{10.24}$$

where Ω represents the face to be distinguished and Ω_k represents someone's face in the training set, both are represented by the weights of Eigenface. Formula (10.24) is to solve the Euclidean distance between them. When the distance is less than the threshold, the distinguished face and the k th face in the training set belong to the same person. When traversing all the training set and the distance is always larger than the threshold, the distinguished face can be divided into two situations according to the size of the distance: a new face or not a face. The threshold setting is not fixed according to the different training sets.

The MATLAB code of facial expression recognition is implemented in PROGRAMME 10.2:

PROGRAMME 10.2: Facial expression recognition

```
clear all
close all
clc
xunlian=[];% All training images
for i=1:10
    a=imread(strcat('happiness\',num2str(i),'.bmp'));
    subplot(121);
    imshow(a);title('training data set');drawnow;
    title('training data set');
    b=a(1:100*100);%The b is the row vector of 1*N, where N is 10000
    b=double(b);
xunlian=[xunlian;b];% The xunlian is a matrix of M*N, and each line of data in it is a picture, in which
    M is
%10.
end;
for i=1:10
    a=imread(strcat('sadness \',num2str(i),'.bmp'));
```

```
        imshow(a);title('training data set');drawnow;
        title('traning data set\');
        b=a(1:100*100);% The b is the row vector of 1*N, where N is 10000
        b=double(b);
        xunlian=[xunlian;b];% The xunlian is a matrix of 20*10000, and each line of data in it is a picture.
end;
for i=1:10
        a=imread(strcat(' surprise \',num2str(i),'.bmp'));
        imshow(a);title('training data set');drawnow;
        imshow([]);title(' The training is over !');
        b=a(1:100*100);% The b is the row vector of 1*N, where N is 10000
        b=double(b);
        xunlian=[xunlian; b];% The xunlian is a matrix of 30*1000, and each line of data in it is a picture.
end;
xunlian=xunlian';% Each column is a picture, and the xunlian'dimension is 10000*30.
for i=1:1000
        cy(i,1:30)=xunlian(10*i,:);
end
pmetrix=cy*cy';
[vet vetvalue t1]=pcacov(pmetrix);% Finding eigenvalues and eigenvectors by principal component
        analysis, and
%find eigenvalue vetvalue and corresponding eigenvector vet from sorted covariance matrix which is in
        ascending
%order
sum2=sum(vetvalue);
temp=0;
con=0;
m=0;
for i=1:1000
        if(con<0.99)
                temp=temp+vetvalue(i);
                con=temp/sum2;
                m=m+1;
        else
                break;
        end
end
A=vet(:,1:m);
y=A'*cy;%m*10000*10000*30 Each kind of emotion is projected into the eigenface space
f=imread('s146.bmp');
```

```
ff=f;
f=f(1:100*100);%The f is the row vector of 1*N, where N is 10000
f=double(f);
f=f';
for i=1:1000
        cs(i)=f(10*i);
end;
xl=cs';
sum1=zeros(18,1);
sum2=zeros(18,1);
sum3=zeros(18,1);
zbceshi=A'* xl; %18*1 Find the coordinates of it in the eigenface space.
%size(zbceshi)
for i=1:10
        %size(y:i)
        %size(sum1)
        sum1=y(:,i)+sum1;
end
for i=11:20
        sum2=y(:,i)+sum2;
end
for i=21:30
        sum3=y(:,i)+sum3;
end
a=zeros(18,3);
a(:,1)=sum1./10;
a(:,2)=sum2./10;
a(:,3)=sum3./10;
for i=1:3
        wucha(i)=norm(zbceshi-a(:,i));
end
[h,I]=min(wucha);% Using nearest neighbor method to do face recognition.
if I==1
        subplot(122);
        imshow(ff);
        title('The emotion which is identified is happiness!');
elseif I==3
        subplot(122);
        imshow(ff);
         title(' The emotion which is identified is suprise!');
```

else

    subplot(122);

    imshow(ff);

      title(' The emotion which is identified is sadness!');

end

We use images in the YALE database as the training set and test set. 10 images are selected for each kind of emotional expression. After the end of the training, the image of the known category was tested, which realize the recognition of happiness, sadness, and surprise.

The YALE database contains 165 grayscale images of 15 people which is in the size of 100 * 100. Each person has 11 different images, which show the characteristics in positive light irradiation, the existing eyes, the happy expression, left side irradiation, the non-existing eyes, neutral expression, neutral light, right side irradiation, the sad expression, the sleepy expression, the surprised expression, and nictation. We selected images with three kinds of emotions as the training set, which are happiness, sadness, surprise, and each kind takes 10 images. After reducing the dimension of the face by the PCA method, the least nearest neighbor method is used to identify an unknown facial emotion image.

10.3   Extraction and Recognition of Characters in Pictures

The information of character in the image contain rich semantic information of the high level, and extracting these characters is very helpful for the understanding, indexing and retrieval of the high level semantics of the image. Image character extraction is divided into two types: dynamic image character extraction and static image character extraction. Static image character extraction is the basis of dynamic image character extraction, whose application range is more extensive, and its research is fundamental. The characters in the static image can be divided into two categories: one is the characters contained in the scene itself in the image, which are called the scene characters; the other is the characters added to the post production of the image, called the artificial characters. The scene characters are generally difficult to detect and extract because of the randomness of their location, color, and shape. While the artificial characters are more standard and easy to identify. Moreover, the size of the characters has a certain limitation; The color is monochromatic and is more easy to be detected and extracted than the former. The general identification method of artificial character extraction is as follows (Fig. 10.8):

The input color image contains a lot of color information, which takes up more storage space and reduces the execution speed of the system. Thus, when performing the image recognition and other processing, the color image is often converted to the grayscale image to speed up the processing speed. The image is processed by using grayscale processing, edge extraction, and morphological method to locate the character region. Image binarization has many mature algorithms, and we can use the adaptive threshold algorithm, or the given threshold algorithm.

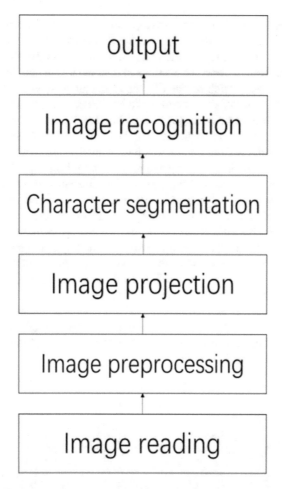

Fig. 10.8 The artificial character extraction system block diagram

The image after the morphological filtering is very close to the correct character position, and in the automatic recognition process, the character segmentation has the function of carrying forward. The character segmentation is based on the previous location of the region, and then the character recognition is carried out by using the segmentation results. Usually, the image to be identified contains a lot of characters, which should be judged according to the features of each character. Firstly, the image is scanned progressively from bottom to top until the first black pixel is encountered and recorded. And then continue to scan the image to find the next black pixel. Repeat the above process, so that the range of the maximum height of each line of the image will be found. Then, we continue to scan the image until there is a column without a black pixel, which means that the character segmentation is completed. Then continue to scan to the right end of the image according to the above method, which will give a more precise range of the width of each character. In order to obtain the result from coarse to fine, in the range of the known width of each character, scan the image progressively from bottom to top until the first black pixel is encountered and recorded. Next, the image is scanned progressively from top to bottom until the first black pixel is encountered and recorded. Thus, the approximate height range of the image can be found in this way. In the end, a top-down and bottom-up scan is conducted to obtain the precise height range of each character.

Because of the large difference in the size of the characters in the scanned images, the higher the size of the character recognition is, the higher the recognition rate is. The standardization of images is to unify the different original characters to the same size. Each rectangle after normalization is arranged at the same height, and there is a certain interval between these character rectangles.

Static image character extraction is generally divided into: character region detection and location, character segmentation and character extraction, character post-processing functions and so on. The code for static image character extraction is shown in PROGRAMME 10.3 (Fig. 10.9).

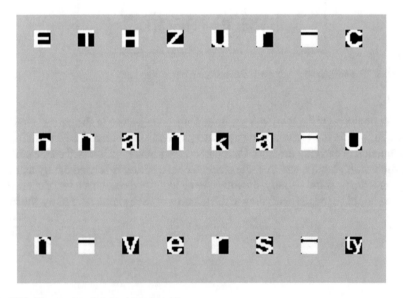

Fig. 10.9 The results of character recognition

PROGRAMME 10.3: Static image character extraction

```
function getPicChar %% The character extraction function is set up and can be run directly on the
%MATLAB platform
% Using MATLAB UI, directly open the character pictures you need to extract
[filename,pathname,~]=uigetfile({'*.jpg';'*.bmp';'*.png'},'Chose a picture');
picstr=[pathname filename];
if ~ischar(picstr)
    return;
end
pic=imread(picstr);%Open image
if length(size(pic))==3 % Determine the image of the dimension, unified for the gray image
    pic=rgb2gray(pic);
end
pic=(pic<127);% Convert to binary image
pic=xylimit(pic);% The first border of the image region is defined
%%%%%% The first stage %%%%%
m=size(pic,1);
Ycount=zeros(1,m);
for i=1:m
    Ycount(i)=sum(pic(i,:));
end
lenYcount=length(Ycount);
Yflag=zeros(1,lenYcount);
for k=1:lenYcount-2
    if Ycount(k)<3 && Ycount(k+1)<3 && Ycount(k+2)<3
        Yflag(k)=1;
    end
end
for k=lenYcount:1+2
    if Ycount(k)<3 && Ycount(k-1)<3 && Ycount(k-2)<3
        Yflag(k)=1;
    end
end
Yflag2=[0 Yflag(1:end-1)];
Yflag3=abs(Yflag-Yflag2); % Differential operations
[~,row]=find(Yflag3==1); % Find the mutated position
row=[1 row m]; % Adjust the mutation point
row1=zeros(1,length(row)/2);% Capture the initial position vector of the image
row2=row1;% Capture the ending position vector of the image
for k=1:length(row)
```

```
        if mod(k,2)==1;% The odd as the beginning
            row1((k+1)/2)=row(k);
        else%The even as the ending
            row2(k/2)=row(k);
        end
end
pic2=pic(row1(1):row2(1),:);%Capture the first coloum of characters
alpha=1024/size(pic2,2);% Calculate the scaling ratio
pic2=imresize(pic2,alpha);% Adjust the size of the first column of characters as a benchmark
for k=2:length(row)/2
        pictemp=imresize(pic(row1(k):row2(k),:),[size(pic2,1) size(pic2,2)]);
        pic2=cat(2,pic2,pictemp);% Horizontally connect image blocks
end
pic=xylimit(pic2);% Limit the image region
%%%%%%The second stage%%%%%
[~,n]=size(pic);
Xcount=zeros(1,n);
for j=1:n
        Xcount(j)=sum(pic(:,j));
end
lenXcount=length(Xcount);
Xflag=zeros(1,lenXcount);
for k=1:lenXcount-2
        if Xcount(k)<3 && Xcount(k+1)<3 && Xcount(k+2)<3
            Xflag(k)=1;
        end
end
for k=lenXcount:1+2
        if Xcount(k)<3 && Xcount(k-1)<3 && Xcount(k-2)<3
            Xflag(k)=1;
        end
end
Xflag2=[0 Xflag(1:end-1)];
Xflag3=abs(Xflag-Xflag2);
[~,col]=find(Xflag3==1);
col=[1 col size(pic,2)];
coltemp=col(2:end)-col(1:end-1);
[~,ind]=find(coltemp<3);
col(ind)=0;
col(ind+1)=0;
```

```
col=col(col>0);
col1=zeros(1,length(col)/2);
col2=col1;
for k=1:length(col)
    if mod(k,2)==1
        col1((k+1)/2)=col(k);
    else
        col2(k/2)=col(k);
    end
end
picnum2=length(col)/2;
piccell2=cell(1,picnum2);
for k=1:picnum2
    piccell2{k}=pic(:,col1(k):col2(k));
    piccell2{k}=xylimit(piccell2{k});
    piccell2{k}=imresize(piccell2{k},[128 128]);
end
% Show extracted characters, up to 8 characters per line
if mod(picnum2,8)
    rownum=ceil(picnum2/8)+1;
else
    rownum=picnum2/8;
end
for k=1:picnum2
    subplot(rownum,8,k);
    imshow(piccell2{k});
end
%%The function xylimit is as follows:
function newpic=xylimit(pic)
%function name:XYLIMIT
%   Input pic:binary image
%   Output newpic:binary image
% Uses: The boundary of the binary image is limited, requiring the image to be black background and
    white piexls
%example:
%%pic=imread(' Numeric characters.jpg');
%%pic=rgb2gray(pic);
%%pic=(pic<127);
%%pic=xylimit(pic);
%%imshow(pic);
[m,n]=size(pic);
```

```
[m,n]=size(pic);
%%% Vertical scanning %%%
Ycount=zeros(1,m);
for i=1:m
    Ycount(i)=sum(pic(i,:));% Gets the number of pixels per row
end
Ybottom=m;% Bottom boundary
Yvalue=Ycount(Ybottom);
while(Yvalue<3)
    Ybottom=Ybottom-1;
    Yvalue=Ycount(Ybottom);
end

Yceil=1;%      Top    boundary
Yvalue=Ycount(Yceil);

while(Yvalue<3)

    Yceil=Yceil+1;

    Yvalue=Ycount(Yceil);

end

%%% Horizontal scanning %%%

Xcount=zeros(1,n);

for j=1:n

    Xcount(j)=sum(pic(:,j));% Gets the number of pixels per column

end

Xleft=1;% Left border

Xvalue=Xcount(Xleft);

while(Xvalue<2)

    Xleft=Xleft+1;

    Xvalue=Xcount(Xleft);

end

Xright=n;%      R ight border

Xvalue   =Xcount(Xright);

while(Xvalue<2)
```

```
Xright=Xright-1;

Xvalue=Xcount(Xright);
```

end

%%%Capture images%%%

newpic=pic(Yceil:Ybottom,Xleft:Xright);

References

1. Turk M, Pentland A (1991) Eigenfaces for recognition. J Cogn Neurosci 3(1):71
2. Vapnik VN (1998) Statistical learning theory. Encycl Sci of Learn 41(4):3185–3185
3. Phillips PJ, Flynn PJ, Scruggs T et al (2005) Overview of the face recognition grand challenge
4. Song F, Song F, Feng G et al (2010) Short communication: a novel local preserving projection scheme for use with face recognition. Exp Syst Appl Int J 37(9):6718–6721
5. Dasarathy BV (1991) Nearest neighbor (NN) norms: NN pattern classification techniques. Los Alamitos IEEE Comput Soc Press 13(100):21–27
6. Xu Y, Zhong A, Yang J et al (2010) LPP solution schemes for use with face recognition. Pattern Recogn 43(12):4165–4176
7. Hanmandlu M, Singhal S (2017) Face recognition under pose and illumination variations using the combination of Information set and PLPP features. Appl Soft Comput 53:396–406
8. Jain D, Shikkenawis G, Mitra SK et al (2013) Face and facial expression recognition using extended locality preserving projection. Comput Vis Pattern Recogn Image Process Graph. IEEE. 1–4
9. Zhu Lei, Zhu Shanan (2007) Face recognition based on orthogonal discriminant locality preserving projections. Neurocomputing 70(7):1543–1546
10. Guo G, Dyer CR (2005) Learning from examples in the small sample case: face expression recognition. IEEE Transactions on Systems Man & Cybernetics Part B Cybernetics A Publication of the IEEE Systems Man & Cybernetics Society 35(3):477–88

Part III
Advances in Video Processing and then Associated Chapters

Chapter 11
Visual Object Tracking

Moving object tracking is to find out the candidate object region which is the most similar area in the image sequence through the effective expression of the object, that is to locate the target in the sequence image so as to obtain the complete motion trajectory of the moving target. In this chapter, we first introduce the moving object detection method in static background. We also present the Adaptive background modeling method by using a mixture Gaussians. In the next three sections, there are three methods for object tracking: Ransac, Meanshift and Particle Filter. In the last section, we introduce the multi-object tracking method.

11.1 Adaptive Background Modeling by Using a Mixture of Gaussians

The main idea of the Gaussian mixture model is to characterize the pixels in each frame of the video sequence by the weighted sum of finite Gaussian models. Usually, the more the number of pixels in the Gaussian model is, the more complete the feature is. However, with the increase number of Gaussian model, the calculation will be more complex and increase. When a new image arrives, the background model needs to be updated. For each pixel, define K Gaussian models, taking into account the speed and effectiveness of the algorithm, the value usually takes between 3 and 5.

The implementation of Gaussian mixture model includes three parts: model definition, model update and foreground detection.

(1) Model Definition

In the Gaussian mixture model, the color values of a pixel in a video frame (or image sequence) form the corresponding pixel process:

$$\{X_1, X_2 \ldots, X_t\} = \{I(x, y, i) : 1 \leq i \leq t\} \tag{11.1}$$

© Springer International Publishing AG, part of Springer Nature 2019
S. Gong et al., *Advanced Image and Video Processing Using MATLAB*,
Modeling and Optimization in Science and Technologies 12,
https://doi.org/10.1007/978-3-319-77223-3_11

where $I(x, y, i)$ represents the color value of pixel (x, y) at time i. Modeling the background by a mixture of Gaussians assumes that the pixel process satisfies the mixed Gaussian distribution, that is, a Gaussian mixture model is composed by K single Gaussian models for each pixel:

$$P(X_t) = \sum_{k=1}^{K} \omega_{k,t} \eta(X_t, \mu_{k,t}, \Sigma_{k,t}) \tag{11.2}$$

where $\omega_{k,t}$ is the weight of the kth Gaussian component at time t, which also means probability density function, then $P(X_t)$ represents the probability observed by the observed pixel value X at time t. In order to avoid cumbersome matrix operations, it is common to assume that the components of the pixel value X, such as red, green and blue components of the RGB color model, are independent of each other and have the same covariance. It could speed up the computation and have little effect on the results.

(2)   Model Update

The mixed Gaussian background modeling first calculates the match between the current pixel value and K Gaussian distributions in the model, and then matches the distribution if the pixel value is within the $k\sigma$ (usually $k = 2.5$) range of a Gaussian distribution average. If the current pixel value does not match K Gaussian distributions, a new Gaussian distribution will replace the distribution with the smallest weight value, and the new distribution average is the current pixel value. If there is a Gaussian distribution that matches, the weight values for each distribution are adjusted as follows:

$$\omega_{k,t+1} = (1 - \alpha)\omega_{k,t} + \alpha M_{k,t+1} \tag{11.3}$$

where σ is the learning rate and its value is between $(0, 1)$; for the Gaussian distribution matched to the current pixel $M_{k,t+1} = 1$ otherwise $M_{k,t+1} = 0$ by Formula (11.3), the Gaussian distribution weight value resulting in the matching is increased while Match the Gaussian distribution weight value.

For a Gaussian distribution that matches the current pixel value, adjust its parameters as follows:

$$\mu_{k,t+1} = (1 - \alpha)\mu_{k,t} + \rho X_{k,t+1} \tag{11.4}$$

$$\sigma_{k,t+1}^2 = (1 - \alpha)\sigma_{k,t}^2 + \rho(X_{k,t+1} - \mu_{k,t+1})^T (X_{t+1} - \mu_{k,t+1}) \tag{11.5}$$

where ρ is the other learning rate, its value is $\rho = \alpha\eta(X_t|\mu_k, \sigma_k)$, and for the Gaussian distribution without matching, its parameters remain unchanged.

(3)   Foreground Detection

According to the model update method described earlier, the Gaussian distribution with smaller covariance and larger weight has more possibility to be the distribution of background pixels. Therefore, in order to determine the specific background model, we will arrange the K Gaussian distributions according to the order of the ω/σ value for each pixel in the image. For the first B Gaussian distributions satisfying Formula (11.6) as a description of the background:

$$B = \arg\min_{b}\left(\sum_{k=1}^{b}\omega_{k,y} > T\right) \qquad (11.6)$$

T is the background model proportional threshold. If the T value is small, the Gaussian mixture model will degrade into a single Gaussian distribution model. If the T value is large, it can be a complex dynamic background, such as shaking leaves and fluctuating lakes, many Gaussian distribution of the mixed models will be built to simulate.

If at least one Gaussian distribution matches the current pixel value in the B Gaussian distributions described in the background after the current sorted by the ω/σ value, the current pixel is a background pixel, otherwise it is determined as the foreground pixel.

The MATLAB code is shown in PROGRAMME 11.1.

PROGRAMME 11.1: Gaussian mixture model for background detection

```
video = VideoReader('pedestrian.avi');
frame_num = video.NumberOfFrames;
height = video.Height;
width = video.Width;
fg = zeros(height, width); % define foreground and background matrix
bg_bw = zeros(height, width);
C = 3; % the number of single Gaussian model (usually 3-5)
M = 3; % the number of background model
D = 2.5; % deviation threshold
alpha = 0.01; % learning rate
thresh = 0.25; % foreground threshold
sd_init = 15; % initialize the standard deviation
w = zeros(height,width,C); % initialize the weight matrix
mean = zeros(height,width,C); % pixel value
sd = zeros(height,width,C); % pixel standard deviation
u_diff = zeros(height,width,C); % the absolute distance between the pixel and the mean of a Gaussian
    model
p = alpha/(1/C); % initialize the p variable to update the mean and standard deviation
rank = zeros(1,C); % the priority of each Gaussian distribution (w/sd)
pixel_depth = 8; % each pixel 8bit resolution
pixel_range = 2^pixel_depth -1; % pixel value range [0,255]
for i=1:height
  for j=1:width
    for k=1:C
```

```
mean(i,j,k) = rand*pixel_range; % initialize the mean of the kth Gaussian distribution
w(i,j,k) = 1/C; % initialize the weight of the kth Gaussian distribution
sd(i,j,k) = sd_init; % initialize the standard deviation of the kth Gaussian distribution

      end
   end
end
for n = 1:frame_num
   I1 = read(video, n);
   I1 = rgb2gray(I1);
   fr_bw =I1;
   % calculate the absolute distance between the new pixel and the mean of the mth Gaussian model
   for m=1:C
      u_diff(:,:,m) = abs(double(fr_bw)-double(mean(:,:,m)));
   end
   % update the parameters of the Gaussian model
   for i=1:height
      for j=1:width
         match = 0; % match mark
         for k=1:C
            % pixels match the kth Gaussian model
            if (abs(u_diff(i,j,k)) <= D*sd(i,j,k))
               match = 1; % set the match flag to 1
               % update weight, mean, standard deviation, p
               w(i,j,k) = (1-alpha)*w(i,j,k) + alpha;
               p = alpha/w(i,j,k);
               mean(i,j,k) = (1-p)*mean(i,j,k) + p*double(fr_bw(i,j));
               sd(i,j,k)=sqrt((1-p)*(sd(i,j,k)^2)+p*((double(fr_bw(i,j))-mean(i,j,k)))^2);
            else % pixels do not match the kth Gaussian model
               w(i,j,k) = (1-alpha)*w(i,j,k); % slightly reduce the weight
            end
         end
         bg_bw(i,j)=0;
         for k=1:C
            bg_bw(i,j) = bg_bw(i,j)+ mean(i,j,k)*w(i,j,k);
         end
         % if the pixel value does not match any Gaussian model, a new model is created
         if (match == 0)
            [min_w, min_w_index] = min(w(i,j,:)); % find the minimum weight
            % the initialized mean is the mean of the current observed pixel
            mean(i,j,min_w_index) = double(fr_bw(i,j));
            sd(i,j,min_w_index) = sd_init; % the initial standard deviation is 6
         end
         rank = w(i,j,:)./sd(i,j,:); % calculate model priority
```

```
        rank_ind = [1:1:C];% priority index
        % calculate the foreground
        fg(i,j) = 0;
        while ((match == 0)&&(k<=M))
            % pixels match the kth Gaussian model
            if (abs(u_diff(i,j,rank_ind(k))) <= D*sd(i,j,rank_ind(k)))
                fg(i,j) = 0; % the pixel is the background, set to black
            else
                fg(i,j) = 255; % Otherwise it is foreground and white
            end
            k = k+1;
        end
    end
  end
  imshow(uint8(fg));
end
figure(2)
subplot(1,3,1),imshow(fr_bw); % displays the last frame image
subplot(1,3,2),imshow(uint8(bg_bw)) % show background
subplot(1,3,3),imshow(uint8(fg)) % show foreground
```

See Fig. 11.1.

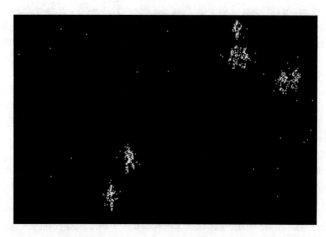

Fig. 11.1 GMM method to detect the prospects

11.2 Object Tracking Based on Ransac

Feature-based tracking method matches and traces a set of feature points (such as boundary line, centroid, corner, etc.) in successive frame images, including feature extraction [1, 2] and matching. The main advantage of this type of tracking method is that even if the object in the scene is partially occluded, the object can be continuously tracked as long as the feature points are visible. SIFT, SURF, Harris, SUSAN and many other algorithms can be applied to the feature extraction. After the feature extraction [3–6] of the moving object, the similarity metric algorithm needs to be matched with the frame image to achieve the object tracking. Common similarity measures are Euclidean distance, city-block distance, chessboard distance, weighted distance, Hausdorff distance, and so on. On the basis of rough matching, the random sampling consistency (RANSAC) algorithm can be further refined to filter the noise error data and reduce the deviation.

Random sample consensus algorithm can estimate the parameters of a mathematical model from an array of observations that contain "external points". The basic assumptions of the random sampling are: (1) data consists of "internal point", such as the distribution of data can be explained by some parameters; (2) "external point" is not able to adapt to the model data. The other data is noise. Random sampling consistent made the following assumptions: Given a set of (usually small) points, there is a process that can be used to estimate model parameters, which can be interpreted or applied to local points.

Figure 11.2 shows an example of finding the appropriate two-dimensional line from a set of observations. Assume that the observed data contain the local points and the external points, where the local points are approximated by a straight line and the outright points are far from the straight line. The simple least squares method cannot find a straight line that adapts to the internal point, since the least squares method tries to adapt to all points including the external points. Instead, RANSAC can derive a model that is calculated using only the internal point with the high enough probability. However, RANSAC cannot guarantee that the results must be correct. In order to ensure that the algorithm has a high enough reasonable probability, we must carefully select the algorithm parameters.

The input of the RANSAC algorithm is a set of observation data, a parametric model adapted to the observed data, and some trusted parameters to achieve the object by repeatedly selecting a set of random subsets in the data. The selected subset is assumed to be an internal point and verified by the following method.

(1) A model is adapted to the assumed internal point, that is, all unknown parameters can be calculated from the hypothetical central point.
(2) Test all other data with the model obtained in step 1, and if a point applies to the estimated model, it is also considered an internal point.
(3) If there are enough points to be classified as hypothetical local points, then the estimated model is justified.
(4) Then, all assumptions are used to re-estimate the model, because it is only estimated by the initial hypothesis.

(a) Contains a lot of external points of the data set (b) RANSAC found the line

Fig. 11.2   Find the right line from a set of observations by RANSAC

(5)   Finally, the model is evaluated by estimating the error rate between the interior point and the model.

This process is repeated a fixed number of times, that the model is either discarded because of too few points, or because it is better than the existing model to be selected.

The flow chart of the object tracking algorithm based on RANSAC is as follows in Fig. 11.3.

The MATLAB + VLFeat source code is shown in PROGRAMME 11.2:

PROGRAMME 11.2: SIFT operator and Ransac algorithm

```
function mosaic = sift_mosaic(im1, im2)
if nargin == 0
  im1 = imread(fullfile(vl_root, 'data', 'picture1.jpg')) ;
  im2 = imread(fullfile(vl_root, 'data', 'picture2.jpg')) ;
end
% make single
im1 = im2single(im1) ;
im2 = im2single(im2) ;
% make grayscale
if size(im1,3) > 1, im1g = rgb2gray(im1) ; else im1g = im1 ; end
if size(im2,3) > 1, im2g = rgb2gray(im2) ; else im2g = im2 ; end
% SIFT matches
[f1,d1] = vl_sift(im1g) ;
[f2,d2] = vl_sift(im2g) ;
[matches, scores] = vl_ubcmatch(d1,d2) ;
numMatches = size(matches,2) ;
X1 = f1(1:2,matches(1,:)) ; X1(3,:) = 1 ;
X2 = f2(1:2,matches(2,:)) ; X2(3,:) = 1 ;
% RANSAC with homography model
```

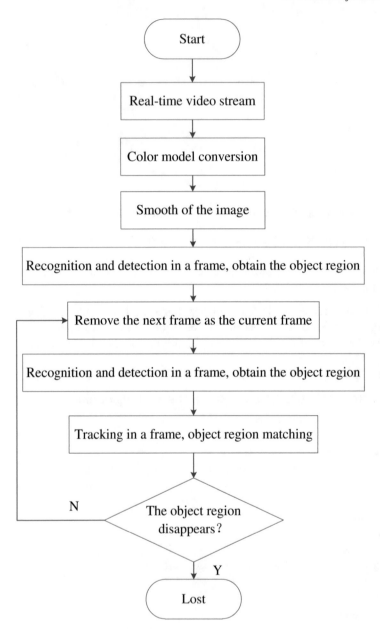

Fig. 11.3 RANSAC algorithm flow chart

```
clear H score ok ;
for t = 1:100
  % estimate homograpyh
  subset = vl_colsubset(1:numMatches, 4) ;
  A = [] ;
  for i = subset
    A = cat(1, A, kron(X1(:,i)', vl_hat(X2(:,i)))) ;
  end
  [U,S,V] = svd(A) ;
  H{t} = reshape(V(:,9),3,3) ;
  % score homography
  X2_ = H{t} * X1 ;
  du = X2_(1,:)./X2_(3,:) - X2(1,:)./X2(3,:) ;
  dv = X2_(2,:)./X2_(3,:) - X2(2,:)./X2(3,:) ;
  ok{t} = (du.*du + dv.*dv) < 6*6 ;
  score(t) = sum(ok{t}) ;
end
[score, best] = max(score) ;
H = H{best} ;
ok = ok{best} ;
% Optional refinement
function err = residual(H)
  u = H(1) * X1(1,ok) + H(4) * X1(2,ok) + H(7) ;
  v = H(2) * X1(1,ok) + H(5) * X1(2,ok) + H(8) ;
  d = H(3) * X1(1,ok) + H(6) * X1(2,ok) + 1 ;
  du = X2(1,ok) - u ./ d ;
  dv = X2(2,ok) - v ./ d ;
  err = sum(du.*du + dv.*dv) ;
end
if exist('fminsearch') == 2
  H = H / H(3,3) ;
  opts = optimset('Display', 'none', 'TolFun', 1e-8, 'TolX', 1e-8) ;
  H(1:8) = fminsearch(@residual, H(1:8)', opts) ;
else
  warning('Refinement disabled as fminsearch was not found.') ;
end
% Show matches
dh1 = max(size(im2,1)-size(im1,1),0) ;
dh2 = max(size(im1,1)-size(im2,1),0) ;
figure(1) ; clf ;
subplot(2,1,1) ;
```

```matlab
imagesc([padarray(im1,dh1,'post') padarray(im2,dh2,'post')]) ;
o = size(im1,2) ;
line([f1(1,matches(1,:));f2(1,matches(2,:))+o], ...
     [f1(2,matches(1,:));f2(2,matches(2,:))]) ;
title(sprintf('%d tentative matches', numMatches)) ;
axis image off ;
subplot(2,1,2) ;
imagesc([padarray(im1,dh1,'post') padarray(im2,dh2,'post')]) ;
o = size(im1,2) ;
line([f1(1,matches(1,ok));f2(1,matches(2,ok))+o], ...
     [f1(2,matches(1,ok));f2(2,matches(2,ok))]) ;
title(sprintf('%d (%.2f%%) inliner matches out of %d', ...
                sum(ok), ...
                100*sum(ok)/numMatches, ...
                numMatches)) ;
axis image off ;
drawnow ;
% Mosaic
box2 = [1    size(im2,2) size(im2,2)   1 ;
        1    1                size(im2,1)   size(im2,1) ;
        1    1                1                    1 ] ;
box2_ = inv(H) * box2 ;
box2_(1,:) = box2_(1,:) ./ box2_(3,:) ;
box2_(2,:) = box2_(2,:) ./ box2_(3,:) ;
ur = min([1 box2_(1,:)]):max([size(im1,2) box2_(1,:)]) ;
vr = min([1 box2_(2,:)]):max([size(im1,1) box2_(2,:)]) ;
[u,v] = meshgrid(ur,vr) ;
im1_ = vl_imwbackward(im2double(im1),u,v) ;
z_ = H(3,1) * u + H(3,2) * v + H(3,3) ;

u_ = (H(1,1) * u + H(1,2) * v + H(1,3)) ./ z_ ;
v_ = (H(2,1) * u + H(2,2) * v + H(2,3)) ./ z_ ;
im2_ = vl_imwbackward(im2double(im2),u_,v_) ;
mass = ~isnan(im1_) + ~isnan(im2_) ;
im1_(isnan(im1_)) = 0 ;
im2_(isnan(im2_)) = 0 ;
mosaic = (im1_ + im2_) ./ mass ;
figure(2) ; clf ;
imagesc(mosaic) ; axis image off ;
title('Mosaic') ;
if nargout == 0, clear mosaic ; end
end
img1 = imread('1bmpfile.bmp');
img2 = imread('2bmpfile.bmp');
sift_mosaic(img1, img2);
```

Fig. 11.4   Results of the SIFT operator and the Ransac algorithm

See Fig. 11.4.

11.3   Object Tracking Based on MeanShift

In the process of object tracking, if you match all the content directly in the scene to look for the best match position, you need to deal with a lot of redundant information, so that the amount of computing is relatively large. It is meaningfully to estimate the position state of the object in the future and narrow the object search range by the search algorithm. The commonly used algorithms to reduce the search range include the mean shift algorithm (Meanshift algorithm), the continuous adaptive mean shift algorithm (Camshift algorithm) and the confidence region algorithm. They all use the nonparametric method to optimize the object template and candidate object distance iterative convergence process to achieve the purpose of narrowing the scope of the search.

Meanshift algorithm is a method of gradient optimization to achieve fast object location, real-time tracking of non-rigid objects, suitable for tracking non-linear moving objects, and have a good applicability with the object deformation, rotation and other conditions. However, the Meanshift algorithm does not use the moving direction and velocity information of the object in the object tracking process, and it is easy to lose the object when there is interference (such as light, occlusion) in the surrounding environment. The Camshift algorithm is based on the Meanshift algorithm and has been extended to an improved mean shift algorithm based on the object color information. Since the histogram of the target image records the probability of the appearance of the color, this method is not affected by the change of the object shape, it can effectively solve the problem of object deformation and partial occlusion with the higher operation efficiency, but the algorithm needs to manually specify the object before it starts.

MeanShift algorithm is a nonparametric probability density estimation algorithm that can converge quickly to the local maximum of the probability density function by iteration. The tracking process of the algorithm is to find the process of local maximum of probability density.

11.3.1 Description of the Object Model

The description of the object model is, above all, the initialization of the object, that is, the object area to be tracked in the first frame image. The object area can be determined by manual selection, or the object area can be automatically selected based on the result of motion detection. If the center of the object area is x_0, then the object model can be described as the probability value for all eigenvalues on the object area. The probability density estimated by the eigenvalue of the object model $u = 1, \ldots, m$ is:

$$\hat{q}_u = C \sum_{i=1}^{n} K \left[\left\| \frac{x_0 - x_i}{h} \right\|^2 \right] \delta[b(x_i) - u] \tag{11.7}$$

where $K(x)$ is the contour function of the kernel function. Since the pixels near the center of the object model are more reliable than the external pixels, $K(x)$ gives a large weight to the center pixel and a small weight for the pixel away from the center. $b(x_i)$ is the characteristic value of pixel x_i, $\delta(x)$ is Delta function, $\delta[b(x_i) - u]$ is used to determine whether the object area of any pixel x_i eigenvalue is equal to the u-th eigenvalue, if it is equal then 1, otherwise 0. C is a normalized constant coefficient.

11.3.2 A Description of the Candidate Model

Moving object in each frame and later, the area that may contain the object is called the candidate region, the center coordinate is y, and the probability density of the pixel eigenvalue a of the candidate model $u = 1, \ldots, m$ is

$$\hat{p}_u(y) = C_h \sum_{i=1}^{n_h} K \left[\left\| \frac{y - x_i}{h} \right\|^2 \right] \delta[b(x_i) - u] \tag{11.8}$$

where h is the bandwidth parameter, MeanShift's tracking window size depends on bandwidth h, where C_h is the normalization constant, which is

$$C_h = \left[\sum_{i=1}^{n_h} K\left[\left\| \frac{y - x_i}{h} \right\|^2 \right] \right]^{-1}.$$

11.3.3 Similarity Function

The similarity function is used to describe the degree of similarity between the object model and the candidate object. The Bhattacharyya coefficient can be used as a similarity function:

$$\hat{\rho}(y) = \rho\big(\hat{p}(y), \hat{q}\big) = \sum_{u=1}^{m} \sqrt{\hat{p}_u(y)\hat{q}_u} \tag{11.9}$$

Its value is between 0 and 1. The larger the value $\hat{\rho}(y)$, the more similar the two models.

11.3.4 Object Location

In order to maximize $\hat{\rho}(y)$, we should first locate the object center of the current frame as the position y_0 of the object center in the previous frame, and then start looking for the best matching object from this point y_1. When locating, the Taylor series expansion is performing at $\hat{\rho}(y_u)$, and the similarity function can be approximated as:

$$\rho\big(\hat{p}(y), \hat{q}\big) = \frac{1}{2} \sum_{u=1}^{m} \sqrt{\hat{p}_u(y_0)\hat{q}_u} + \frac{C_h}{2} \sum_{i=1}^{n_h} w_i k\left[\left\| \frac{y - x_i}{h} \right\|^2 \right] \tag{11.10}$$

where $w_i = \sum_{u=1}^{m} \sqrt{\frac{\hat{q}_u}{\hat{p}_u(y_0)}} \delta[b(x_i) - u]$, and $\tilde{O}_{n,k} = \sum_{i=1}^{n_h} \frac{C_h}{2} w_i k\left[\left\| \frac{y - x_i}{h} \right\|^2 \right]$.

That is the kernel density estimate existing weight w_i. It shows that computing the maximum value of the similarity function is equal to computing the maximum of the Formula (11.10), and the MeanShift vector $y_1 - y_0$ can be calculated by maximizing the similarity function. In each MeanShift iteration, if $y_1 - y_0 < E$, the iteration is stopped, and the center position of the object area is moved from y_0 to the new position y_1.

$$y_1 - y_0 = \frac{\sum_{i=1}^{n_h} x_i w_i g\left[\left\|\frac{y-x_i}{h}\right\|^2\right]}{\sum_{i=1}^{n_h} w_i g\left[\left\|\frac{y-x_i}{h}\right\|^2\right]} - y_0 \qquad (11.11)$$

where the object area $g(x) = -kc(x)$ can be moved from the initial position to the real object position step by step.

According to the similarity function $\hat{\rho}(y_0)$, Taylor series expansion is required to start in the neighborhood, which limits the distance between starting point y_0 and y_1 cannot be too large, if the movement is too fast, the MeanShift algorithm tracking effect is not good.

The steps of MeanShift tracking algorithm are as follows:

Step 1: In the initial frame, the object area is first selected by the user and the object model is constructed, and the center position of the object is initialized;
Step 2: Selecting a candidate object area in the current frame, constructing a candidate object model with the object center of the previous frame as the center of the candidate object area;
Step 3: Estimate the similarity function, and calculate the weight coefficient, initialize the number of iterations, and then calculate the new candidate area center;
Step 4: And then re-estimate the similarity function by constructing a new candidate object model with a new candidate regional center;
Step 5: The similarity function is compared and then estimated again;
Step 6: Set the iteration threshold and the maximum number of iterations, and if the condition is satisfied, the iteration is terminated. Otherwise, return to Step 2 to continue iterating.

The code for MATLAB is shown in PROGRAMME 11.3.

PROGRAMME 11.3: Object Tracking Based on MeanShift Method

```
function [] = select()
close all;
clear all;
% Track the object based on a goal-wide image
I=imread('pedestrian.bmp');
figure(1);
imshow(I);
[temp,rect]=imcrop(I);
```

```
[a,b,c]=size(temp); %a:row,b:col
% Calculate the weight matrix of the object image
y(1)=a/2; y(2)=b/2;
tic_x=rect(1)+rect(3)/2;
tic_y=rect(2)+rect(4)/2;
m_wei=zeros(a,b);% Weight matrix
h=y(1)^2+y(2)^2 ;% bandwidth
for i=1:a
    for j=1:b
        dist=(i-y(1))^2+(j-y(2))^2;
        %epanechnikov profile
        m_wei(i,j)=1-dist/h;
    end
end
C=1/sum(sum(m_wei));% Normalization coefficient
% Calculate the object weight histogram
%object model
%hist1=C*wei_hist(temp,m_wei,a,b);
hist1=zeros(1,4096);
for i=1:a
    for j=1:b
        %The rgb color space is quantified as16*16*16 bins
        %fix is rounding function that approximation 0
        q_r=fix(double(temp(i,j,1))/16);
        q_g=fix(double(temp(i,j,2))/16);
        q_b=fix(double(temp(i,j,3))/16);
        % Set the proportion of red, green, and blue components for each pixel
        q_temp=q_r*256+q_g*16+q_b;
        % Calculates the weight of each pixel in the histogram statistics
        hist1(q_temp+1)= hist1(q_temp+1)+m_wei(i,j);
    end
end
hist1=hist1*C;
rect(3)=ceil(rect(3));
rect(4)=ceil(rect(4));
% Read the sequence image
myfile=dir('*.bmp');
lengthfile=length(myfile);
for l=1:lengthfile
    Im=imread(myfile(l).name);
    num=0;
    Y=[2,2];
    %mean shift iteration
    while((Y(1)^2+Y(2)^2>0.5)&num<20) % Iterative condition
```

```
num=num+1;
temp1=imcrop(Im,rect);
% Calculate the candidate area histogram
%object candidates pu
%hist2=C*wei_hist(temp1,m_wei,a,b);
hist2=zeros(1,4096);
for i=1:a
   for j=1:b
      q_r=fix(double(temp1(i,j,1))/16);
      q_g=fix(double(temp1(i,j,2))/16);
      q_b=fix(double(temp1(i,j,3))/16);
      q_temp1(i,j)=q_r*256+q_g*16+q_b;
      hist2(q_temp1(i,j)+1)=hist2(q_temp1(i,j)+1)+m_wei(i,j);
   end
end
hist2=hist2*C;
figure(2);
subplot(1,2,1);
plot(hist2);
hold on;
w=zeros(1,4096);
for i=1:4096
   if(hist2(i)~=0) % not equal
      w(i)=sqrt(hist1(i)/hist2(i));
   else
      w(i)=0;
   end
end
% Variable initialization
sum_w=0;
xw=[0,0];
for i=1:a;
   for j=1:b
      sum_w=sum_w+w(uint32(q_temp1(i,j))+1);
      xw=xw+w(uint32(q_temp1(i,j))+1)*[i-y(1)-0.5,j-y(2)-0.5];
   end
end
Y=xw/sum_w;
% Center point location update
rect(1)=rect(1)+Y(2);
rect(2)=rect(2)+Y(1);
end
```

```
% Tracking trajectory matrix
tic_x=[tic_x;rect(1)+rect(3)/2];
tic_y=[tic_y;rect(2)+rect(4)/2];
v1=rect(1);
v2=rect(2);
v3=rect(3);
v4=rect(4);
% Show trace results
subplot(1,2,2);
imshow(uint8(Im));
title(' Object tracking result and its trajectory ');
hold on;
plot([v1,v1+v3],[v2,v2],[v1,v1],[v2,v2+v4],
        [v1,v1+v3],[v2+v4,v2+v4],[v1+v3,v1+v3],
        [v2,v2+v4],'LineWidth',2,'Color','r');
plot(tic_x,tic_y,'LineWidth',2,'Color','b');
end
```

The MeanShift tracking algorithm is implemented and the experimental results are analyzed. For the sake of convenience, the two sets of video sequences are named $S1$ and $S2$, respectively. In the experiment, the RGB color model is used and 16 parts of each component are quantized, and the color histogram is used as the object model. Figure 11.5 shows the partial tracking results.

In the video sequence S_1, the object is the white human body, the separability of the object and the background is relatively high. From the video sequence tracking results, it can be seen that although there is a partial occlusion in the tracking process, the positioning of the object is somewhat biased in the 95th frame, but in the tracking of all the 120 frames, the object can always be tracked successfully and never lose the object. It can be seen from the experimental results, MeanShift algorithm for the object and background differencing in the scene, showing a good tracking performance (Fig. 11.6).

In the sequence S2, the green player and the court color is very similar, that is, the object and the background is difficult to distinguish. From the tracking results can be seen, the tracking effect is unsatisfied, tracking box gradually deviate from the center of the object, resulting in a larger tracking bias, and ultimately lost the object, resulting in tracking failure.

Comparing with the results of the tracking in S_1 and S_2, we can see that the separability of the object and the background is critical, it determines whether the MeanShift algorithm can always track the object effectively throughout the process. The classic MeanShift tracking algorithm is used to track the object of color information. It can achieve good tracking effect on the obvious difference between the object and the background. However, if the object is similar to the background, the tracking effect is not ideal and will cause the tracking failure. Therefore, it is very important to improve the performance of the tracking algorithm by choosing a distinguishing

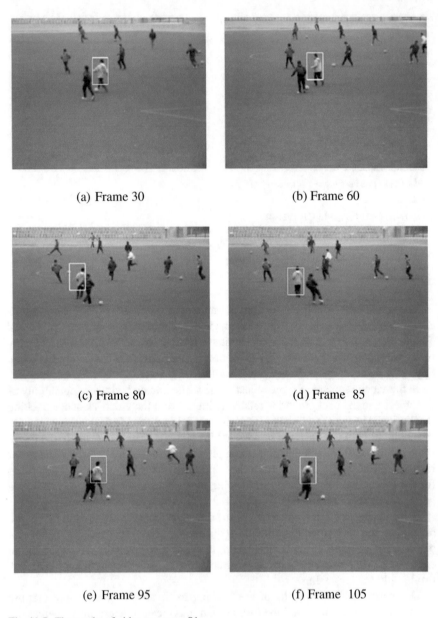

(a) Frame 30   (b) Frame 60

(c) Frame 80   (d) Frame 85

(e) Frame 95   (f) Frame 105

Fig. 11.5 The results of video sequence S1

feature to construct the object model, so that the characteristics of the object and the background are obviously different.

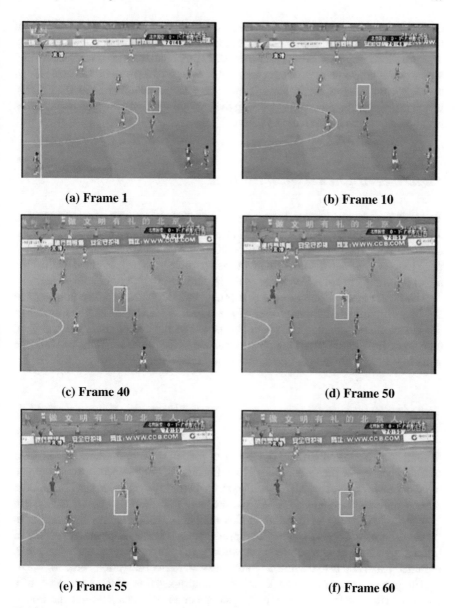

(a) Frame 1

(b) Frame 10

(c) Frame 40

(d) Frame 50

(e) Frame 55

(f) Frame 60

Fig. 11.6 The results of the video sequence S2 tracks

11.4 Object Tracking Based on Particle Filter

Particle filter theory describes an effective object tracking framework, which uses particle weighting to represent the posterior probability of the object state. Starting from $k = 0$, the system is initialized to determine the prior probability representation of the object state, giving the initial weights to each particle. At the next moment, the state prediction transfer is carried out first, and each particle follows the state transition equation to carry out its own state propagation. Then, the observation amount of the new state is obtained, and the weight of each particle is calculated by the system measurement phase (similarity with the actual state of the object), which is actually the process of updating the particle state, and then the particles are resampled to continue the state transition.

11.4.1   Prior Knowledge of the Goal

The goal has a certain a priori characteristics, which is generally considered to be a distinguishing feature of other goals, in other words, it is a specified descriptive character with a certain semantics. Different feature descriptions determine different prior probabilities and the initial state of each particle is also determined by this. We select the weighted color histogram as the feature description of the object here.

The object area is found in the first frame and an object template is generated to obtain the initial state parameter of the object, which represents the center of the object area, indicating the width and height of the object area. The weighted color histogram of the object region is calculated as the initial template, and then the particle set is distributed near the initial state of the object.

11.4.2   System State Transition

System state transition is the propagation of particles, which refers to the process of time updating of the state of the object. Because of the independence movement trend of moving object is generally obvious, the particle propagation can be a random motion process. It should be noted that the state transition process of the system is independent of the observation at this moment. That is to say, this step is to "assume" how the object state will be propagated. It is the propagation process of priori probability. It is also unknown whether the propagation of each particle is reasonable and needs to be verified in the next "system observation" process.

The propagation of particles is actually the propagation of the parameters of a particle. In the first frame, a set of particles is generated within a certain range around the object. When reading subsequent frames, these particles will pass through the state.

11.4.3 System Observation

After the "hypothesis" of the propagation of the object state, it is necessary to validate it with the acquired observations (time), which is the systematic observation. The so-called observation is the resulting k-th frame image intuitive, accurately it is the color characteristic extracted after the processing of the k-th frame image. The verification of system state transition results using observations is, in fact, a process of similarity measurement. Bhattacharrryya similarity coefficients are used to calculate the distance between the color histogram of each particle and the color histogram of the object model.

q_u and $p_u(y)$ are used to represent the color histogram of the known object template and the object image to be selected, respectively, with Bhattacharrryya similarity coefficients:

$$\rho(y) = \rho[p(y), q] = \sum_{u=1}^{m} \sqrt{p_u(y)q_u} \tag{11.12}$$

It represents the similarity between q_u and $p_u(y)$. Here, the degree of similarity between the object template and the candidate region is measured using the following distance:

$$d = \sqrt{1 - \rho(y)} \tag{11.13}$$

The smaller the distance d in the above equation, the closer the object to be selected is to the actual situation. Since each particle represents a possibility of the object state, the purpose of the system observation is to give a larger weight to the particles that are close to the actual situation, and to give a smaller weight to the particles that differ greatly from the actual situation.

After measuring the similarity distance, the weights of the particles are distributed by the Gaussian function, then the following equation is obtained.

$$p(y_k|x_k^i) = \frac{1}{\sqrt{2\pi}\sigma} \exp\{1\frac{1}{2\sigma^2}d^i\} \tag{11.14}$$

where σ is a constant and d is a Bhattacharyya distance, the weight of the particle can be calculated as follows:

$$w_k^i = w_{k-1}^i p(y_k|x_k^i) \tag{11.15}$$

11.4.4 Posterior Probability Calculation

The posterior probabilities can be calculated in two general criteria. One is the maximum posteriori criterion. That is, the state of a particle of maximum weight is the final form of the posterior probability. This method is very intuitive, generally speaking, the most similar one has the highest probability. Another is a weighted criterion, meaning that each particle depends on its weight size to determine its proportion in the posterior probability. This method can better reflect the superiority of particle filter tracking. The final result, which is the most similar share of the largest proportion, is determined by many particles, so the posterior probability is smoother. From this point of view, the weighted criterion is superior to the maximum criterion, so the weighting criterion is adopted in the process of implementing particle filter tracking.

After the weight of each particle is updated, the state estimate at time k is represented by the weighted sum of the respective particles. As is shown below:

$$cx_k = \sum_{i=1}^{n} cx_k^i w_k^i \tag{11.16}$$

$$cy_k = \sum_{i=1}^{n} cy_k^i w_k^i \tag{11.17}$$

where (cx_k, cy_k) represents the center position of the object estimate in k-th frame.

11.4.5 Particle Resampling

In the process of particle propagation, some of the particles deviate from the actual state of the object to obtain the smaller and smaller weights, so that only a few particles have a large weight, resulting in a large number of calculations wasted on these small weight particles. Although these small weight particles also represent the possibility of the object state, the possibility is too small. So, it should be ignored and the focus will be on the weight of some of the larger particles.

Resampling can alleviate this problem to some extent. In the resampled particles, the larger particles produce more "offspring" particles, and the weight of the particles corresponding to the "descendants" particles are less, and "offspring" particles are reset to the same weight. This process can be described as a black box, just need to define a threshold. When the weight of some particles below the threshold, the process will be executed. And the weight of the particle "offspring" will be reset to ensure that the number of particles remains constant during the tracking process.

11.4.6 Implementation Steps

Step 1: Initialization, in time of $k = 0$, sampling N evenly distributed particle set S_{k-1} and establish the object model:

$$\hat{q}_u = f \sum_{i=1}^{l} k\left(\left\|\frac{x_0 - x_i}{a}\right\|^2\right)\delta[h(x_i) - u] \qquad (11.18)$$

Step 2: Observe the color distribution

(a) Calculate the color distribution of each particle in the particle set S_k:

$$\hat{P}_u(y) = f \sum_{i=1}^{l} k\left(\left\|\frac{y - x_i}{a}\right\|^2\right)\delta[h(x_i) - u] \qquad (11.19)$$

(b) Calculate the similarity of each particle of particle set S_k to the object template, which represents Bhattacharyya coefficient:

$$\rho(s) = \rho[p(s), q] = \sum_{u=1}^{m} \sqrt{\hat{p}_u(s)\hat{q}_u} \qquad (11.20)$$

(c) Calculate the probability density of the observed values:

$$p\left(z_k | x_k^i\right) = \frac{1}{\sqrt{2\pi}\sigma}e^{-\frac{(1-\rho[p(s),q])}{2\sigma^2}} \qquad (11.21)$$

(d) Calculate the weight of each particle:

$$w_k^i = w_{k-1}^i p\left(z_k | x_k^i\right) \qquad (11.22)$$

(e) Normalized weights:

$$w_k^i = \frac{w_k^i}{\sum_{i=1}^{N} w_k^i} \qquad (11.23)$$

Step 3: Resampling according to the weight w_k^i of each particle. The resampling method is as follows:

(a) Produce a uniformly distributed random number $\mu_1 \in [0, 1]$ to find the smallest m that satisfies the following formula:

$$\sum_{j=0}^{m-1} w_{k-1}^i < \mu_l \le \sum_{j=0}^{m} w_k^i \qquad (11.24)$$

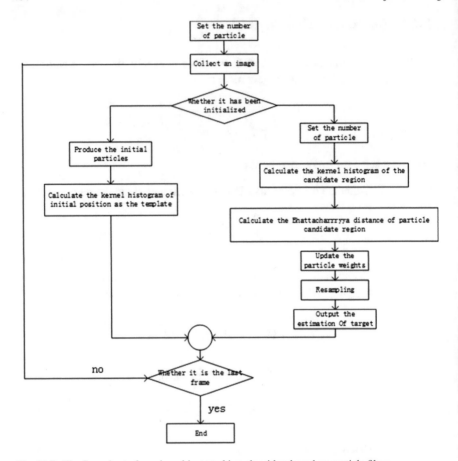

Fig. 11.7 The flow chart of moving object tracking algorithm based on particle filter

    (b) Copy the sample x_k^m.

Step 4: State Estimation

    Calculate the weighted average state $S_k(E)$

$$S_k(E) = \sum_{n=1}^{N} w_k^i S_k^i \tag{11.25}$$

Figure 11.7 gives the algorithm flow chart.

The code based on importance sampling is shown in PROGRAMME 11.4.

PROGRAMME 11.4: Main Program of Moving Object Tracking Based on Importance Sampling

```
% The video is uncompressed in avi format
% Enter the full path of the video file
clear all
close all
M = VideoReader('pedestrian.avi');
% Get the number of frames in the video sequence
numberofframes=M.numberOfFrames; % Number of video frames
% Define object area
samplesnum=30; % Number of sampling particles
% The storage matrix of the location coordinates of the sampled particle
  particles=zeros(samplesnum,2);
center(1,1)=108; % The center y of the object window (representing rows)
center(1,2)=280; % X (representing columns)
w_halfsize(1) =15; % Half of the window's height
w_halfsize(2) =15; % Half of the window's width
%Draw the first frame of the tracking box
cmin=center(1,2)-w_halfsize(2);
cmax=center(1,2)+w_halfsize(2);
rmin=center(1,1)-w_halfsize(1);
rmax = center(1,1)+w_halfsize(1);
trackim=read(M,1);%Frames{1};
for r= rmin:rmax
   trackim(r, cmin,:) = 0;
   trackim(r, cmax,:) = 0;
end
for c= cmin:cmax
   trackim(rmin, c,:) = 0;
   trackim(rmax, c,:) = 0;
end
% Get the histogram of the object window
q_u=rgbPDF(double(read(M,1)),center,w_halfsize);
%Sampling particle distribution
halfwidth=10; %The half width of the area allocated for the sampling particles
%Sampling the abscissa of the particle
particles(:,1)=round(center(1,1)-halfwidth+
                         2*halfwidth*rand(samplesnum,1));
% Sampling the ordinate of the particle
particles(:,2)=round(center(1,2)-halfwidth+
                         2*halfwidth*rand(samplesnum,1));
figure(1);hold on;
% Start tracking
for i = 2:numberofframes
% Sampling the transfer of the particle
```

```
% Sampling the abscissa transfer of the particle
particles(:,1)=round(particles(:,1)+
                        2*rand(samplesnum,1)-2*rand(samplesnum,1));
%Sampling the ordinate transfer of the particle
particles(:,2)=round(particles(:,2)-
                        6*rand(samplesnum,1)+rand(samplesnum,1));
%Calculate the Bhattacharyya distance and importance sampling (weighted particle)
    for k=1:samplesnum
        center_p(1,1)= particles(k,1);
        center_p(1,2)= particles(k,2);
p_halfsize(1) =10; % Half of the window's height
p_halfsize(2) =10; % Half of the window's width
% The histogram statistics of the pixel points of the surrounding area of the particle
    hist=rgbPDF(double(read(M,i)),center_p,p_halfsize);
        sumwi=0;
        sum=0;
        for t=1:4096
            suma= hist(t) * q_u(t);
            sum=sum+sqrt(suma) ;% Bhattacharyya coefficient
    end
sum=1-sum;% Similarity distance
wi(k)=exp(-20*sum); %Sampling the weight of the particle
sumwi=sumwi+wi(k); %Sampling the sum of weights of the particle
end
% Resampling
wi(1)= wi(1)/sumwi;
    st=2;
    while(st<samplesnum+1)
        wi(st)=wi(st)/sumwi;
        wi(st)=wi(st-1)+wi(st);
        st=st+1;
    end
    resample=rand(1,samplesnum);% A random number is generated between 0-1
for ii=1:1:samplesnum
    jj=1;
    while( resample(1,ii)>wi(jj) )
        jj=jj+1;
    end
if(jj<=samplesnum)% Resampled particles
wi(ii)= wi(jj);
```

```
        particles(ii,1)= particles(jj,1);
        particles(ii,2)= particles(jj,2);
   end
   end

% The location of the new center point of the object window
locationx=0;
   locationy=0;
   for m=1:1:samplesnum
      locationx=particles(m,1)+locationx;
   end
locationx=round(locationx/samplesnum); % The abscissa of the center point
for n=1:1:samplesnum
      locationy=particles(n,2)+ locationy;
   end
   locationy=round( locationy/samplesnum); % The ordinate of the center point
%Draw the tracking box
rmin=locationx-w_halfsize(1);
   rmax=locationx+w_halfsize(1);
   cmin=locationy-w_halfsize(2);
   cmax=locationy+w_halfsize(2);
   trackim=read(M,i);
   for r= rmin:rmax
      trackim(r, cmin,:) = 0;
      trackim(r, cmax,:) = 0;
   end
   for c=cmin:cmax
      trackim(rmin, c,:) = 0;
      trackim(rmax, c,:) = 0;
   end
% display the tracking results dynamically
imshow(trackim);title([num2str(i),'/',num2str(numberofframes)]);drawnow;
end
```

Subroutine rgbPDF.m is shown in PROGRAMME 11.5:

PROGRAMME 11.5: Subroutine rgbPDF

```
function q_u=rgbPDF(image,center,w_halfsize) % Nuclear histogram
sum_q=0;
histo=zeros(16,16,16);
% Calculate the location of the object window
rmin=center(1)-w_halfsize(1);
rmax=center(1)+w_halfsize(1);
cmin=center(2)-w_halfsize(2);
cmax=center(2)+w_halfsize(2);
```

```
% The maximum distance from the object window to the center
wmax=(rmin-center(1)).^2+(cmin-center(2)).^2+0.001;
for i=rmin:rmax    % Calculate its eigenvector space
for j=cmin:cmax
d=(i-center(1)).^2+(j-center(2)).^2; %Kernel-based tracking(Weighting of the kernel function)
w=wmax-d; % The larger the distance, the smaller the weight
    R=floor(image(i,j,1)/16)+1;
      G=floor(image(i,j,2)/16)+1;
      B=floor(image(i,j,3)/16)+1;
      histo(R,G,B)=histo(R,G,B)+w;
  end
end
for i=1:16
    for j=1:16
        for k=1:16
            index=(i-1)*256+(j-1)*16+k;
            q_u(index)=histo(i,j,k);
            sum_q=sum_q+q_u(index);
        end
    end
end
q_u=q_u/sum_q; % Normalized
```

In order to verify the effectiveness of the proposed algorithm, the 240 * 360, 70-frame Sam video is employed in the laboratory pedestrian video library. It shows in Fig. 11.8 that the tracking results are more accurate in most cases. But in the second frame and the 55th frame, the position of the tracking object is slightly deviated as shown in Fig. 11.8a, f. At last, the algorithm can automatically retrieve the object area and recover the effective tracking in the subsequent video frame. The results of the experiment show that this method can track the pedestrian object effectively, also it can recover the object and recover the tracking when the tracking is incorrect or lost. But the algorithm has high complexity and costs more time.

The object tracking experiment using this method has a high accuracy and stability and the loss of tracking is less likely to occur. But the complexity of the algorithm is relatively high. Also, it did not consider the case of moving objects' occlusion, resulting in that it is only suitable for tracking pedestrians without occlusion.

11.5   Multiple Object Tracking

Multiple Object Tracking (MOT) [7] plays an important role in computer vision. MOT is to locate identify and yield the individual trajectories in an input video. The objects can be, for example, pedestrians on the street, vehicles in the road, sport players on the court, or of animals in group. Multiple "objects" could also be viewed

Fig. 11.8 Experimental effect drawings of the algorithm

as different parts of a single object. In this section, we mainly focus on the research of pedestrian tracking. The underlying reasons for this specification are threefold.

First, by comparing to other common objects in our environment, pedestrians are typical nonrigid objects, which are the ideal examples to study the MOT problem. Second, videos of pedestrians arise in a huge number of practical applications, which further result in great commercial potential. Third, according to all collected data by the author, at least 70% of current MOT research efforts are devoted to pedestrians.

There are three steps in MOT procedure: detection, prediction, and data association.

Detection: Selecting the appropriate approach to detect objects of interest depends on what you want to track and whether the camera is stationary.

Prediction: To track an object over time means that you must predict its location in the next frame. The simplest method of prediction is to assume that the object will move to the area near the previous location. In other words, the previous detection serves as the next prediction. This method is especially effective for high frame rates. However, it may fail the prediction for those objects which move at varying speeds, or when the frame rate is low relative to the speed of the object in motion.

Data association: It is the process of associating detections corresponding to the same physical object across frames. The temporal tracking of a particular object consists of multiple detections, and is called a track. A track representation can include the entire history of the previous locations of the object. Alternatively, it can consist only of the object's last known location and its current velocity.

This example shows how to perform automatic detection and motion-based tracking of moving objects in a video from a stationary camera.

Detection of moving objects and motion-based tracking are important components of many computer vision applications, including activity recognition, traffic monitoring, and automotive safety. The solution of motion-based object tracking can be divided into two parts:

Detecting moving objects in each frame.

Associating the detections corresponding to the same object over time.

The detection of moving objects uses a background subtraction algorithm based on Gaussian mixture models. Morphological operations are applied to the resulting foreground mask to eliminate noise. Finally, blob analysis detects groups of connected pixels, which are likely to correspond to moving objects.

The association of detections to the same object is solely based on motion. The motion of each track is estimated by a Kalman filter [8]. The filter is used to predict the track's location in each frame, and determine the likelihood of each detection being assigned to each track.

Track maintenance becomes an important aspect of this example. In any given frame, some detections may be assigned to tracks, while other detections and tracks may remain unassigned. The assigned tracks are updated using the corresponding detections. The unassigned tracks are marked invisible. An unassigned detection begins a new track.

Each track keeps count of the number of consecutive frames, where it remained unassigned. If the count exceeds a specified threshold, it can be assumed that the object has left the field of view and the track will be deleted.

```
% Create System objects used for reading video, detecting moving objects,
% and displaying the results.
obj = setupSystemObjects();
tracks = initializeTracks(); % Create an empty array of tracks.
nextId = 1; % ID of the next track
% Detect moving objects, and track them across video frames.
while ~isDone(obj.reader)
  frame = readFrame();
  [centroids, bboxes, mask] = detectObjects(frame);
  predictNewLocationsOfTracks();
  [assignments, unassignedTracks, unassignedDetections] = ...
  detectionToTrackAssignment();
  updateAssignedTracks();
  updateUnassignedTracks();
  deleteLostTracks();
  createNewTracks();
  displayTrackingResults();
end


% Create System Objects
% Create System objects used for reading the video frames, detecting foreground
% objects, and displaying results.
function obj = setupSystemObjects()
        % Initialize Video I/O
        % Create objects for reading a video from a file, drawing the tracked
        % objects in each frame, and playing the video.
        % Create a video file reader.
        obj.reader = vision.VideoFileReader('atrium.mp4');
        % Create two video players, one to display the video,
        % and one to display the foreground mask.
        obj.maskPlayer = vision.VideoPlayer('Position', [740, 400, 700, 400]);
        obj.videoPlayer = vision.VideoPlayer('Position', [20, 400, 700, 400]);
        % Create System objects for foreground detection and blob analysis
        % The foreground detector is used to segment moving objects from
        % the background. It outputs a binary mask, where the pixel value
        % of 1 corresponds to the foreground and the value of 0 corresponds
        % to the background.
        obj.detector = vision.ForegroundDetector('NumGaussians', 3, ...
                'NumTrainingFrames', 40, 'MinimumBackgroundRatio', 0.7);
        % Connected groups of foreground pixels are likely to correspond to moving
        % objects.   The blob analysis System object is used to find such groups
        % (called 'blobs' or 'connected components'), and compute their
        % characteristics, such as area, centroid, and the bounding box.
        obj.blobAnalyser = vision.BlobAnalysis('BoundingBoxOutputPort', true, ...
                'AreaOutputPort', true, 'CentroidOutputPort', true, ...
```

```
                    'MinimumBlobArea', 400);
end


% Initialize Tracks
% The function creates an array of tracks, where each track is a structure representing
% a moving object in the video.

function tracks = initializeTracks()
          % create an empty array of tracks
          tracks = struct(...
               'id', { }, ...
               'bbox', { }, ...
               'kalmanFilter', { }, ...
               'age', { }, ...
               'totalVisibleCount', { }, ...
               'consecutiveInvisibleCount', { });
end


% Read a Video Frame
% Read the next video frame from the video file.
function frame = readFrame()
          frame = obj.reader.step();
end


% Detect Objects
% The function returns the centroids and the bounding boxes of the detected objects.
function [centroids, bboxes, mask] = detectObjects(frame)
          % Detect foreground.
          mask = obj.detector.step(frame);
          % Apply morphological operations to remove noise and fill in holes.
          mask = imopen(mask, strel('rectangle', [3,3]));
          mask = imclose(mask, strel('rectangle', [15, 15]));
          mask = imfill(mask, 'holes');
          % Perform blob analysis to find connected components.
          [~, centroids, bboxes] = obj.blobAnalyser.step(mask);
end


% Predict New Locations of Existing Tracks
% Use the Kalman filter to predict the centroid of each track in the current frame,
% and update its bounding box accordingly.
function predictNewLocationsOfTracks()
          for i = 1:length(tracks)
               bbox = tracks(i).bbox;
               % Predict the current location of the track.
```

```
            predictedCentroid = predict(tracks(i).kalmanFilter);
            % Shift the bounding box so that its center is at
            % the predicted location.
            predictedCentroid = int32(predictedCentroid) - bbox(3:4) / 2;
            tracks(i).bbox = [predictedCentroid, bbox(3:4)];
        end
end


% Assign Detections to Tracks
% Assigning object detections in the current frame to existing tracks is done by
% minimizing cost.
function [assignments, unassignedTracks, unassignedDetections] = ...
            detectionToTrackAssignment()
        nTracks = length(tracks);
        nDetections = size(centroids, 1);
        % Compute the cost of assigning each detection to each track.
        cost = zeros(nTracks, nDetections);
        for i = 1:nTracks
            cost(i, :) = distance(tracks(i).kalmanFilter, centroids);
        end
        % Solve the assignment problem.
        costOfNonAssignment = 20;
        [assignments, unassignedTracks, unassignedDetections] = ...
            assignDetectionsToTracks(cost, costOfNonAssignment);
    end


% Update Assigned Tracks
% The updateAssignedTracks function updates each assigned track with the
% corresponding detection.
function updateAssignedTracks()
        numAssignedTracks = size(assignments, 1);
        for i = 1:numAssignedTracks
            trackIdx = assignments(i, 1);
            detectionIdx = assignments(i, 2);
            centroid = centroids(detectionIdx, :);
            bbox = bboxes(detectionIdx, :);
            % Correct the estimate of the object's location
            % using the new detection.
            correct(tracks(trackIdx).kalmanFilter, centroid);
            % Replace predicted bounding box with detected
            % bounding box.
            tracks(trackIdx).bbox = bbox;
```

```
% Update track's age.
tracks(trackIdx).age = tracks(trackIdx).age + 1;
% Update visibility.
tracks(trackIdx).totalVisibleCount = ...
        tracks(trackIdx).totalVisibleCount + 1;
tracks(trackIdx).consecutiveInvisibleCount = 0;
            end
end


% Update Unassigned Tracks
% Mark each unassigned track as invisible, and increase its age by 1.
function updateUnassignedTracks()
        for i = 1:length(unassignedTracks)
            ind = unassignedTracks(i);
            tracks(ind).age = tracks(ind).age + 1;
            tracks(ind).consecutiveInvisibleCount = ...
                    tracks(ind).consecutiveInvisibleCount + 1;
        end
end


% Delete Lost Tracks
% The function deletes tracks that have been invisible for too many consecutive
% frames.
function deleteLostTracks()
        if isempty(tracks)
                return;
        end
        invisibleForTooLong = 20;
        ageThreshold = 8;
        % Compute the fraction of the track's age for which it was visible.
        ages = [tracks(:).age];
        totalVisibleCounts = [tracks(:).totalVisibleCount];
        visibility = totalVisibleCounts ./ ages;
        % Find the indices of 'lost' tracks.
        lostInds = (ages < ageThreshold & visibility < 0.6) | ...
                [tracks(:).consecutiveInvisibleCount] >= invisibleForTooLong;
        % Delete lost tracks.
        tracks = tracks(~lostInds);
end


% Create New Tracks
% Create new tracks from unassigned detections.
function createNewTracks()
        centroids = centroids(unassignedDetections, :);
        bboxes = bboxes(unassignedDetections, :);
```

```
        for i = 1:size(centroids, 1)
            centroid = centroids(i,:);
            bbox = bboxes(i, :);
            % Create a Kalman filter object.
            kalmanFilter = configureKalmanFilter('ConstantVelocity', ...
                centroid, [200, 50], [100, 25], 100);
            % Create a new track.
            newTrack = struct(...
                'id', nextId, ...
                'bbox', bbox, ...
                'kalmanFilter', kalmanFilter, ...
                'age', 1, ...
                'totalVisibleCount', 1, ...
                'consecutiveInvisibleCount', 0);
            % Add it to the array of tracks.
            tracks(end + 1) = newTrack;
            % Increment the next id.
            nextId = nextId + 1;
        end
end


% Display Tracking Results
% The function draws a bounding box and label ID for each track on the video frame
% and the foreground mask.
function displayTrackingResults()
        % Convert the frame and the mask to uint8 RGB.
        frame = im2uint8(frame);
        mask = uint8(repmat(mask, [1, 1, 3])) .* 255;
        minVisibleCount = 8;
        if ~isempty(tracks)
            % Noisy detections tend to result in short-lived tracks.
            % Only display tracks that have been visible for more than
            % a minimum number of frames.
            reliableTrackInds = ...
                [tracks(:).totalVisibleCount] > minVisibleCount;
            reliableTracks = tracks(reliableTrackInds);
            % Display the objects. If an object has not been detected
            % in this frame, display its predicted bounding box.
            if ~isempty(reliableTracks)
                % Get bounding boxes.
                bboxes = cat(1, reliableTracks.bbox);
                % Get ids.
```

```
ids = int32([reliableTracks(:).id]);
% Create labels for objects indicating the ones for
% which we display the predicted rather than the actual
% location.
labels = cellstr(int2str(ids'));
predictedTrackInds = ...
        [reliableTracks(:).consecutiveInvisibleCount] > 0;
isPredicted = cell(size(labels));
isPredicted(predictedTrackInds) = {' predicted'};
labels = strcat(labels, isPredicted);
% Draw the objects on the frame.
frame = insertObjectAnnotation(frame, 'rectangle', ...
        bboxes, labels);
% Draw the objects on the mask.
mask = insertObjectAnnotation(mask, 'rectangle', ...
        bboxes, labels);
        end
    end
    % Display the mask and the frame.
    obj.maskPlayer.step(mask);
    obj.videoPlayer.step(frame);
end
```

This example [9] creates a motion-based system for detecting and tracking multiple moving objects. Try using a different video to see if you are able to detect and track objects. Or modifying the parameters for the detection, assignment, and deletion steps (Fig. 11.9).

The tracking in this example is solely based on motion with the assumption that all objects move in a straight line with constant speed. When the motion of an object significantly deviates from this model, the example may produce tracking errors. Notice the mistake in tracking the person who is occluded by the tree.

The likelihood of tracking errors can be reduced by using a more complex motion model, such as constant acceleration, or by using multiple Kalman filters for each object. Also, you can incorporate other cues for associating detections over time, such as size, shape, or color.

(a) Motion-based pedestrians tracking.

(b) Binary result of pedestrians tracking.

Fig. 11.9 The results of motion-based multiple object tracking, **a** result of pedestrians tracking; **b** binary result of pedestrians tracking

References

1. Guyon I, Elisseeff A, Jankowski N, Grabczewski K, Dreyfus G, Guyon I et al (2006) Feature extraction. Stud Fuzziness Soft Comput 31(7):1737–1744
2. Sonka M, Hlavac V, Ceng RBDM (2008) Image processing, analysis and machine vision. J Electron Imaging xix(82):685–686
3. Kenny T, Chawla R, Pacsai E et al (2004) Object tracking: US, US 20040036595 A1

4. Wu Y, Lim J, Yang MH (2013) Online object tracking: a benchmark. In: IEEE conference on computer vision and pattern recognition. IEEE Computer Society, pp 2411–2418
5. Babenko B, Yang M-H, Belongie S (2011) Robust object tracking with online multiple instance learning. IEEE Trans Pattern Anal Mach Intell 33(8):1619
6. Lu Y, Wu T, Zhu SC (2014) Online object tracking, learning, and parsing with and-or graphs. In: IEEE conference on computer vision and pattern recognition. IEEE Computer Society, pp 3462–3469
7. Luo W, Xing J, Zhang X et al (2015) Multiple object tracking: a literature review
8. Berclaz J, Fleuret F, Turetken E et al (2011) Multiple object tracking using K-shortest paths optimization. IEEE Trans Pattern Anal Mach Intell 33(9):1806–1819
9. http://ww2.mathworks.cn/help/vision/ug/multiple-object-tracking.html

Chapter 12
Dynamic Scene Classification Based on Topic Models

This chapter briefly introduces the background reading of scene classification and two topic models: LDA model and Topic Model using Belief Propagation (TMBP). In Sect. 12.2, there are two TMBP based on factor graph and fusion porior knowledge. Moreover, we present the dynamic scene classification based on TMBP and the Behavior Recognition based on LDA topic model.

12.1 Overview

Due to the massive deployment of video surveillance system, the contents of the dynamic scene get more complex which brings a challenge for the manual management of the video scene. In other words, it is impossible to classify and label millions of the video manually due to the high-cost of the required labor force. Therefore, it is necessary to classify the scene automatically depending on the video contents by using computer science.

Scene classification refers to the specific meaning of the image data which is set for automatic labeling. Fast and accurate classification of dynamic scenes has become a keen topic in the unsupervised model. The use of dynamic scene classification which can assist in manual labeling and the management of digital image data provide support for a deeper level of digital video analysis. This chapter focuses on the dynamic scene and takes the visual lexicon—semantic theme and modeling—dynamic scene semantic classification as the main line. It includes the construction of dynamic scene visual dictionary, the subject model modeling of message passing based on prior knowledge, and the realization of dynamic scene semantic classification.There are two main types of dynamic scene classification: tracking-based and feature extraction based classification.

The basic idea of the tracking method is to track the moving objects in the dynamic scene and get its trajectory. Dynamic scene classification is realized by analyzing the trajectory. The method first performs target detection and tracking on the video, and

© Springer International Publishing AG, part of Springer Nature 2019
S. Gong et al., *Advanced Image and Video Processing Using MATLAB*,
Modeling and Optimization in Science and Technologies 12,
https://doi.org/10.1007/978-3-319-77223-3_12

the detection result triggers the tracking by forcibly detects the tracking trajectory; with the passage of time, the tracking trajectory effectively update the tracking route to improve the detection effect; finally, implement the dynamic scene classification through the analysis of the trajectory. The dynamic scene classification algorithm based on feature extraction can be divided into two levels according to the feature extraction strategy: scene classification using low-level visual features and scene classification using middle-level semantics. Scene classification using low-level visual features: first, extract the underlying features of the dynamic scene, such as color, texture and shape, and then combine these features with supervised training methods, such as the quantized feature as the input of the probability statistical model, complete the classification of dynamic scenes. Commonly used probably statistical models are LDA, HDP and so on. The dynamic scene classification algorithm based on feature extraction is not concerned with the single moving target in the scene comparing with the dynamic scene classification method on tracking, but it focuses on the movement trend in the whole scene. The dynamic scene classification algorithm based on feature extraction shows a better classification effect for complex scenes where contains more motion targets or have occlusion.

12.2   Introduction to the Topic Models

Topic model is a statistical model for analyzing large-scale data, its ideas originating potential semantic analysis presented by Deerwester et al. in 1990, they constructed a new Latent Semantic space using the Singular Value Decomposition (SVD) method so as to achieve the effect of reducing dimension. In 1999, Hofmann et al. proposed a Probabilistic Latent Semantic Analysis (PLSA) model based on LSA. The model introduces the representation of probability and simulates the generation of words in documents by a probabilistic model. In 2003, D. M. Blei et al. Extended the PLSA based on a random implicit variable satisfying the Dirichlet distribution to represent the subject probabilistic distribution of the document, resulting in a more complete probability generation model LDA (Latent Dirichelet Allocation, LDA). LDA model parameters are all random variables, only two external control parameters achieve a fully complete probability. It is an unsupervised learning model. At present, the mainstream algorithms for solving LDA model are Variational Bayes (VB), Gibbs Sampling (GS) and Belief Propagation (BP).

12.2.1   LDA Model

The LDA model generates the probability model of the sample, and then classifies by the probability model of the sample. It is based on the Bag-of-Words (BOW) model assumption that the text is regarded as a set of unordered words, ignoring the syntax

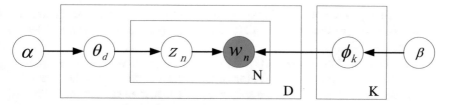

Fig. 12.1 Image representation of LDA model

Table 12.1 Variables defined in the LDA model

Variables	Definition
K	Number of topics
D	Number of documents
$N_{d=1\ldots D}$	The number of vocabularies in the document D
$w_{d=1\ldots D, w=1\ldots N_d}$	Word w in the document D
α	Superparine of Dirichlet distribution
β	Superparine of Dirichlet distribution
$z_{w=1\ldots N_d, d=1\ldots D}$	The subject tag of the word w in the document D
$\phi_{k=1\ldots K}$	Probability distribution of vocabulary in topic K
$\theta_{d=1\ldots D}$	The probability distribution of the subject in the document D

and the order of the words. The graph representation of the LDA model is shown in Fig. 12.1.

The LDA model is a three-level Bayesian model. The black nodes in the figure represent observable variables, other nodes are potential variables, K is the number of topics, N is the number of vocabularies in the current document, and D is the number of documents. At the word layer, there are two variables, w_n and z_n, representing the word w_n of the nth document and the subject tag z_n of the word. At the document level, T has Φ_k and two variables, T is a matrix of $k \times v$ (v is the dimension of the word), and each row represents the vocabulary distribution of a topic. θ_d is a matrix of $D \times K$, each row representing the subject probability distribution of a document; at the corpus level, α and β are the two Dirichlet distributions of the superparinehyperparameter. As can be seen from Fig. 12.1, there is only one observable variable w_n in the model, others are potential variables. The variables defined in the model are given in Table 12.1.

In this generation model, the document is treated as a potential mixture of subjects, each of which is determined by the characteristics of the lexical distribution. The basic idea of the LDA model is described as follows:

1. Determine the subjective distribution of the document;
2. Select the theme according to the theme distribution;
3. According to the selected theme to determine the theme of the word distribution;
4. Select the word according to the theme distribution and word distribution.

This is the process of a word generation. Therefore, that is, the completion of a word generation process, repeat step 2 to step 4 to generate the document with length of the repeated times.

θ_d is a $1 \times K$ random line vector $(\theta_1, \theta_2, \ldots, \theta_K), \theta_K \geq 0$ and $\theta_1 + \theta_2 + \cdots + \theta_K = 1$, the physical meaning is the distribution of the subject of the current document, i.e. θ_k indicates the probability that the subject z_k will appear on the current document. Thus θ satisfies the Dirichlet distribution, where α is the hyperparameter of the distribution, i.e. $\theta \sim Dir(\alpha)$. This corresponds to step 1.

$z_{w,d}$ represents the subject of the current word, in the subject set T to take K discrete values. $p(z|\theta)$ is given θ when the conditional distribution of z, with the function of the expression is relatively simple, the direct use of θ as a probability value: $p(z = k|\theta) = \theta_k$, i.e. the probability that z is the $1, 2, \ldots, K$-topic is $\theta_1, \theta_2, \ldots, \theta_K$. This corresponds to step 2.

If V is used to represent the number of words in the vocabulary, φ_k is the row vector $(\varphi_1, \varphi_2, \ldots, \varphi_V)$ of $1 \times V$, $\varphi_v \geq 0$ and $\varphi_1 + \varphi_2 + \cdots + \varphi_v = 1$, the physical meaning is the word distribution of the current topic, that is, φ_V represents the probability that the word W appears on the current subject. So it satisfies the Dirichlet distribution, β is the super parameter of the distribution, i.e. $\varphi \sim Dir(\beta)$. This corresponds to step 3.

W represents the word, it is a discrete random variable, in the vocabulary V to take $|V|$ discrete values. $p(w|z, \varphi)$ represents the occurrence of W after the subject z is determined and given the probability distribution of the word appearing under the subject. This corresponds to step 4.

It is clear from the above description that the generation process of the LDA generation model is the theme of the first generation, and then the specific word is generated according to the probability distribution of the word under the subject. The probability of generating an LDA model can be expressed as:

$$p(\theta, z, w|\alpha, \beta) = p(\theta|\alpha) \prod_{n=1}^{N} p(z_n|\theta)p(w_n|z_n, \beta) \tag{12.1}$$

The process of LDA modeling of the obtained word document data is based on the final word distribution information, and the parameters in the LDA model are deduced. All the parameters in the model are obtained by iterative learning of the training data, especially θ and φ, these two parameters are the key to infer the subject of the test document after modeling.

The parameters learning of LDA model mainly includes Variational Bayesian (VB) and Gibbs Sampling (GS). The basic idea of the VB algorithm is to use an approximate lower bound function to continually approximate the posterior probability of the solution. In theory, the VB algorithm is more accurate than the GS algorithm, but in the actual operation, the VB algorithm introduces a more complex function operation significantly resulting in the time complexity of the algorithm increases, even sometimes as the GS algorithm. The basic idea of the GS algorithm is to scan each word W and then sample a subject tag z from the posterior probabil-

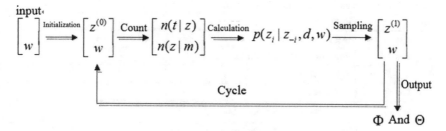

Fig. 12.2 The process of Gibbs sampling in LDA model

ity $p(z)$ and multiply the iteration until it converges to output the parameters to be estimated. In theory $p(z)$ will converge to the true posterior probability distribution. The process of Gibbs sampling in the LDA model is shown in Fig. 12.2.

GS algorithm convergence is very slow; usually, in practice, there is need to scan the document data 500–1000 times to reach the convergence. In addition, GS needs to scan each word. In the text classification, the scanning time cost of the words will increase which depends on the quantity of the words of the document.

12.2.2 TMBP Model Based on Factor Graph

The TMBP model belongs to the LDA model. However, in order to facilitate the reasoning and parameter learning, the LDA model needs to be transformed into the equivalent factor graph, and then the Belief Propagation (BP) algorithm is used to reason, this approach greatly improves the learning speed of the model. The factor diagram is shown in Fig. 12.3. The definition of the variables in the factor graph is shown in Table 12.2.

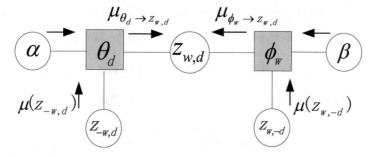

Fig. 12.3 The TMBP model is represented by a factor graph

Table 12.2 Variables defined in the TMBP model

Variable	Definition
$1 \leq d \leq D$	Document index
$1 \leq w \leq W$	Word index in the word list
$1 \leq k \leq K$	Subject index
$w = \{w, d\}$	Word bag
$z = \{z_{w,d}\}$	The subject of the word label
$z_{-w,d}$	The subject tag of all the word index in document d except the word w
$z_{w,-d}$	The subject label of word w for all documents except the document d
α	Superparine of Dirichlet distribution
β	Superparine of Dirichlet distribution
ϕ_w	The factor of word w
θ_d	The factor of document d
$\mu(z_{.,d})$	$\sum_w \mu(z_{w,d})$
$\mu(z_{w,.})$	$\sum_d \mu(z_{w,d})$

In Fig. 12.3, the factors θ_d and ϕ_w are represented by boxes, and their connection variables $Z_{w,d}$ are represented by circles. In contrast to Fig. 12.1, at the word layer, the original variables w_n and z_n are merged into a variable $Z_{w,d}$; $Z_{w,d}$ indicates the subject tag where the word w in document d is located. At the document level, there are factor variables θ_d and ϕ_k, which represent the distribution of the theme distribution corresponding to the specified document and the corresponding word list, and their neighbor variables are $z_{-w,d}$ and $z_{w,-d}$, respectively, as shown in Fig. 12.1. $z_{-w,d}$ is the subject label of all the word indexes except for the word w in the document d, $z_{w,-d}$ is the subject tag of the word w except for the word in the document d; in the corpus layer to retain α and β two super-parameters, used to control the document layer θ_d and ϕ_k variables, and Fig. 12.1 is also the same.

Figures 12.1 and 12.3 equivalent mainly for the following two reasons:

1. Have the same neighbor system. Because the hidden variables of the connection are the same.
2. In the factor diagram, the corresponding penalty function or set potential function is defined to implement the three essential assumptions of the subject model:

    (1) The same document in the same word index tends to give them the same theme;
    (2) The same word index in different documents also tends to give the same subject;
    (3) All word indexes cannot be given the same subject.

The confidence-based reasoning BP algorithm in the TMBP model does not provide an accurate solution to the problem-based reasoning problem of the structural tree factor, rather than the approximate reasoning process of the factor graph through the cyclic process. It does not directly calculate the posterior distribution $p(z|w)$, but rather calculates its joint probability $p(z_{w,d})$, also known as message $\mu(z_{w,d})$. The message is passed from the variable which is connected to the factor to the corresponding factor node. All the incoming messages are localized by the factor node, and the message is passed to the relevant variable by the factor node. Repeat this until the convergence or iteration is terminated. According to Fig. 12.3, the message is obtained by the neighbor node:

$$\mu(z_{w,d}) \propto \mu_{\theta_d \rightarrow z_{w,d}}(z_{w,d}) * \mu_{\phi_w \rightarrow z_{w,d}}(z_{w,d}) \tag{12.2}$$

The arrows in Fig. 12.3 indicate the direction of transmission between messages, where $\mu(z_{-w,d}) = \sum_{v \neq w} \mu(z_{v,d})$. Similarly, $\mu(z_{w,-d}) = \sum_{s \neq d} \mu(z_{w,s})$. The message passed from the factor to the variable is a stack of messages that are passed into all the neighbor variables and multiplied by the corresponding setpoint function:

$$\mu_{\theta_d \rightarrow z_{w,d}}(z_{w,d}) = \sum_{Z_{-w,d}} f_d \prod_{-w} \mu(z_{-w,d}) \alpha \tag{12.3}$$

$$\mu_{\varphi_w \rightarrow z_{w,d}}(z_{w,d}) = \sum_{Z_{w,-d}} f_d \prod_{-d} \mu(z_{w,-d}) \beta \tag{12.4}$$

In practical applications, the Eqs. (12.3) and (12.4) cause the calculated incoming message value to be close to zero. In order to avoid this phenomenon, the incoming message is generally used and the calculated operation is substituted for the calculation of the product. Since the sum of the corresponding sum increases as the product of the multiplication of the two numbers increases. Equations (12.3) and (12.4) are transformed into Eqs. (12.5) and (12.6):

$$\prod_{-w} \mu(z_{-w,d}) \alpha \propto \sum_{-w} \mu(z_{-w,d}) + \alpha \tag{12.5}$$

$$\prod_{-d} \mu(z_{w,-d}) \beta \propto \sum_{-d} \mu(z_{w,-d}) + \beta \tag{12.6}$$

In the Markov model, it is usually based on the current all the prior knowledge of the local topic of the label, and set the corresponding set of potential function. The TMBP model is defined as follows:

$$f_d = \frac{1}{\sum_k [\mu(z_{-w,d} = k) + \alpha]} \tag{12.7}$$

$$f_w = \frac{1}{\sum_w [\mu(z_{w,-d} = k) + \beta]} \tag{12.8}$$

According to Eq. (12.7), the incoming message will normalize all the $\mu(z_{w,d})$ on the document. The normalized operation eliminates the independence of the documents. The incoming message is normalized according to the words in the word list by Eq. (12.8). To simplify the formula: $\sum_{z_{w,d}=k} \mu(z_{w,d} = k) = \sum_k \mu(z_{w,d} = k)$. The messages from the Eqs. (12.2) to (12.8) are updated as follows:

$$\mu(z_{w,d} = k) \propto \frac{\mu(z_{-w,d} = k) + \alpha}{\sum_k \mu(z_{-w,d} = k) + \alpha} \times \frac{\mu(z_{w,-d} = k) + \beta}{\sum_w \mu(z_{w,-d} = k) + \beta} \qquad (12.9)$$

For the updated message, it is also necessary to normalize the dimension of the subject, that is: $\sum_k \mu(z_{w,d} = k) = 1$. And then fixed the updated message parameters $\mu(z_{w,d})$, The parameters θ_d and φ_w are updated using Eqs. (12.10) and (12.11) respectively until the iteration ends:

$$\theta_d \propto \frac{\mu(z_{\cdot,d} = k) + \alpha}{\sum_k \mu(z_{\cdot,d} = k) + \alpha} \qquad (12.10)$$

$$\varphi_w \propto \frac{\mu(z_{w,\cdot} = k) + \beta}{\sum_w \mu(z_{w,\cdot} = k) + \beta} \qquad (12.11)$$

The flow of the entire BP algorithm is described in Table 12.3. In the input parameter, K is the number of classification subjects, T is the maximum number of iterations, α and β are calculated as the known quantity. The output parameter θ_d is the matrix of $D \times K$, and records the probability that all documents appear on the corresponding subject. φ_w is the matrix of $K \times W$, and stores the probability that all words appear on the corresponding subject.

In general, hyperparameters α and β determine the sparseness of θ and φ, and have a certain effect on the results of the model. However, in order to simplify the model, it is assumed that the hyperparameters are symmetric in the original LDA model, which is still used in the TMBP model and its value is being provided as a priori knowledge. In order to reduce the complexity of the reasoning in the entire LDA model, the hyperparameters of the symmetric Dirichlet distribution are fixed.

12.2.3 TMBP Model Fusing Prior Knowledge

In the previous LDA model, the information of the only observable variable is not preprocessed, and it only used the frequency of the word in the document as the input of the model. In the dynamic scene, there is meaningless or redundant in visual word, if we apply LDA or TMBP to dynamic scene classification, we need to consider that the visual word on the expression of the subject is meaningful. Therefore, we extend the original TMBP model using the metric of the visual word prior knowledge.

In the TMBP model, if the prior knowledge of the word is added, the reasoning result of the model is more in line with human thinking. The subject matter of the document is derived from the frequency of the word. If the word can be given the weight according to the importance of the word.For example, for the meaningless or irrelevant words of the subject, a lower weight is given, and accordingly, the larger

Table 12.3 BP algorithm

BP algorithm
Input parameters: w, K, T, α, β
Output parameters: θ_d, φ_w
Step 1: initialization $\mu_{w,d}^1(k)$;
Step 2 : for k=1 to T
$$\mu_{w,d}^{t+1}(k) \propto \frac{\mu_{-w,d}^t(k)+\alpha}{\sum_k[\mu_{-w,d}^t(k)+\alpha]} \times \frac{\mu_{w,-d}^t(k)+\beta}{\sum_w[\mu_{w,-d}^t(k)+\beta]}$$;
end
Step 3 : $\theta_d(k) = [\mu_{.,d}(k)+\alpha]/\sum_k[\mu_{.,d}(k)+\alpha]$;
Step 4 : $\varphi_d(k) = [\mu_{w,.}(k)+\beta]/\sum_w[\mu_{w,.}(k)+\beta]$;

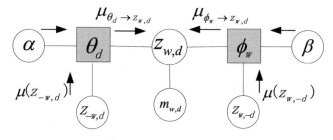

Fig. 12.4 Knowledge-TMBP graph model

word that contributes to the topic gives them a larger weight, which will provide the reasoning for the next word an effective prior knowledge. Therefore, we try to add the knowledge of Term Frequency Inverse Document Frequency (TF-IDF) as a priori knowledge of visual words to the TMBP model. After the TMBP model is expanded, the Knowledge-TMBP model is shown in Fig. 12.4.

Compared to the TMBP model, The only changes of the model is one $m_{w,d}$ node. The node represents a priori knowledge of the word weight. The prior knowledge is calculated from the TF-IDF, i.e., the inverse frequency of the word. TF-IDF is used to determine the probability that a word is in a particular document compared to all document libraries. In short, this calculation determines the relevance of a given word in a particular document. If a word appears in a document or a small part of the document, the word tends to be given a higher TF-IDF value, and accordingly, the words that appear in most or all of the documents tend to be given a Low TF-IDF value. Of course, TF-IDF has many methods of calculation, but all the methods are calculated by the following method. Given a document library D, one of the documents as d, TF-IDF is calculated as follows:

$$m_w = f_{w,d} * \log(\frac{D}{n_w} + 0.1) \tag{12.12}$$

where $f_{w,d}$ is the frequency at which the word w appears in document d, D is the number of documents in the entire document library, and n_w represents the number of documents in which the word w appears. The meaning of each variable in the new model is shown in Table 12.4.

Table 12.4 The variables defined in knowledge-TMBP model

Variable	Definition
$1 \leq d \leq D$	Document index
$1 \leq w \leq W$	Word index in the word list
$1 \leq k \leq K$	Subject index
$w = \{w, d\}$	Word bag
$z = \{z_{w,d}\}$	The subject of the word label
$m_{w,d}$	The prior knowledge of the word weight
$z_{-w,d}$	The subject of all the word index in document d except the word w
$z_{w,-d}$	The word w is the subject label for all documents except the document d
α	Superparine of Dirichlet distribution
β	Superparine of Dirichlet distribution
ϕ_w	The factor of word w
θ_d	The factor of document d
$\mu(z_{.,d})$	$\sum_w \mu(z_{w,d})$

12.3 Dynamic Scene Classification Based on TMBP

The process of dynamic scene classification based on the topic model is actually the subject of the maximum probability of finding the video file. The main steps are as follows:

(1) The video is processed as an image sequence

For the input video, first of all, dealing with a single frame of the image sequence, in which key frames can be appropriate to intercept. On the one hand, it can help reduce the amount of data to facilitate the later calculation; on the other hand, if the difference between the two frames is small, the movement will be too small, the motion information is more likely to be extracted, so the selection of the key frame also helps to extract the motion information between the two frames. There are many ways to select keyframes. There are common methods of selecting a frame directly in the video time series, and other adaptive key frame extraction methods. By extracting the key frame, the input video can be transformed into an image with a time series.

(2) Extract the gray difference feature of adjacent frames

Paired two adjacent frames, the next frame minus the previous frame, get a differential image. Then a 100-dimensional eigenvector is formed by dividing the difference image into a 100-dimensional eigenvector, plus the average gray value of the image block to obtain a 101-dimensional eigenvector. This vector is used as motion information to describe the dynamic scene.

(3) Visual word generation

Each feature map is described as a 72-dimensional feature vector. And then cluster the feature vector using the K-means clustering method to generate the visual word dictionary, and the clustering center is the visual word.

(4) Modeling with the topic model

Doing statistics of each training video frequency according to the visual dictionary, representing each dynamic scene file with the word frequency, and then using the topic model for dynamic scene modelling. After training, we can get the probability distribution of the corresponding visual word corresponding subject and the probability distribution of the corresponding subject of the video.

(5) Test

For the test data, the video is first processed as a key frame image sequence. After obtaining the key frame, extract the grayscale difference feature of the adjacent key frame calculate the Euclidean distance between the extracted feature and the gray scale difference feature corresponding to each visual word in the visual dictionary,

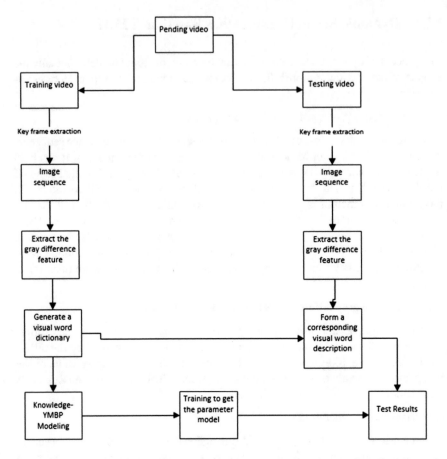

Fig. 12.5 Dynamic scene classification based on the topic model implementation flow chart

and then use the nearest visual word to represent the image frame in the video. Finally, represent dynamic scene video of the test is expressed as the word frequency table of the visual word into the model, and test the probability distribution of the corresponding word of the visual word obtained in the training is tested.

(6) The processing of the model output results

Through the test, the model will output the probability distribution of each test data for each subject, select the topic with the maximum probability as the subject category of the dynamic scene.

Figure 12.5 shows the flow chart of dynamic scene classification based on TMBP algorithm.

The MATLAB source program for dynamic scene classification based on TMBP algorithm is shown in PROGRAMME 12.1 to PROGRAMME 12.4.

PROGRAMME 12.1: **Extract keyframes**

```
VideoFolder='Video';   % The folder where the video file is located
filenames=strcat(VideoFolder,'/*.avi');
filenames1=dir(filenames);
VideoNum=size(filenames1,1);
letter1='a';
cont=0;
for m=1:VideoNum
    VideoName=strcat(VideoFolder,'/');
    VideoName=strcat(VideoName,filenames1(m).name);% In turn get a video
    xyloObj = VideoReader(VideoName);% Read the video
    nFrames = xyloObj.NumberOfFrames;      % Total number of frames
    wd = xyloObj.Width;% Get the image size
    hd = xyloObj.Height;
    count=0;       % Record the number of frames intercepted
    for k = 1 : 2: nFrames-1
        mov(k).cdata =   read(xyloObj,k);      % Picture color data
        strtemp = strcat('addStrain/Data/');
        count=count+1;
        cont=cont+1;
        imwrite(mov(k).cdata,strtemp1);      % Save as jpg picture named strtemp
        if rem(k,2)==1
            cont1 = fix(count/2);
            mov1 = mov(k).cdata;
            mov1 =im2double(rgb2gray(mov1));
            mov2 = read(xyloObj,k+1);
            mov2 =im2double(rgb2gray(mov2));
            mov3 = mov2-mov1;     % Get the difference image
            movi.diffimg{cont1+1}=mov3;   % Differential image data
        end
    end
    len=length(xyloObj.Name);
    strtemp3 = xyloObj.Name(1:len-4);
    strtemp3 = strcat(strtemp,strtemp3 );
    save(strtemp3,'movi');     % Save the difference image
    clear('movi');
end
```

PROGRAMME 12.2: Extract the gray difference feature

```
space=10;
n=0;
filename='Data/';   % Save the path of the differential image
filename1=strcat(filename,'*.mat');
filenames1=dir(filename1);
fileNum=size(filenames1,1);
for aa=1:fileNum
fm=strcat(filename,filenames1(aa).name);
load(fm);
GraySet=[];GrayNumOfDoc=[];
imgNum=size(movi.diffimg,2);
  for aaa=1:imgNum    % Gets the difference image in the same video
          img=movi.diffimg{aaa};
          img2=movi.meancolor{aaa};
          [hd,wd] = size(img);

          numr=fix(hd/space);
          numc=fix(wd/space);   % The difference image is divided by the size of 10 * 10
          t1 = (0:numr-1)*space + 1;
          t2 = (1:numr)*space;
          t3 = (0:numc-1)*space + 1;
          t4 = (1:numc)*space;
          for i = 1 : numr
          for j = 1 : numc
             x = t1(i):t2(i);
             y = t3(j):t4(j);
             B=img(x,y);B1=img2(x,y);
             vmean=mean(B1(:));   % The average grayscale feature of the recorded image block
             C=reshape(B,1,space*space);
             C=[C,vmean];
             proSet((i-1)*numc+j,:)=C;
          end
          end
          proSet(find(sum(abs(proSet),2)==0),:)=[];% The image block is characterized by an all-zero
    discard
          [r,c]=size(proSet);
          GrayNumOfDoc(aaa,:)=r;   % Record the number of image blocks contained in the video
```

frame image

```
        GraySet=[GraySet;proSet];
        img=[];proSet=[];
    end
        len=length(fm);
        fm1=fm(1:len-4);
        strtemp=strcat(fm1,'_FeaSet');
        save(strtemp,'GraySet','GrayNumOfDoc');
end
```

PROGRAMME 12.3: All the extracted gray-scale differential features are clustered to form visual words

```
openpath=strcat("Data/*_FeaSet.mat');
filenames=dir(openpath);
fileNum=size(filenames,1);
for bb=1:fileNum
loadname=openpath;
load(loadname);
    totGreySet=[totGreySet; GraySet];
    [numofthis,n]=size(GraySet);
    NumofDoc=[NumofDoc,numofthis];
End
save('Data/totGraySet.mat',' totGreySet',' NumofDoc'); %Save all the gray-scale difference features
    and the
% number of words contained in the file
for k=200:50:800
[wordnum,center]=kMeansCluster(totGraySet,k,1);
filename = sprintf('Data/word_%d', k);
save(filename,'wordnum','center');
end
% Subfunction:
function [y,c]=kMeansCluster(m,k,isRand)
%K-means Clustering method
% Input: m Characteristic matrix, k Clustering center number, isRand random number (1 for random
    initial
% clustering center, 0 for the first n of the feature cluster center)
%Output: Clustering result matrix y, Clustering center c
[maxRow, maxCol]=size(m);
if maxRow<=k,
    y=[m, 1:maxRow];
else
```

```
if isRand,
    p = randperm(size(m,1)); % Random initialization of clustering centers
    for i=1:k
    c(i,:)=m(p(i),:);
    end
else
    for i=1:k
        c(i,:)=m(i,:);              % Sequential initialization of clustering centers
    end
end
temp=zeros(maxRow,1);              % Initialize with a 0 matrix
while 1,
    bj=[];
    whole=fix(maxRow/2000);
    remainder=mod(maxRow,2000);
% Automatically assigns the acquired vector group to the initialization category
    for i=1:whole
        te=(i-1)*2000+1:i*2000;
        d=DistMatrix(m(te,:),c);          % c Calculate the distance to the center of the cluster
        [z,g]=min(d,[],2);        % Find the smallest cluster g
        bj=[bj;g];
    end
    te=2000*whole+1:2000*whole+remainder;
    d=DistMatrix(m(te,:),c);
    [z,g]=min(d,[],2);
    bj=[bj;g];
    if bj==temp,
        break;                 % Stop iteration
    else
        temp=bj;                 % Assign new clustering results to temp
    end
    for i=1:k
        f=find(bj==i);
        if f                      % Recalculate the cluster center
            c(i,:)=mean(m(find(bj==i),:),1);
        end
    end
end
y=bj;
end
```

```
function d=DistMatrix(A,B)
% Calculate the Euclidean distance of the two matrices
% Input: Matrix A, B
% Output: A, B distance
[hA,wA]=size(A);
[hB,wB]=size(B);
if wA ~= wB,
    error('second dimension of A and B must be the same');
end
for k=1:wA
    C{k}= repmat(A(:,k),1,hB);
    D{k}= repmat(B(:,k),1,hA);
end
S=zeros(hA,hB);
for k=1:wA
    S=S+(C{k}-D{k}').^2;
end
d=sqrt(S);
```

PROGRAMME 12.4: Count the word frequency matrix of all video files

```
load('Data\word_400.mat')
load('Data\totGraySet.mat')
word=countWord(wordnum,center,NumofDoc);
save('Data\wordtrain.mat','word');
Subfunction:
function wd = countWord( wordnum,totwCent,totteNofD )
% Calculate the word frequency table of the training video
% Input: trained visual word matrix, visual word center matrix, the number of words contained in each
    file
% Output: The word corresponding to the word word matrix
SCol=size(totwCent',2);
totteNofD=totteNofD';
    start=1;
    wd=[];
    a=size(totteNofD,1);
    for i=1:a
        add=totteNofD(i,1);
        if start==1
        over=start+add-1;
```

```
        else
        over=start+add;
        end
        doc=start:over;
        x=wordnum(doc,:);
        x=sort(x);
        d=diff([x;max(x)+1]);
        count = diff(find([1;d])) ;
        y =[x(find(d)) count];
        for j=1:size(y,1)
             wd(y(j,1),i)=y(j,2);
        end
          FCol=size(wd,1);
        if SCol>FCol
            wd(FCol+1:SCol,:)=0;
        end
        start=over;
    end
end
```

For the test video, the process of key frame extraction is same as the grayscale difference feature extraction. The process of calculating the visual word does not use K-means clustering, but find the corresponding visual word with visual dictionary after corresponding training.The code is shown in PROGRAMME 12.5.

PROGRAMME 12.5: Testing process

```
load('Data\word_400.mat')
load('Data\totGraySet.mat')
wordtest=calculateWord(totFlow,center,NumofDoc);
save('Data\wordtest.mat','wordtest');
% Subfunction:
function    wd= calculateWord( totteFeat,totwCent,totteNofD)
% Calculate the visual word in which the feature in the test video belongs and output the frequency band
    of the test
%video.
% Input: Test the extracted feature matrix, the visual word center matrix, the number of words contained
    in each
```

```
%file
% Output: The word corresponding to the word word matrix
SCol=size(totwCent,2);
c=totwCent';
[maxRow, maxCol]=size(totteFeat);
    bj=[];
    whole=fix(maxRow/200);
    remainder=mod(maxRow,200);
    for i=1:whole
            te=(i-1)*200+1:i*200;
            d=DistMatrix(totteFeat(te,:),c);          % calculate objcets-centroid distances
            [z,g]=min(d,[],2);          % find group matrix g
            bj=[bj;g];
    end
    te=200*whole+1:200*whole+remainder;
    d=DistMatrix(totteFeat(te,:),c);          % calculate objcets-centroid distances
    [z,g]=min(d,[],2);          % find group matrix g
    bj=[bj;g];
    k=bj;

    start=1;
    wd=[];
    a=size(totteNofD,1);
    for i=1:a
        add=totteNofD(i,1);
        if start==1
        over=start+add-1;
        else
        over=start+add;
        end
        doc=start:over;
        x=k(doc,:);
        x=sort(x);
        d=diff([x;max(x)+1]);
        count = diff(find([1;d])) ;
        y =[x(find(d)) count];
        for j=1:size(y,1)
                wd(y(j,1),i)=y(j,2);
```

```
            end
            start=over;
        end
    FCol=size(wd,1);
    if SCol>FCol
        wd(FCol+1:SCol,:)=0;
    end
end
```

PROGRAMME 12.6: Visual word weight calculation

```
load('Data\wordtrain.mat');    % Read the training word frequency matrix
  count = sum(wordtrain>0,2);
  tot_d=size(wordtrain,2);


for a=1:size(count)
w_tfidf(a)=log(tot_d/count(a));
    wd_tfidf(a,:)=wd_train(a,:)/sum(wd_train(a,:))*(w_tfidf(a)+0.1);
end
save('Data\w_tfidf.mat',' w_tfidf");
```

PROGRAMME 12.7: Training and testing of thematic models

```
ALPHA = 1e-2;
BETA = 1e-2;
N = 6000;
SEED = 1;
OUTPUT = 1;
% %%
load('Data\wordtrain.mat')    % Read the training frequency
load('Data\wordtest.mat')     % Read the test word frequency
load('Data\w_tfidf.mat')      % Read the word weight information
wd=[wordtrain,wordtest];
mm=size(wd,2);
for aa=1:mm
    wdt=wd(:,aa);
    wdt=w_tfidf'.*wdt;
    wd(:,aa)=wdt;
end
```

```
wd=ceil(wd);
tot_wd=sparse(wd);
[r,c]=size(wordtrain);
load('Data\w_tfidf.mat')
wd_train=sparse(wd_train);
wd_test=sparse(wd_test);
clear dp wp dp_te perplexity t
J=7;
tr=1:c;
te = setdiff(1:size(wd,2),tr);
tic
[wp, dp, z] = tfidfLDABPtrain(tot_wd(:,tr), J, N, ALPHA, BETA, SEED, OUTPUT); % Training process
t = toc;
[dp_te, z] = tfidfLDABPpredict(tot_wd(:,te), wp, N, ALPHA, BETA, SEED, OUTPUT); % Testing
    process
```

We conducted an experiment of 14 categories of videos with dynamic image library Dynamic_Scenes, namely ocean, sky-clouds, snowing, waterfall, fountain, forest fire, beach, highway, elevator, lighting, storm, railway windmill farm, rushing river. Each category contains 30 videos, the location of the camera in the video is fixed, the dynamic information in the video is mainly the scene content of the movement. An example of a partial dynamic scenario is shown in Fig. 12.6.

In order to verify the effect of Knowledge-TMBP and other thematic models on the dynamic scene classification results, 7 categories of scenarios are selected, namely Sky-clouds, Waterfall, Fountain, Forest Fire, Beach, Highway, Elevator. Experimental hardware environment including: Windows 7, Pentium 4 processor, clocked at 2.8G, memory for 4G. The code runtime environment is: MATLAB 2013a. The visual words are established by using the grayscale features of the difference graphs of simple video frames, and experiments are carried out in PLSA, GS-LDA, TMBP and Knowledge-TMBP models respectively. The training time is shown in Table 12.5. Although the prior knowledge of the word is added, the training time will be slightly longer than that of the TMBP model, but it is better than the training time of the PLSA model and the GS-LDA model. The classification performance of the four models is evaluated by four evaluation criteria: precision(P), accuracy(ACC), recall(R), and F-measure(F). The classification performance of the four models is shown in Fig. 12.7a. The graph is the classification precision(P) of the four models; Fig. 12.7b is the classification of four types of recall rate(R) comparison; Fig. 12.7c is the classification of the four models F-measure(F) comparison; Fig. 12.7d is the classification accuracy of the four models ACC comparison; Fig. 12.7e is the comparison of the four standard evaluation criteria. It can be seen from the data in Fig. 12.7 that the TMBP model with a priori knowledge is 5% higher than the original LDA model, the recall rate is increased by 6%, the F-measure is improved by 6%, and the classification accuracy is improved 2%.

Fig. 12.6 Dynamic_scenes dynamic scene example

Table 12.5 Comparison of training time for four models

	PLSA	GS-LDA	TMBP	Knowledge-TMBP
Training time t/s	15513.7405	651.0008	355.4706	367.7502

12.4 Behavior Recognition Based on LDA Topic Model

The main process of behavior recognition includes: 1. Detect points of interest in the image or video. 2. Use some features to describe the information around the points of interest. 3. Use the clustering algorithm to cluster the generated features and then take the clustering center as the visual word. 4. Use the classification model to classify the generated visual words. The process is shown in Fig. 12.8.

For the input video, first calculate the significant figure to obtain the person foreground area, then calculate the threshold matrix according to the saliency value and the foreground region, and carry out the point detection according to the threshold matrix; After extracting the points of interest, the surrounding 3D-SIFT feature and the HOOF feature of the whole frame image are calculated, then the two features are merged and the visual word dictionary is generated by spectral clustering. Finally, the TMBP model is used to classify the generated visual words and identify the characters in the video. The implementation of the behavioral recognition based on the topic model is as follows:

1. GBVS significant graph generation

A significant image is actually a simulation of human visual behavior to find out the image of the observer attention to the target. Compared with the original image, a significant figure highlights the target, weakening the background. The Itti method is a more classic visual attention model, which is applied to the analysis of real scene images and obtains experimental results closer to human visual perception. However, in the case of complex scenes, there are still some gaps between the results of the method and the human eye to observe the actual goal. Graph-Based Visual Saliency (GBVS) method is an improved model of Itti, which is simpler and more bionic.

For a given input image GBVS model, the corresponding feature map is firstly calculated and then consider each pixel of the feature map on each pixel (which can be patch) as a node of the map. The edge between nodes represents the difference between any two nodes, and the difference is defined as follows:

$$d((i, j)\|(p, q)) \triangleq |\log \frac{M(i, j)}{M(p, q)}| \tag{12.13}$$

$$\omega_1((i, j)\|(p, q)) \triangleq d((i, j)\|(p, q))F(i - p, j - q) \tag{12.14}$$

$$F(a, b) \triangleq \exp(-\frac{a^2 + b^2}{2\sigma^2}) \tag{12.15}$$

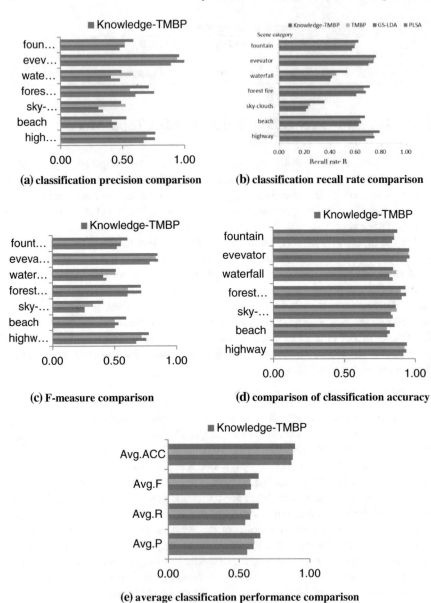

Fig. 12.7 Comparison of classification performance of four models

where $M(i, j)$ indicates the eigenvalue represented by pixel (i, j), and $M(p, q)$ indicates the eigenvalue represented by pixel (p, q). $d((i, j)||(p, q))$ is the distance between two points, given by Eq. 12.13, F is given by Eq. 12.15 and ω_1 is the difference between the two nodes given by Eq. 12.14. According to the calculation

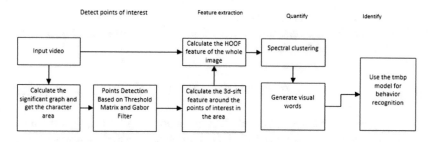

Fig. 12.8 Character behavior identification framework

of Eq. 12.14, we can get the matrix of the difference between each node and all other nodes, then normalize each row of the matrix. Last, we get an adjacency matrix A of the graph. The GBVS method treats this matrix as a corresponding Markov chain, and each node on the chain corresponds to the node of the graph. According to Markov's thought, any state can be continuously updated to enter a final steady state, which means that the state of the system has not changed after the next jump. The update of the adjacency matrix is defined by Eq. 12.16:

$$\omega_1((i, j)\|(p, q)) \triangleq A(p, q)F(i - p, j - q) \tag{12.16}$$

After normalizing each row of ω_1, the final state is obtained. With this steady state, you can analyze the probability that each node is accessed per unit time. If a small cluster node differs greatly from its surroundings, the probability of reaching these nodes from any state is very small, so that this small cluster node is significant.

2. Spatio—Temporal Point of Interest Detection Based on Dynamic Threshold

Taking the point of interest to describe the action behavior without the need for the image before the background segmentation and target tracking, then you can extract the sparse representation of the video. Common point of interest detection methods are the corner-based method, the LOG-based method and the filter based method. Corner point method is extending the two-dimensional corner to the three-dimensional space and take the calculation of the corner of the video as a point of interest. The LOG-based approach uses Gaussian Laplacian as a response function to detect points of interest based on this. The filtering method uses a three-dimensional convolution window to convolve the entire video, and then find the local maximum as a point of interest. When the first two methods detect the points of interest, it is found that the number of points of interest is too small, which is not conducive to the extraction of video features. The method based on filtering increases the number of points of interest detection, and because of the use of convolution operations, the first two methods are simpler and easier to implement, and the time complexity is lower. Therefore, this section uses Gabor-based points of interest to detect the position of the local response value in the sub-search image as the point of interest.

The steps of Gabor filter based on the point of interest detection are as follows:

(1) Using the Gaussian filter in space for each frame image filtering.
(2) Using two orthogonal one-dimensional Gabor filters to filter in time, and then define the response function:

$$R = (S * g * h_{ev})^2 + (S * g * h_{od})^2 \qquad (12.17)$$

where $g(x, y; \sigma)$ is a two-dimensional Gaussian-smoothing kernel, S is the input image of each frame, h_{ev} and h_{od} are a pair of orthogonal one-dimensional Gabor filters:

$$h_{ev}(t; \tau, \omega) = -\cos(2\pi t \omega)e^{-t^2/\tau^2} \qquad (12.18)$$

$$h_{od}(t; \tau, \omega) = -\sin(2\pi t \omega)e^{-t^2/\tau^2} \qquad (12.19)$$

where σ and τ are the filter space and time on the two scale parameters, $\omega = 4/\tau$.

(3) For each pixel, calculate its corresponding response value to find out the local maximum value as the time and space points of interest of the entire video.

When calculating the local maximum, we first use the GBVS significant graph to determine the approximate region of the character with different thresholds inside and outside the region, and then calculate the threshold matrix of each pixel, and then find the local maximum as the point of interest. Define the threshold for each pixel in the space:

$$w_i = \begin{cases} (\frac{S+\delta}{S_{in}})^{-1} \times \varepsilon_{in}, \; pixel \; in \; plane \\ (\frac{S+\delta}{S_{out}})^{-1} \times \varepsilon_{out}, \; pixel \; out \; plane \end{cases} \qquad (12.20)$$

where $S_{in} = \sum_{1}^{n} S_i$ and S_i are the saliency values corresponding to the pixels, and S_{in} represents the sum of the saliency values of all the pixels in the region. S_{out} is the sum of the saliency values of all the pixels outside the region. δ is a small value to prevent the denominator to 0. ε_{in} and ε_{in} are two weight factors, so that the weight in the region is always smaller than the weight outside the region. We calculate the average of the weight order of a continuous $2 \times \xi$:

$$\overline{w_t} = avg\left(\sum_{i=t-\xi}^{t+\xi} w_i\right) \qquad (12.21)$$

After the calculation of Eq. 12.21, we get a three-dimensional threshold matrix. In the calculation of the subsequent local maximum, we use this three-dimensional threshold matrix instead of a single threshold.

3. Visual word generation

Visual words are often considered to be local information extracted from images or some of the regions of the video, and this extracted information is often able to describe the features around the area to characterize the entire image or video. Visual words are different from ordinary low-level features, which simulate the cognitive process of human brain and transform the high-level semantic information into a variety of low-level features. Therefore, the feature expression of the behaviors of the visual word depends on the low-level visual characteristics. SIFT feature, as a traditional feature descriptor, has the characteristics of scale invariance, rotation invariance, light invariance and so on. The 3D-SIFT feature descriptor is a three-dimensional gradient direction histogram operator proposed by Scovanner et al. It is an extension of the two-dimensional SIFT descriptor from image to video, which can better reflect the gradient information around the point of interest. The HOOF feature uses the optical flow histogram to describe the global motion information of the whole frame image, which can make up for the shortcomings of SIFT as a lack of motion information as a local feature. So we use the 3D-SIFT feature and the HOOF feature to describe the local and global information of the image points of interest.

The 3D-SIFT feature can be calculated as follows:

In the two-dimensional space, the gradient size and direction of each pixel can be calculated from Eqs. 12.22 and 12.23:

$$m_{2D}(x, y) = \sqrt{L_x^2 + L_y^2} \tag{12.22}$$

$$\theta(x, y) = \tan^{-1}(\frac{L_y}{L_x}) \tag{12.23}$$

Since each pixel in the image is discrete and cannot calculate the continuous partial derivative function, a discrete approximation algorithm is used to calculate the specific values when calculating L_x and L_y. For L_x, use $L(x+1, y) - L(x-1, y)$ to approximate, for L_y, with $L(x, y+1) - L(x, y-1)$ to approximate. After extending the two-dimensional gradient to three dimensions, the gradient can be obtained by the following formula:

$$m_{3D}(x, y, t) = \sqrt{L_x^2 + L_y^2 + L_t^2} \tag{12.24}$$

$$\theta(x, y, t) = \tan^{-1}(L_y/L_x) \tag{12.25}$$

$$\phi(x, y, t) = \tan^{-1}(\frac{L_t}{\sqrt{L_x^2 + L_y^2}}) \tag{12.26}$$

where ϕ represents the angle in the two-dimensional plane gradient direction, in the range of $(-\frac{\pi}{2}, \frac{\pi}{2})$. The gradient direction of each point is represented by a unique point pair (θ, ϕ). In the calculation, as with the two-dimensional gradient calculation, the discrete-difference method is used to approximate the value of the partial derivative function. In terms of a candidate point, calculate the gradient value and direction of each pixel around it, and then statistical the gradient direction histogram to get a main direction, then use the Eq. 12.27:

$$\begin{bmatrix} \cos\theta\cos\phi & -\sin\theta & -\cos\theta\sin\phi \\ \sin\theta\cos\phi & \cos\theta & -\sin\theta\sin\phi \\ \sin\theta & 0 & \cos\theta \end{bmatrix} \tag{12.27}$$

Rotate the gradient direction of all the pixels to the main direction, re-count the bin size of the histogram with the formulas 12.28 and 12.29:

$$hist(i_\theta, i_\varphi)+ = \frac{1}{\omega}m_{3D}(x', y', t')e^{\frac{-((x-x')^2+(y-y')^2+(t-t')^2)}{2\sigma^2}} \tag{12.28}$$

$$\omega = \int_\varphi^{\varphi+\Delta\varphi}\int_\theta^{\theta+\Delta\theta}\sin\theta d\theta d\varphi = \Delta\varphi\int_\theta^{\theta+\Delta\theta}\sin\theta d\theta$$

$$= \Delta\varphi[-\cos\theta]_\theta^{\theta+\Delta\theta} = \Delta\varphi(\cos\theta - \cos(\theta + \Delta\theta)) \tag{12.29}$$

To get the final bin value by weighting. Extend all bin values into vectors as the final SIFT feature. When calculating the 3D-SIFT feature, we can choose two methods of 8 or 64 regions when selecting the surrounding area of the candidate point. We use the area around the candidate point to calculate the SIFT feature. For the direction of the gradient direction using twenty regular triangles to build a positive icosahedron. In order to improve the representation of the feature, each side of the triangle is subdivided into four regular triangles constitute a regular octahedron. The direction of the vector to the center of each triangles is the bin of each direction of the histogram, so the final feature length is dimension.

The HOOF feature can be calculated as follows:

There is one problem with the use of the SIFT feature which is the movement direction information of the object cannot be fully reflected. And the SIFT feature is a local feature that cannot obtain information about the entire frame image, so we use the Histograms of Oriented Optical Flow (HOOF) feature to represent the motion information of a person, which is a histogram of all the optical flows in a frame as a global feature of the frame image.

Optical flow field through the image of the distribution of different gray levels describes the movement of space in the information. The optical flow field reflects the trend of gray scale of each pixel on the image. This trend can be regarded as the instantaneous velocity field generated by the motion of the pixel with gray scale on the plane, and it is also a kind of real- Approximate estimates.

For an image, suppose $E(x, y, t)$ is the gray scale of point (x, y) at time t. Let $t + \Delta t$ point the point of movement of the point of $(x + \Delta x, y + \Delta y)$, then its gray scale $E(x+\Delta x, y+\Delta y, t+\Delta t)$. Since the two points are corresponding to each other, according to the optical flow constraint equation, we can get the formula 12.30:

$$E(x, y, t) = E(x + \Delta x, y + \Delta y, t + \Delta t) \tag{12.30}$$

And then by calculating the Taylor expansion on the right, and let $\Delta t \to 0$, then the formula 12.31:

$$E_{xu} + E_{yv} + E_t = 0 \qquad (12.31)$$

Among them: $E_x = \frac{dE}{dx}$, $E_y = \frac{dE}{dy}$, $E_t = \frac{dE}{dt}$, $u = \frac{dx}{dt}$, $v = \frac{dy}{dt}$, by using the discrete difference method to approximate the partial derivative function, u and v are calculated as two dimension values of the optical flow feature.

For a video stream, the optical flow characteristics of each frame are first calculated, and for each optical stream vector, it is assigned to each histogram according to its angle with the horizontal direction and the weight of the size. Assuming the optical flow vector $v = [x, y]^T$, its direction $\theta = \tan^{-1}(\frac{y}{x})$ is in the range $(-\frac{\pi}{2} + \pi \frac{b-1}{B}, -\frac{\pi}{2} + \pi \frac{b}{B})$, according to its angle, we divide it into the bth histogram component. Finally, the histogram is normalized so that the sum of all the components is 1.

The 3D-SIFT and HOOF joint features can be generated as follows.

We combine the 3D-SIFT feature of each point of interest with the HOOF feature of the frame image corresponding to that point of interest, combining the local and global features of each frame of the image. For each point of interest, we can compute a 640-dimensional 3D-SIFT feature $(x_1, x_2, \ldots, x_{640})$, assuming that the global HOOF feature of the frame of interest is $(y_1, y_2, \ldots, y_t)$, where t is the bin value of the histogram. By splicing the two features together to get a new feature $(x_1, x_2, \ldots, x_{640}, y_1, y_2, \ldots, y_t)$. In the subsequent clustering generation of visual words, the larger t value will make the clustering center more biased towards the HOOF feature. Therefore, we do not directly add the weighting factor to the feature fusion, but rather adjust the histogram bin value to adjust the global feature in the proportion of the overall feature.

For the behavior with larger difference, only use the 3D-SIFT feature can be well recognized, and for the more close to the behavior, you need to increase the HOOF characteristics on this basis. In the comparison of multiple experiments, we selected the value 150 of 120 to 200 which can get a higher recognition accuracy.

Visual word dictionary can be generated by using spectral clustering.

The core idea of spectral clustering is to use a graph-based Laplace matrix, so it is only necessary to have a similarity matrix between the data, rather than requiring the data to be a vector in the Euclidean space as K-means does. In video processing, the visual word corresponds to the text word, the video corresponds to the article. The difference is not big in the cluster operation, so we use spectral clustering as a visual word clustering method.

Given a data set $X_1, \ldots, X_n$, define the similarity matrix S, where S_{ij} is the similarity between X_i and X_j. The non-normalized Laplace matrix is defined as follows, where is a diagonal matrix. The non-normalized Laplace matrix is defined as $L = D - S$, where D is a diagonal matrix $D_{ii} = \sum_{j=1}^{n} S_{ij}$.

Step1: Calculate the similarity matrix $S \in R^{n \times n}$;
Step2: Calculate the non-normalized Laplacian matrix L;

Step3: Calculate the first k eigenvectors $u_1, \ldots, u_2$ of the L matrix;

Step4: Constructs a matrix $U \in R^{n \times k}$, where each column is a vector $u_1, \ldots, u_2$;

Step5: The clustering algorithm is obtained by clustering the matrix U using the K-means clustering algorithm.

In the experiment, we use the Euclidean distance as the measurement of similarity to construct the similarity matrix. After clustering the generated features by spectral clustering, we use the clustering center as the visual word.

When using TMBBP for behavior classification, K in input parameter is the total number of classes, T is the maximum number of iterations set, and α and β are iterated as known parameters. The output parameter θ_d is a matrix of $K \times D$, the probability of recording the video on the corresponding behavior category; φ_w is a matrix of $K \times W$, recording the probability that visual words appear on the corresponding behavior category.

Based on the LDA topic model of the behavior classification MATLAB code as shown in PROGRAMME 12.8 to PROGRAMME 12.15.

PROGRAMME 12.8: Read the picture and return gbvs significant graphs

```
    out.inputimg = img;
e function [out,motionInfo] = gbvs(img,param,prevMotionInfo)
%%%%%%%%%%%%%%%%%%%%%%%%%%%%%%%%%%%%%%%%%%%%%%
  %%%%%%%%%%%%%%%%%%%%%%%%%%%%%%%%%%%%%%%%%%%%%
  %%%%%
% This computes the GBVS map for an image and puts it in master_map.
%
% If this image is part of a video sequence, motion.Info needs to be recycled in a loop, and information
    from the previous frame/image will be used if "flicker" or "motion" channels are employed.
% You need to initialize prevMotionInfo to [] for the first frame   (see demo/flicker_motion_demo.m)
%input
%-img can be a filename, or image array (double or uint8, grayscale or rgb)
%-(optional)param contains parameters for the algorithm (see makeGBVSParams.m)
%output structure 'out'.fields:
%-master_map is the GBVS map for img. (.._resized is the same size as img)
%-feat_maps contains the final individual feature maps, normalized
%-map_types contains a string description of each map in feat_map (resp. for each index)
%-intermed_maps contains all the intermediate maps computed along the way (act. & norm.)which are
    used to compute feat_maps, which is then combined into master_map
%-rawfeatmaps contains all the feature maps computed at the various scales
%%%%%%%%%%%%%%%%%%%%%%%%%%%%%%%%%%%%%%%%%%%%%%
  %%%%%%%%%%%%%%%%%%%%%%%%%%%%%%%%%%%%%%%%%%%%%
  %%%%%
```

```
if ( strcmp(class(img),'char') == 1 ) img = imread(img); end
if ( strcmp(class(img),'uint8') == 1 ) img = double(img)/255; end
if ( (size(img,1) < 120) || (size(img,2) < 120) )
     fprintf(2,'GBVS Error: gbvs() meant to be used with images >= 128x128\n');
     out = [];
     return;
end
if ( (nargin == 1) || (~exist('param')) || isempty(param) ) param = makeGBVSParams; end
[grframe,param] = initGBVS(param,size(img));
if ( (nargin < 3) || (~exist('prevMotionInfo')) )
     prevMotionInfo = [];
end
if ( param.useIttiKochInsteadOfGBVS )
     mymessage(param,'NOTE: Computing STANDARD Itti/Koch instead of Graph-Based Visual
     Saliency (GBVS)\n\n');
end
%%%%
%%%% STEP 1 : compute raw feature maps from image
%%%%
mymessage(param,'computing feature maps...\n');
if ( size(img,3) == 3 ) imgcolortype = 1; else, imgcolortype = 2; end
[rawfeatmaps motionInfo] = getFeatureMaps( img , param , prevMotionInfo );
%%%%
%%%% STEP 2 : compute activation maps from feature maps
%%%%
mapnames = fieldnames(rawfeatmaps);
mapweights = zeros(1,length(mapnames));
map_types = {};
allmaps = {};
i = 0;
mymessage(param,'computing activation maps...\n');
for fmapi=1:length(mapnames)
     mapsobj = eval( [ 'rawfeatmaps.' mapnames{fmapi} ';'] );
     numtypes = mapsobj.info.numtypes;
     mapweights(fmapi) = mapsobj.info.weight;
     map_types{fmapi} = mapsobj.description;
     for typei = 1 : numtypes
```

```
        if ( param.activationType == 1 )
            for lev = param.levels
                mymessage(param,'making    a    graph-based    activation    (%s)    feature
map.\n',mapnames{fmapi});
                i = i + 1;
                [allmaps{i}.map,tmp] = graphsalapply( mapsobj.maps.val{typei}{lev} , ...
                    grframe, param.sigma_frac_act , 1 , 2 , param.tol );
                allmaps{i}.maptype = [ fmapi typei lev ];
            end
        else
            for centerLevel = param.ittiCenterLevels
                for deltaLevel = param.ittiDeltaLevels
                    mymessage(param,'making a itti-style activation (%s) feature map using center-
surround subtraction.\n',mapnames{fmapi});
                    i = i + 1;
                    center_ = mapsobj.maps.origval{typei}{centerLevel};
                    sz_ = size(center_);
                    surround_ = imresize( mapsobj.maps.origval{typei}{centerLevel+deltaLevel},
sz_ , 'bicubic' );

                    allmaps{i}.map = (center_ - surround_).^2;
                    allmaps{i}.maptype = [ fmapi centerLevel deltaLevel ];
                end
            end
        end
    end
end
%%%%
%%%% STEP 3 : normalize activation maps
%%%%
mymessage(param,'normalizing activation maps...\n');
norm_maps = {};
for i=1:length(allmaps)
    mymessage(param,'normalizing a feature map (%d)... ', i);
    if ( param.normalizationType == 1 )
        mymessage(param,' using fast raise to power scheme\n ', i);
        algtype = 4;
        [norm_maps{i}.map,tmp]    =    graphsalapply(    allmaps{i}.map    ,    grframe,
param.sigma_frac_norm, param.num_norm_iters, algtype , param.tol );
    elseif ( param.normalizationType == 2 )
        mymessage(param,' using graph-based scheme\n');
```

```
        algtype = 1;
        [norm_maps{i}.map,tmp]        =        graphsalapply(        allmaps{i}.map        ,        grframe,
    param.sigma_frac_norm, param.num_norm_iters, algtype , param.tol );
    else
        mymessage(param,' using global - mean local maxima scheme.\n');
        norm_maps{i}.map                                                                     =
    maxNormalizeStdGBVS( mat2gray(imresize(allmaps{i}.map,param.salmapsize, 'bicubic')) );
    end
    norm_maps{i}.maptype = allmaps{i}.maptype;
end
%%%%
%%%% STEP 4 : average across maps within each feature channel
%%%%
comb_norm_maps = {};
cmaps = {};
for i=1:length(mapnames), cmaps{i}=0; end
Nfmap = cmaps;
mymessage(param,'summing across maps within each feature channel.\n');
for j=1:length(norm_maps)
    map = norm_maps{j}.map;
    fmapi = norm_maps{j}.maptype(1);
    Nfmap{fmapi} = Nfmap{fmapi} + 1;
    cmaps{fmapi} = cmaps{fmapi} + map;
end
%%% divide each feature channel by number of maps in that channel
for fmapi = 1 : length(mapnames)
    if ( param.normalizeTopChannelMaps)
        mymessage(param,'Performing additional top-level feature map normalization.\n');
        if ( param.normalizationType == 1 )
            algtype = 4;
            [cmaps{fmapi},tmp] = graphsalapply( cmaps{fmapi} , grframe, param.sigma_frac_norm,
    param.num_norm_iters, algtype , param.tol );
        elseif ( param.normalizationType == 2 )
            algtype = 1;
            [cmaps{fmapi},tmp] = graphsalapply( cmaps{fmapi} , grframe, param.sigma_frac_norm,
    param.num_norm_iters, algtype , param.tol );
        else
            cmaps{fmapi} = maxNormalizeStdGBVS( cmaps{fmapi} );
        end
    end
    comb_norm_maps{fmapi} = cmaps{fmapi};
end
```

```
%%%%
%%%% STEP 5 : sum across feature channels
%%%%
mymessage(param,'summing across feature channels into master saliency map.\n');
master_idx = length(mapnames) + 1;
comb_norm_maps{master_idx} = 0;
for fmapi = 1 : length(mapnames)
    mymessage(param,'adding in %s map with weight %0.3g (max = %0.3g)\n', map_types{fmapi},
        mapweights(fmapi) , max( cmaps{fmapi}(:) ) );
    comb_norm_maps{master_idx}  =  comb_norm_maps{master_idx}  +  cmaps{fmapi}  *
        mapweights(fmapi);
end
master_map = comb_norm_maps{master_idx};
master_map = attenuateBordersGBVS(master_map,4);
master_map = mat2gray(master_map);
%%%%
%%%% STEP 6: blur for better results
%%%%
blurfrac = param.blurfrac;
if ( param.useIttiKochInsteadOfGBVS )
    blurfrac = param.ittiblurfrac;
end
if ( blurfrac > 0 )
    mymessage(param,'applying final blur with with = %0.3g\n', blurfrac);
    k = mygausskernel( max(size(master_map)) * blurfrac , 2 );
    master_map = myconv2(myconv2( master_map , k ),k');
    master_map = mat2gray(master_map);
end
if ( param.unCenterBias )
    invCB = load('invCenterBias');
    invCB = invCB.invCenterBias;
    centerNewWeight = 0.5;
    invCB = centerNewWeight + (1-centerNewWeight) * invCB;
    invCB = imresize( invCB , size( master_map ) );
    master_map = master_map .* invCB;
    master_map = mat2gray(master_map);
end
%%%%
```

```
%%% save descriptive, rescaled (0-255) output for user
%%%
feat_maps = {};
for i = 1 : length(mapnames)
    feat_maps{i} = mat2gray(comb_norm_maps{i});
end
intermed_maps = {};
for i = 1 : length(allmaps)
  allmaps{i}.map = mat2gray( allmaps{i}.map );
  norm_maps{i}.map = mat2gray( norm_maps{i}.map );
end
intermed_maps.featureActivationMaps = allmaps;
intermed_maps.normalizedActivationMaps = norm_maps;
master_map_resized = mat2gray(imresize(master_map,[size(img,1) size(img,2)]));
out = {};
out.master_map = master_map;
out.master_map_resized = master_map_resized;
out.top_level_feat_maps = feat_maps;
out.map_types = map_types;
out.intermed_maps = intermed_maps;
out.rawfeatmaps = rawfeatmaps;
out.paramsUsed = param;
if ( param.saveInputImage )
end
```

The stfeatures function is responsible for reading the three-dimensional video matrix and returning subs to indicate the point of interest, as shown by PRO-GRAMME 12.9.

PROGRAMME 12.9: Read the three-dimensional video matrix and return subs to indicate the point of interest

```
% Interface to spatio-temporal feature detectors.
%
% Wrapper for stfeatures detectors that first shrinks I and then resizes outputs
% appropriately.   It also applies nonmaximal supression and finds local maxima of detector
% output.
%
```

```
% INPUTS
%    I              - 3D uint (or double) array - input video
%    sigma          - spatial scale
%    tau            - temporal scale
%    periodic       - if 1 uses periodic detector else uses harris detector
%    thresh         - [optional] abolute threshold for features strength
%    maxn            - [optional] maximum number of featrues (alternative to thresh)
%    overlap_r      - [optional] controls amount of overlap between cuboids, range: [0,2)
%                        (0 -> no overlap, 2 -> minimal restriction on overlap; 1.7 -> default )
%    show           - [optional] figure to use for display (no display if == 0)
%    shr_spt        - [optional] spatial shrink factor.   Must be integer>1
%    tau_spt        - [optional] temporal shrink factor.   Must be integer>1
%
% OUTPUTS
%    R          - detector strength response at each image location
%    subs       - detected features locations
%    vals       - relative feature strengths
%    cuboids - the detected cuboids (size depends on sigma/tau)
%    V          - location of responses visualization
%
% EXAMPLE
%    load example;
%    [R,subs,vals,cuboids,V] = stfeatures( I, 2, 3, 1, 2e-4, [], 1.85, 2, 1, 1 );
%    [R,subs,vals,cuboids,V] = stfeatures( I, 2, 3, 0, eps, [], 1.85, 2, 1, 5 );
%
% See also STFEATURES_HARRIS, STFEATURES_PERIODIC, NONMAXSUPR_WINDOW
function [R,subs,vals,cuboids,V] = stfeatures( I, sigma, tau, periodic, thresh, maxn,overlap_r, shr_spt,
    shr_tmp, show )
if( ndims(I)~=3 ) error('I must a MxNxK array'); end;
if( nargin<5 || isempty(thresh)) thresh=eps; end;
if( nargin<6 || isempty(maxn)) maxn=-1; end;
if( nargin<7 || isempty(overlap_r)) overlap_r=1.7; end;
if( nargin<8 || isempty(shr_spt)) shr_spt=1; end;
if( nargin<9 || isempty(shr_tmp)) shr_tmp=1; end;
if( nargin<10 || isempty(show)) show=0; end;
thresh = max( thresh, eps );
```

```
%if( isempty(cuboids_rs) ) % pad a bit for filtering etc.
cuboids_rs = ceil( [sigma*3 sigma*3 tau*3] );    %#P
%%% shrink I [creat I_sm]
if( show ) disp('Shrink Image'); end;
shrink_all = [shr_spt shr_spt shr_tmp];
I_sm = imshrink( I, shrink_all );
sigma_sm = sigma / shr_spt;
tau_sm = tau / shr_tmp;
cuboids_rs_sm = ceil([cuboids_rs(1:2)/shr_spt, cuboids_rs(3)/shr_tmp]);
%%% convert I_sm to double in range [-1,1]
if( isa(I_sm,'uint8') ) I_sm = double(I_sm)/255 * 2 -1; end
if( ~isa(I_sm,'double') ) I_sm = double(I_sm); end
if( max(abs(I_sm(:)))>1 ) I_sm=I_sm-min(I_sm(:)); I_sm=I_sm/(max(I_sm(:))/2) - 1; end;
%%% apply feature extraction
if( show ) disp('Apply feature detector'); end;
if( periodic )
      R = stfeatures_periodic( I_sm, sigma_sm, tau_sm );
else
      R = stfeatures_harris( I_sm, sigma_sm, tau_sm );
end;
%%% Apply nonmaximal suppression, resize subs appropriately.   Note that
%%% cuboids_rs_sm is an approximation, so we use cuboids_rs for actual suppression
if( nargout<3 && show==0 ) return; end;
if( show ) disp( 'Apply nonmaximal supression'); end
suprradii_sm = max(1,ceil(cuboids_rs_sm*(2-overlap_r)));
[subs_sm, vals] = nonmaxsupr( R, suprradii_sm, thresh );
subs = round( imsubs_resize( subs_sm, shrink_all ) );
[subs, vals] = nonmaxsupr_window( subs,vals,1+cuboids_rs,size(I)-cuboids_rs,[],maxn);
subs_sm = round( imsubs_resize( subs, 1./shrink_all ) );
%%% extract cuboids
if( nargout<4 && show==0 ) return; end;
if( length(subs)==0 ) cuboids=[]; return; end;
cuboids = cell2array( cuboid_extract( I, cuboids_rs, subs, 0 ) );
if( nargout<5 && show==0 ) return; end;
V = cuboid_display( I_sm, cuboids_rs_sm, subs_sm );
%%% optionally display
```

```
if( show )
    disp( ['Create visualization; ' num2str(size(vals,1)) ' features' ]);
    figure(show); clf; montage2( I_sm,1 );
    figure(show+1); clf; montage2( R,1,1 ); colormap jet;
    figure(show+2); clf; montage2( V,1,1   );
    figure(show+3); clf; montages2( cuboids, {1} );
end;
```

The thresh_matrix function is responsible for reading the 3D video matrix and the foreground area, and returns thresh to represent the threshold matrix, as shown by PROGRAMME 12.10.

PROGRAMME 12.10: Read three-dimensional video matrix and foreground area, return thresh to indicate threshold matrix

```
%% Calculate the threshold matrix
function thresh = thresh_matrix(I,rect,factorIn,factorOut,delta_frame)
    [row,column,frame] = size(I);
    thresh = [];
    temp = [];
    thresh = zeros(row,column,frame);
    io_Matrix = [];
    S_in = zeros(1,frame);
    S_out = zeros(1,frame);


    [io_Matrix,S_in,S_out] = IO_Matrix(I,rect);


    for t = 1:frame
        for i =1 :row
            for j=1:column
                if (io_Matrix(i,j,t) == 1)
                    thresh(i,j,t) = (S_in(1,t)/I(i,j,t)) * factorIn;
                else
                    thresh(i,j,t) = (S_out(1,t)/I(i,j,t)) * factorOut;
                end
            end
        end
    end
```

```
    temp = thresh;

    for t = 1+delta_frame : frame - delta_frame-1
        for m = t - delta_frame:t+delta_frame
            thresh(:,:,t) = temp(:,:,t) + temp(:,:,m);
        end
        thresh(:,:,t) = thresh(:,:,t)./(2*delta_frame+1);
    end
end
```

The Create_Descriptor function is responsible for reading the three-dimensional video matrix and returns the keypoint to represent the 3D-SIFT descriptor, as shown by PROGRAMME 12.11.

PROGRAMME 12.11: Read the 3D video matrix and return the keypoint to represent the 3D-SIFT descriptor

```
function [keypoint reRun] = Create_Descriptor(pix, xyScale, tScale, x, y, z)
% Main function of 3DSIFT Program from http://www.cs.ucf.edu/~pscovann/
%
% Inputs:
% pix - a 3 dimensional matrix of uint8
% xyScale and tScale - affects both the scale and the resolution, these are
% usually set to 1 and scaling is done before calling this function
% x, y, and z - the location of the center of the keypoint where a descriptor is requested
%
% Outputs:
% keypoint - the descriptor, varies in size depending on values in LoadParams.m
% reRun - a flag (0 or 1) which is set if the data at (x,y,z) is not
% descriptive enough for a good keypoint
%
% Example:
% See Demo.m
LoadParams;
reRun = 0;
radius = int16(xyScale * 3.0);
fv = sphere_tri('ico',Tessellation_levels,1);
myhist = buildOriHists(x,y,z,radius,pix,fv);
```

```
[yy ix] = sort(myhist,'descend');

% Dom_Peak = ix(1);

% Sec_Peak = ix(2);

if     (TwoPeak_Flag    ==    1    &&    dot(fv.centers(ix(1),:),fv.centers(ix(2),:))    >    .9    &&
       dot(fv.centers(ix(1),:),fv.centers(ix(3),:)) > .9)
         disp('MISS : Top 3 orientations within ~25 degree range : Returning with reRun flag set.');
         keypoint = 0;
         reRun = 1;
         return;
end

keypoint = MakeKeypoint(pix, xyScale, tScale, x, y, z);
```

HSoptflow function calculates the image corresponding to the optical flow characteristics, return us, vs represent the two directions of optical flow, as shown in PROGRAMME 12.12.

PROGRAMME 12.12: Calculate the corresponding optical flow characteristics of the image, return us, vs represent the two directions of the optical flow

```
function [us,vs] = HSoptflow(Xrgb,n)

%%%%%%%%%%%%%%%%%%%%%%%%%%%%%%%%%%%%%%%%%%%%%%%%%%%%%%%
     %%%%%%%%%%%

% Author: Gregory Power gregory.power@wpafb.af.mil

% This MATLAB code shows a Motion Estimation map created by

% using a Horn and Schunck motion estimation technique on two

% consecutive frames.    Input requires.

%        Xrgb(h,w,d,N) where X is a frame sequence of a certain

%                       height(h), width (w), depth (d=3 for color),

%                       and number of frames (N).

%        n= is the starting frame number which is less than N

%        V= the output variable which is a 2D velocity array

%

% Sample Call: V=HSoptflow(X,3);

%%%%%%%%%%%%%%%%%%%%%%%%%%%%%%%%%%%%%%%%%%%%%%%%%%%%%%%
     %%%%%%%%%%%

[h,w,d,N]=size(Xrgb);

if n>N-1

    error(1,'requested file greater than frame number required');

end;
```

```
%get two image frames
if d==1
    Xn=double(Xrgb(:,:,1,n));
    Xnp1=double(Xrgb(:,:,1,n+1));
elseif d==3
    Xn=double(Xrgb(:,:,1,n)*0.299+Xrgb(:,:,2,n)*0.587+Xrgb(:,:,3,n)*0.114);
Xnp1=double(Xrgb(:,:,1,n+1)*0.299+Xrgb(:,:,2,n+1)*0.587+Xrgb(:,:,3,n+1)*0.114);
else
    error(2,'not an RGB or Monochrome image file');
end;
%get image size and adjust for border
size(Xn);
hm5=h-5; wm5=w-5;
z=zeros(h,w); v1=z; v2=z;
%initialize
dsx2=v1; dsx1=v1; dst=v1;
alpha2=625;
imax=20;
%Calculate gradients
dst(5:hm5,5:wm5) = ( Xnp1(6:hm5+1,6:wm5+1)-Xn(6:hm5+1,6:wm5+1) + Xnp1(6:hm5+1,5:wm5)-
    Xn(6:hm5+1,5:wm5)+    Xnp1(5:hm5,6:wm5+1)-Xn(5:hm5,6:wm5+1)    +Xnp1(5:hm5,5:wm5)-
    Xn(5:hm5,5:wm5))/4;
dsx2(5:hm5,5:wm5) = ( Xnp1(6:hm5+1,6:wm5+1)-Xnp1(5:hm5,6:wm5+1) + Xnp1(6:hm5+1,5:wm5)-
    Xnp1(5:hm5,5:wm5)+    Xn(6:hm5+1,6:wm5+1)-Xn(5:hm5,6:wm5+1)    +Xn(6:hm5+1,5:wm5)-
    Xn(5:hm5,5:wm5))/4;
dsx1(5:hm5,5:wm5) = ( Xnp1(6:hm5+1,6:wm5+1)-Xnp1(6:hm5+1,5:wm5) + Xnp1(5:hm5,6:wm5+1)-
    Xnp1(5:hm5,5:wm5)+    Xn(6:hm5+1,6:wm5+1)-Xn(6:hm5+1,5:wm5)    +Xn(5:hm5,6:wm5+1)-
    Xn(5:hm5,5:wm5))/4;
for i=1:imax
    delta=(dsx1.*v1+dsx2.*v2+dst)./(alpha2+dsx1.^2+dsx2.^2);
    v1=v1-dsx1.*delta;
    v2=v2-dsx2.*delta;
end;
u=z; u(5:hm5,5:wm5)=v1(5:hm5,5:wm5);
v=z; v(5:hm5,5:wm5)=v2(5:hm5,5:wm5);
xskip=round(h/64);
[hs,ws]=size(u(1:xskip:h,1:xskip:w))
us=zeros(hs,ws); vs=us;
N=xskip^2;
```

```
for i=1:hs-1
  for j=1:ws-1
    hk=i*xskip-xskip+1;
    hl=i*xskip;
    wk=j*xskip-xskip+1;
    wl=j*xskip;
    us(i,j)=sum(sum(u(hk:hl,wk:wl)))/N;
    vs(i,j)=sum(sum(v(hk:hl,wk:wl)))/N;
  end;
end;
figure(1);
quiver(us,vs);
colormap('default');
axis ij;
axis tight;
axis equal;
```

The gradient Histogram function is responsible for reading the information of the two dimensions of the optical stream and the number of bins, returning ohog to represent the HOOF feature, as shown by PROGRAMME 12.13.

PROGRAMME 12.13: Read the information of the two dimensions of the optical flow feature and the number of bin, return ohog to represent the HOOF feature

```
function ohog = gradientHistogram(Fx,Fy,binSize)
% Compute HOOF feature
% INPUTS
%    Fx       - X-flow
%    Fy       - Y-flow
%    binSize - number of bins used
%
% OUTPUTS
%    ohog      - output histogram of oriented optical flow
%
% EXAMPLE
%
%% Written by    : Rizwan Chaudhry and Avinash Ravichandran
%% $DATE          : 28-Aug-2008 11:00:58 $
```

```
%% $Revision     : 1.00 $
%% Matlab        : 7.4.0.287 (R2007a)
%% FILENAME      : gradientHistogram.m
%
% (c) Rizwan Chaudhry, Avinash Ravichandran - JHU Vision Lab
%%%%%%%%%%%%%%%%%%%%%%%%%%%%%%%%%%%%%%%%%%%%%%%%%%%%
%gradientHistogram.m
%%%%%%%%%%%%%%%%%%%%%%%%%%%%%%%%%%%%%%%%%%%%%%%%%%%%
magnitudeImage     = (Fx.^2 + Fy.^2 ).^0.5;
orientationImage =    atan2(Fy,Fx);
greaterPiBy2Index = orientationImage > pi/2;
smallerMinusPiBy2Index = orientationImage < -pi/2;
remainingIndex = orientationImage <=pi/2 & orientationImage >= -pi/2;
greaterPiBy2Mat = greaterPiBy2Index.*orientationImage;
smallerMinusPiBy2Mat = smallerMinusPiBy2Index.*orientationImage;
remainingMat = remainingIndex.*orientationImage;
piMat = pi*ones(size(orientationImage));
convertGreaterPiBy2Mat = greaterPiBy2Index.*piMat - greaterPiBy2Mat;
convertSmallerMinusPiBy2Mat = smallerMinusPiBy2Index.*(-piMat) - smallerMinusPiBy2Mat;
newOrientationImage = convertGreaterPiBy2Mat + remainingMat + convertSmallerMinusPiBy2Mat;
% [hog,idx] =
% histc(reshape(orientationImage,1,[]),linspace(-pi,pi,binSize+1) );
[hog,idx] = histc(reshape(newOrientationImage,1,[]),linspace(-pi/2,pi/2,binSize+1) );
values = reshape(magnitudeImage,1,[]);
for k=1:binSize
    bin(k) = sum(values(find(idx==k)));
end
ohog = bin/sum(bin);
ohog = ohog';
```

The spectral_cluster function is responsible for reading the eigenvector and the number of categories, returning IDX to represent each class number of the eigenvector, and C for the clustering center, as shown by PROGRAMME 12.14.

PROGRAMME 12.14: Read the eigenvector and the number of categories return IDX represents each class number of the eigenvector

```
function [IDX, C, SUMD, D] = spectral_cluster(cora_wd,K)

    W =1-squareform(pdist(cora_wd,'euclidean'));
%       W(W< mean(mean(W))) = 0;
    D = diag(sum(W,2));
    L=D-W;
    [V, dummy] = eig(L, D);
    V = V(:, 1:K);
    [IDX, C, SUMD, D] = kmeans(V, K,'emptyaction','drop');
end
```

The LDAGStrain and LDAGSpredict functions are responsible for the training and prediction of the LDA model, returning dp_te to represent the final classification result, as shown in PROGRAMME 12.15.

PROGRAMME 12.15: Training and prediction of the LDA model, returning dp_te to indicate the final classification result

```
clc
clear
ALPHA = 1e-2;
BETA = 1e-2;
N = 500;
SEED = 1;
OUTPUT = 1;
load '/volume1/huangyiqing/workspace/xiefei_temp/UCF/lda/wd/wd.mat'
KTH_wd = sparse(wd);
J=4;
            tic
            [wp, dp, z] = LDAGStrain(KTH_wd, J, N, ALPHA, BETA, SEED, OUTPUT);
            t = toc;
            [dp_te, z] = LDAGSpredict(KTH_wd, wp, N, ALPHA,BETA, SEED, OUTPUT);
dp = full(dp);
dp_te = full(dp_te);
```

(a)   (b)   (c)   (d)   (e)

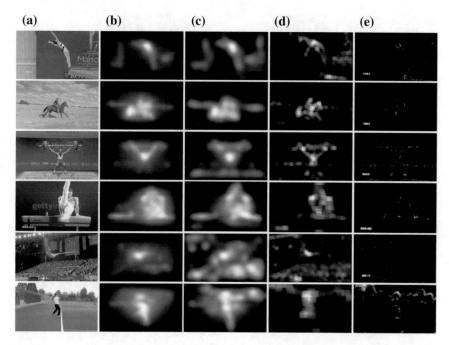

Fig. 12.9 Comparison results for each salience model **a** the original image, **b** GBVS significant figure, **c** itti significant figure, **d** PQFT significant graph, **e** the residual spectrum is significant

Choose a few paragraphs from the UCF library video, then use GBVS function of each frame to calculate the picture. Compare the results of various significant graph, as shown in Fig. 12.9.

Use the function stfeatures to calculate the points of interest for each frame, as shown in Fig. 12.10.

The HOOF feature is calculated using the HSoptflow function and the gradientHistogram function. The results of the optical flow are shown in Fig. 12.11.

After calculating the 3D-SIFT feature and HOOF feature using the functions Create_Descriptor, HSoptflow and gradientHistogram. The spectral_cluster function is used to cluster the features. Finally, the LDAGSpredict function is used to predict the video behavior.

(a) boxing   (b) jogging

(c) diving   (d) swing

(f) riding   (h) benching

Fig. 12.10 Results of the point of interest detection

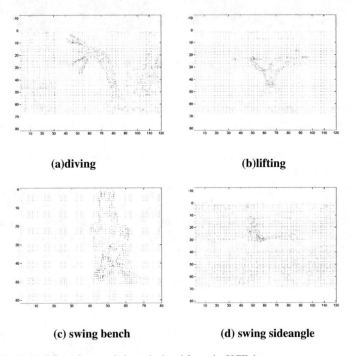

(a)diving   (b)lifting

(c) swing bench   (d) swing sideangle

Fig. 12.11 Optical flow characteristics calculated from the UCF dataset

Chapter 13
Image Understanding-Person Re-identification

Abstract In this chapter, we talk about one of the typical image understanding problems—cross-camera person re-identification. Some classical visual descriptors and metric learning algorithms for person re-identification are detailed.

13.1 Introduction

One important task in image understanding is the cross-view (or cross-modal) retrieval problem. Given one instance as the probe, the objective is to find the most similar (or relevant) instances from a large number of galleries. In this chapter, we talk about a special task in this research field, which is usually called person re-identification (Re-id).

Person re-identification is the task of matching individuals observed from non-overlapping camera views. It is the fundamental task of many applications in video surveillance, such as cross-camera tracking and human retrieval. For example in the cross-camera tracking scenario, when one interested person disappeared from one camera view, we have to identify him/her from another view. The matching task is just the job of Re-id. Another example is the long-term tracking of one specified person in a large-scale camera network. When he/she reappears in the view after a while of occlusion, we need to assign the same label to him/her, then it is also a procedure of Re-id.

To realize the surveillance of a large scope, the cameras are usually set in high locations. The captured pedestrian images are typically of low resolutions and quality. This leads to the non-reliable bio-characteristics, such as face, iris, and gait. As a result, the Re-id task has to rely upon the appearance information of pedestrians. However, due to large variations in image condition caused by viewpoint, illumination, pose, and occlusions, one person's appearance may change significantly in different camera views. This makes the Re-id task inherently rather challenging.

The person re-identification and pedestrian detection are two different concepts but maybe easily confused at the first glance. The pedestrian detection refers to the task of finding out persons in one image or video clip, whereas re-identification focus

© Springer International Publishing AG, part of Springer Nature 2019   475
S. Gong et al., *Advanced Image and Video Processing Using MATLAB*,
Modeling and Optimization in Science and Technologies 12,
https://doi.org/10.1007/978-3-319-77223-3_13

on identifying a specified person from a large number of candidates captured from another camera view. Therefore, they are completely different tasks in computer vision. Nevertheless, they are also closely related. In current literature of person re-identification, it is generally assumed that the persons in the video frames are already detected and cropped out by bounding boxes. Thus, the person re-identification task is somewhat relied on the pedestrian detection first.

Let P be the feature vector representing the pedestrian image used as the probe, the re-identification task can be formulated as:

$$T = \arg\min d(T_i, P), T_i \in T \tag{13.1}$$

where $T = \{T_1, T_2, \ldots, T_N\}$ represents the set of N images to be matched which is usually called the gallery set, and $d(\cdot)$ is a certain distance. From Eq. (13.1), we can find that there are two most important components in person re-identification task:

(1) the extraction of robust feature representations,
(2) a reliable distance measurement.

To achieve efficient re-identification, we have to extract both discriminative and robust descriptors from pedestrian images first, and then choose a certain distance metric to measure the similarity between one image pair.

In Fig. 13.1, we show some example images randomly chosen from public person re-identification datasets. It can be found that even the same person may take strong different appearances in two non-overlapping camera views. In the first pair, the appearances suffer the illumination conditions heavily. In the second pair large portion of the man in the left image is occluded by one woman with long hair. In the third pair, the two images are of different resolutions, especially the right one is of rather low quality. In the fourth pair we can find the appearance of one person may be also heavily affected by different views of two cameras. In the last pair, the images are captured from different persons though their appearances are rather similar due to the same clothes.

Fig. 13.1 Example image pairs from public person re-identification datasets

13.2 Person Re-ID Scenarios

The person re-identification can be classified to different scenarios according to different criteria. For example, according to the corpus form, we can group the methods into image-based or video-based categories. We can also classify them to open set re-identification and closed set re-identification. Here the closed set re-identification means that every gallery image has its correct match in the probe set, while in open set case, there may be some gallery images have no corresponding probe images. As a result, the open set person re-identification task is much more difficult than the closed set scenario due to the existence of many distractors.

(1) Image based and Video based Person Re-identification

The image based person re-identification refers to the task of matching cross-camera pedestrian images, whereas the video based re-identification means the provided material for matching are video clips. Since the video comprises multiple frames which can provide much more robust appearance information, the re-identification is much easier than the image based case. But the computation cost is much higher due to more data is involved.

Since video comprises multiple frames, the video based person re-identification can be simply viewed as the extension of image based scenario with multiple images are provided. Therefore, the image based person re-identification attracts much more attention than video based re-identification. In practice, if there are only one probe image and one gallery image for each pedestrian, the re-identification is called Single-shot versus Single-shot (SvsS) task. In contrast, if there are multiple probe and gallery images for one pedestrian, we call this case the Multiple-shot versus Multiple-shot (MvsM) re-identification. As limited appearance information can be obtained from only single image, the SvsS re-identification is rather challenging. The matching accuracy on rank-1 is generally very low. Nevertheless, it is the basis of other type re-identification tasks. Thus most re-identification researches are carried out to tackle SvsS re-identification problem.

By applying max pooling or average pooling operation to the feature vectors extracted from the images of one person, more stable and robust feature can be obtained in the MvsM re-identification task. In consequence much higher identification accuracy can be achieved. Another way of tackling the MvsM re-identification task is to compute the distances between multiple image pairs and then obtain their mean value for ranking. This can also lead to much higher matching performance than SvsS case.

The video based person re-identification can be simply transformed to the MvsM re-identification by omitting the temporal information. However, this may damage the matching accuracy as the temporal information is abandoned. To make full use of both spatial and temporal information provided in the continuous video frames, the descriptors which can well capture both aspects are commonly employed, e.g., the HOG3D descriptor [1, 2].

(2) Open Set Re-identification and Closed Set Re-identification

Person re-identification in the context of identity retrieval is closer to the classic closed set matching problem, where both probe and gallery sets are fixed. In this case, one person may have multiple observations throughout the network, and his/her images are assumed to be available in every camera view. Thus the gallery set is a set of people IDs seen in selected or all the cameras over a specified period of time. In other words, the gallery set includes many subjects observed by different cameras. After re-identification, there may be multiple observations of the probe retrieved. As a result, the closed set re-identification is a one-to-many matching problem in somewhat ideal condition.

In the context of tracking a specific individual across multiple cameras is a typical open set person re-identification problem. As the gallery evolves over time, there may be no correct matches in the gallery for some probes. Additionally, there might be several subjects that co-exist in time and need to be re-identified simultaneously. In tracking scenario, re-identification provides a means of connecting subjects' tracks that were disconnected due to the subject entering an area not in the field-of-view (FOV) of the camera network. Due to the more comprehensive evolution of gallery and probe sets, open set re-identification is much more difficult, and the research work in this case is more valuable in video surveillance.

There have been some re-identification datasets that simulate the open set re-identification, such as PRID2011 (or PRID) [3] and QMUL GRID [4]. In the PRID2011 dataset, the probe contains 385 images and the gallery images are 749. But there are only 200 images in the probe set have their corresponding correct matches in the gallery set. When we use these 200 images in the probe set to query their correct matches in the gallery set, the extra images will act as the distractors in the real world scenario.

13.3 Methodology

Current person re-identification methods typically contain two main components:

(1) feature representation extraction [5–9],
(2) matching model learning [10–13].

Some works focus on designing feature representations while some others emphasize on learning the matching models.

The re-identification methods based on feature representations aim to design discriminative features to capture the invariance of pedestrian appearances. Since the re-identification task mainly relies on the appearance information, the following aspects of appearance are generally considered in visual features for person re-identification:

(1) color, widely used since the color of clothing constitutes simple but efficient visual signatures, usually encoded within histograms of RGB or HSV values,
(2) shape, e.g. using HOG [14] based signature,
(3) texture, often represented by Gabor filters [15, 16] and some other filters, co-occurrence matrices [17] and Local Binary Pattern (LBP) [18],
(4) interest points, e.g. SURF and SIFT [19],
(5) image regions [6].

Besides these generic representations, there are some more specialized representations, e.g. Epitomic Analysis, Spin Images, Bag-of-Word-based description, and Panoramic Map.

Since different elementary features capture different and complementary aspects of the image, better performance is obtained by combining several signatures. After extracting the visual features, some generic distance metrics without learning procedure are employed to measure the cross-view image pairs, namely L1 or L2 norm, Bhattacharyya distance [6], and χ^2 distance [16].

The re-identification methods based on matching models pays more attention to learn a certain model for matching cross-view image pairs. In current literature, these methods are more prevailing due to higher matching performance, and they can be generally grouped into four categories:

(1) learning SVM models,
(2) learning distance metrics,
(3) learning discriminative dictionaries,
(4) learning deep models.

The idea of learning SVM models lies in that the similarity score of one positive pair (two images from the same person) should be higher than negative ones (images from different persons). This can be formulated as $w^T \phi(x_i, z_j) > w^T \phi(x_i, z_k)$, where (x_i, z_j) represent the same person while (x_i, z_k) represent different persons, ϕ is a mapping function, and w is the parameter to be learned. The metric learning models aim to learn a Mahalanobis distance function $d_M^2(x_i, z_j) = (x_i - z_j)^T M (x_i - z_j)$ parameterized by a positive semi-definite (PSD) matrix M. The basic idea of metric learning is to project the samples into a new space so that the samples of the same class are much closer, and meanwhile the samples of different classes are pushed far away. This is because M can be decomposed as $M = LL^T$ due to its PSD property. Then the Mahalanobis distance function can be reformulated as $d_M^2(x_i, z_j) = \left\| L^T(x_i - z_j) \right\|_2^2$, which means the original Mahalanobis distance is identical to the Euclidean distance in a new subspace. So there are two ways to learn distance metrics, i.e., learning a Mahalanobis form metric and learning a subspace. One advantage of learning Mahalanobis form metric is that the model is always convex, but this needs to project the metric onto convex cone in the optimization procedure. On the contrary, there is no such trouble of ensuring the PSD property in learning a subspace, but the learning model is not convex and only a local optimal solution can be guaranteed. Since the metric learning can explore the second-order information between sample pairs due to the quadratic formulation, better performance is usually yield by metric learning than SVM models in reported results.

With the success of deep learning modes [20] in other research fields of computer vision and machine learning, there are also some works try to use the deep models for matching cross-view image pairs. Since the convolutional neural networks (CNN) have excellent ability to learn features from raw pixel values, the employed deep models are generally the variants of CNN [21–23]. In these models, the procedures of feature extraction and matching model learning are deeply coupled, thus they don't resemble the "two-step" re-identification pipeline. Besides CNN, there are few works utilize the fully connected networks and hand-crafted features to map samples into a new space for matching. Since the Euclidean distance is generally used as the metric to measure the mapped feature vectors, we can view these models as deep non-linear metric learning models due to their non-linear mapping.

13.4   Public Datasets and Evaluation Metrics in Person Re-identification

As has been discussed in aforementioned sections, the visual characteristics of a person vary drastically across cameras in real world scenario, resulting large variances in illumination, poses, view angles, scales and camera resolutions. Factors like occlusions, cluttered background and articulated bodies further add to visual appearance changes. Thus, in order to develop robust re-identification techniques, it is important to build evaluation datasets that capture these factors effectively. Along with high quality data emulating real world conditions, there is also a need to compare re-identification approaches being developed and identify improvements to techniques. There are several available datasets that have been used to test Re-ID models, such as VIPeR [15], PRID450S [24], 3DPeS [25], QMUL GRID [4], CUHK01 [26], CUHK02 [27], CUHK03 [28], Market-1501 [29], PRW [29], etc. Table 13.1 provides a summary of the widely used Re-ID datasets.

Beside public datasets, the evaluation metric also plays an import role in advancing the re-identification research. With years efforts a number of evaluation metrics have been designed to measure the performance of person re-identification techniques, including the cumulative match curve (CMC), top match ranking rate, area under curve (AUC), and so on.

13.4.1   Public Datasets

Currently, one of the most popular and challenging datasets to test people re-identification for image retrieval is VIPeR, which contains 632 pedestrian image pairs taken from arbitrary viewpoints under varying illumination conditions. The

Table 13.1 Summary of public person re-identification datasets

Dataset	Number of persons	Number of cameras	Published year
VIPeR [15]	632	2	2007
ETH1,2,3 [30]	85,35,28	1	2007
i-LIDS MCTS [29]	119	2	2009
GRID [4]	250	8	2009
CAVIAR4REID [5]	72	2	2011
3DPeS [25]	192	8	2011
PRID2011 [3]	934	2	2011
PRID450S [24]	450	2	2014
SAIVT-Softbio [31]	150	8	2012
CUHK01 [26]	972	2	2012
CUHK02 [27]	1816	10 (5 pairs)	2013
CUHK03 [28]	1467	2	2014
iLIDS-VID [32]	300	2	2014
Market1501 [29]	1501	6	2015
PRW [29]	932	6	2016

dataset was collected in an academic setting over the course of several months. Each image is scaled to 128×48 pixels. The images in this dataset are captured from 5 different view angles, including $0°$, $45°$, $90°$, $135°$, and $180°$. Due to complex view angles and the low resolution of images, the published results on this dataset are generally very low. Actually, some matches are hard to identify even by a human. This dataset cannot be fully employed for evaluating methods exploiting multiple shots, video frames, or 3D models, since only one pair of bounding boxes of the same person is collected (Fig. 13.2).

Strictly speaking, the ETHZ dataset is not a standard re-identification dataset, because it was generated from the original ETHZ video dataset captured by only one moving cameras. This dataset is composed of three video sequences which contain 85, 35, and 28 pedestrians respectively. This camera setup provides a range of variations in people's appearances, with strong changes in pose and illumination. As a relatively old dataset, the re-identification accuracy on ETHZ has achieved saturation now.

The i-LIDS Multiple-Camera Tracking Scenario (MCTS) dataset was captured indoor at a busy airport arrival hall. It contains 119 people with a total of 476 shots captured by multiple non-overlapping cameras with an average of four images for each person. Many of these images undergo large illumination changes and subject to heavy occlusions. Most of the people in this dataset are carrying bags or suitcases. These accessories and carried objects can be profitably used to match their owners, but they introduce a lot of occlusions which usually act against the matching. In addition, images have been taken with different qualities (in terms of resolution, zoom level, noise), making very challenging the re-identification over this dataset.

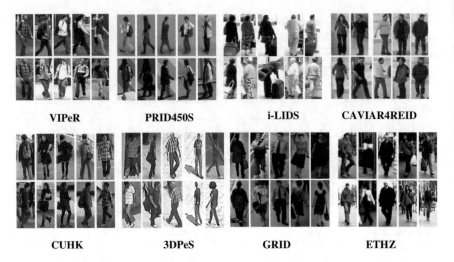

VIPeR	PRID450S	i-LIDS	CAVIAR4REID

CUHK	3DPeS	GRID	ETHZ

Fig. 13.2   Example image pairs from 8 person re-identification datasets

CAVIAR4REID dataset is extracted from a multi-camera tracking dataset captured at an indoor shopping mall by two cameras. It contains multiple images of 72 pedestrians, out of which only 50 appear in both cameras, whereas 22 come from the same camera. The images for each pedestrian take serious appearance variations due to changes of resolution, light, pose, and occlusions. The minimum and maximum size of the images is 17×39 and 72×144, respectively. Due to these challenges, the re-identification on this dataset is rather difficult.

The PRID 2011 dataset consists of person images recorded from two different static cameras. Two scenarios are provided: multi-shot and single-shot. Since we are focusing on single-shot methods in this work, we use only the latter one. Typical challenges on this dataset are viewpoint and pose changes as well as significant differences in illumination, background and camera characteristics. Camera view A contains 385 persons, camera view B contains 749 persons, with 200 of them appearing in both views. Hence, there are 200 person image pairs in the dataset. These image pairs are randomly split into a training and a test set of equal size. For evaluation on the test set, we followed the procedure described in [26], i.e., camera A is used for the probe set and camera B is used for the gallery set. Thus, each of the 100 persons in the probe set is searched in a gallery set of 649 persons (all images of camera view B except the 100 training samples).

The PRID 450S dataset is built on PRID 2011. However, it is arranged in a way similar to VIPeR dataset and more samples than PRID 2011 are included. In particular, the dataset contains 450 single-shot image pairs depicting walking humans captured in two spatially disjoint camera views. For each image instance a binary segmentation mask is provided to separate the foreground from the background. Moreover, it further provides a part-level segmentation 3 describing the following regions: head, torso, legs, carried object at torso level (if any) and carried object below torso (if any). The union of these part segmentations is equivalent to the foreground segment.

The QMUL under GRound Re-IDentification (GRID) dataset is another challenging person re-identification dataset. It was captured from 8 disjoint camera views in a underground station. There are 250 pedestrian image pairs, with each pair contains two images of the same person from different camera views. Besides, there are 775 additional images that do not belong to the 250 persons which can be used to enlarge the gallery set. The images in this dataset have poor image quality and low resolutions, and contain large variations of illumination and viewpoint.

The CUHK01, CUHK02, and CUHK03 person re-identification datasets were collected by the Multimedia Laboratory of Chinese University of Hong Kong. All of them were captured in a campus environment. The CUHK Campus dataset contains 971 persons, and each person has two images in each camera view. Camera A captures the frontal view or back view of pedestrians, while camera B captures the side views. Different from the above datasets, images in this dataset are of higher resolution. All images were scaled to 160×60 pixels. The CUHK02 contains 1816 pedestrians organized in 5 folders. The number of pedestrian images in CUHK03 dataset is much larger. There are total 1360 pedestrians with 13164 images captured from 6 cameras, which makes CUHK03 one of the largest person re-identification datasets. In addition to manually cropped pedestrian images, samples detected with a state-of-the-art pedestrian detector is also provided in CUHK03. Different from CUHK01 and CUHK02, CUHK03 is a more realistic setting with misalignment, occlusions and body part missing.

There are also a line of datasets published in latest years, such as Market1501, PRW, and MARS. Some datasets tried to incorporate the bio-character for re-identification, such as SAIVT-Softbio [31] which provides the gait information to assist the appearance based person re-identification.

Although the public re-identification datasets have greatly promoted the research, there is still a big gap between them and the actual environment. First, the cameras in one city may amounts to tens of thousands, whereas the camera number in above re-identification dataset are all less than 10. Only some larger datasets contain 6 to 8 cameras, even though they cannot simulate the real work scenario. Due to the tedious labeling work, generating a re-identification is rather cost prohibitive. So the re-identification datasets are also much smaller than the datasets for other tasks, such as ImageNet [33] for image classification, and LFW for face recognition. The relatively less instances may also have an impact on the performance of deep learning models.

13.4.2   Evaluation Metrics

The Cumulative Matching Characteristic (CMC) curve is the most widely used evaluation protocol in person re-identification. Because the person re-identification can be treated as a fine-grained recognition and retrieval problem, we can rank the gallery images according to their distances or similarities with the probe image, and then compute the matching accuracies on each rank. This provides a ranking for every image in the gallery w.r.t the probe. This procedure is repeated for every image in the probe set and averaged. By accumulating the accuracies on each rank and plot them, a CMC curve is obtained. The CMC curve is then the expectation of finding the correct match in the top *n* matches.

Synthetic Recognition Rate (SRR) curve is another evaluation protocol based on CMC curve. It measures the probability that any of the *k* best matches is correct. Since the SRR curve is not as intuitive as the CMC curve and it is computed from CMC, few works have reported it for comparison in latest year.

Normalized Area Under Curve (nAUC) is the area under the CMC curve, which is the scalar appraisal of CMC curves and can be used to summarize the overall performance. The higher the nAUC is, the better the performance is.

The Proportion of Uncertainty Removed [34] (PUR) is also a scalar standard for evaluating re-identification algorithms. It measures the entropy reduction of finding the correct matches before and after applying re-identification techniques. The formulation of PUR is as follows:

$$PUR = \frac{\log(S) - \sum_{r=1}^{s} M(r) \log(M(r))}{\log(S)} \qquad (13.2)$$

Rank-1 matching rate and CMC-expectation are two scalar standards obtained from CMC curve. The Rank-1 matching rate is just the matching accuracy on the first rank, which is the most important concern for re-identification operators. A high matching rate on the first rank will greatly ease the human labor in real work applications. But there may be some algorithms that have high re-identification accuracy on the top ranks but not ideal on higher ranks. In this case, the CMC-expectation may be a proper evaluation standard since it computes the mean value of the matching rates on all ranks. The smaller the CMC-expectation is, the higher performance.

13.5   Classic Feature Representations for Person Re-identification

To capture the rich appearance information of pedestrians, a number of feature representations have been designed. With these feature representations, the research on re-identification field has been greatly advanced.

13.5.1   Salient Color Names

Salient Color Names [9] (SCN) is a feature representation specially designed for person re-identification. The SCN uses 16 standard RGB colors as the salient colors,

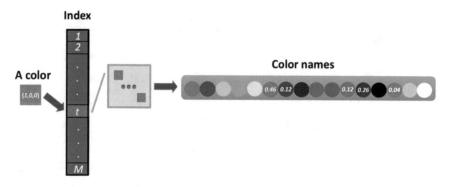

Fig. 13.3 Illustration of the SCN extraction procedure

namely fuchsia, blue, aqua, lime, yellow, red, purple, navy, teal, green, olive, maroon, black, gray, silver, and white. The detailed RGB values can be referenced from http://www.wackerart.de/rgbfarben.html. By building a 16-dimensional vocabulary of the salient colors, we can further compute the statistical distribution of one image's pixel values over them. This just resembles the computation of "bag-of-words" features. The extraction procedure of SCN feature is shown in Fig. 13.3, we detail it in the follows.

The SCN can be viewed as a high-level color distribution based visual descriptor. Although the color histograms have been widely used to describe pedestrian appearance, they are not robust to the variations of illumination and background clutter. In contrast, the SCN only focus on the pixel value distribution over some salient colors, thus it has better photometric invariance and robustness against the illumination changes.

To compute the SCN feature, it is recommended to normalize the pixel values of one image into [0, 1] for all 3 channels of RGB. Then the RGB color space is divided into $32 \times 32 \times 32 = 32,768$ equally spaced cubes. Therefore, each cube contains $8 \times 8 \times 8 = 512$ colors. Let $d = \{w_1, \ldots w_{512}\}$ be a set of the colors in one cube. The most important step is in computing SCN is to calculate the distribution of 512 colors in d over the 16 salient colors.

Let $Z = \{z_1, z_2, \ldots z_{16}\}$ denote the 16 salient colors have been assigned the standard names (i.e., salient color names), then the probability of assigning d to a color name $z(z \in Z)$ is

$$p(z|d) = \sum_{n=1}^{512} p(z|w_n)p(w_n|d) \tag{13.3}$$

where

$$p(z|w_n) = \begin{cases} \dfrac{\exp(-\|z-w_n\|^2 / \frac{1}{K-1} \sum_{z_l \neq z} \|z_l-w_n\|^2)}{\sum \exp(-\|z_p-w_n\|^2 / \frac{1}{K-1} \sum_{z_s \neq z_p} \|z_s-w_n\|^2)}, & if \ z \in KNN(w_n) \\ \\ 0, & otherwise \end{cases} \tag{13.4}$$

and

$$p(w_n|d) = \frac{\exp(-\alpha\|w_n - \mu\|^2)}{\sum_{l=1}^{512} \exp(-\alpha\|w_l - \mu\|^2)} \tag{13.5}$$

the K is the number of nearest neighbors, μ refers to the mean of w_n ($n = (1, \ldots, 512)$). In Eq. (13.4), z_p, z_l, and z_s ($p, l, s = 1, \ldots, K$) belong to K nearest color names of w_n.

To reflect the saliency of the salient color names for d, the Euclidean distances between w_n and the standard colors are first computed to apply KNN algorithms, such that the K nearest color names can be selected. Then, the difference between the one of K nearest color names to the other $K - 1$ color names is utilized to embody the saliency which is calculated as $\phi(x) = \exp(-x)$.

After normalization, the probability distribution of w_n over 16 color names is calculated as in Eq. (13.4), and the final probability of d being assigned to each color name, Eq. (13.5) is employed to weigh the contribution of w_n to d. That is, the nearer of w_n to μ, the more it contributes to d.

With Eqs. (13.3)–(13.5), we can easily obtain the 16-dimensional distribution of the colors in one cube of RGB space. Besides, it is easy to prove that the sum of the distribution of d over all color names is 1, i.e. $\sum_{m=1}^{16} p(z_m|d) = 1$.

Once obtaining the 16-dimensional representations of all 32,678 cubes in the RGB space, they can be used as the dictionary to represent every pixel value in the pedestrian image. By further computing the color names distribution of an image using the dictionary, we can obtain the SCN feature representation of one image. Because the human body is not rigid, the SCN computed from the whole image may not well capture the fine appearance. A part-based model is selected instead of taking a person image as a whole. In practice, we can adopt a simple strategy to partition an image into six horizontal stripes of equal size. Let $H = [h_1, \ldots, h_6]^T$ be the color names distribution of a person image that has been divided into six stripes. Then the mth ($m = 1, \ldots, 16$) element of the distribution in the ith ($i = 1, \ldots, 6$) part $h_i = [h_{i1}, \ldots, h_{i16}]$ is defined as

$$h_{im} = \frac{\sum_{k=1}^{N} p(z_m|x_{ik})}{\sum_{m=1}^{16} \sum_{k=1}^{N} p(z_m|x_{ik})} \tag{13.6}$$

where x_{ik} ($k = 1, \ldots, N$) means the kth pixel in part i, and N denotes the total number of colors in part i. An example of the color names distribution in each part of a person image is shown in Fig. 13.4. The final SCN feature representation of one person image is the concatenation of the color names distribution of all stripes.

The SCN feature representation has the following advantages:

Fig. 13.4 An example of the color names distribution of a person image

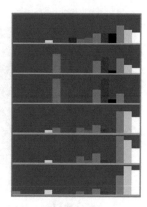

1. Each pixel value in RGB color space is represented by the probability distribution over its salient color names.
2. It can achieve a certain amount of illumination invariance. Because small RGB value changes caused by illumination will have the same color description if only the cubes they belong to are the same.
3. The SCN representation is not restricted to the RGB space. It can also be computed from other color spaces, such as HSV and Lab.
4. It does not rely on complex optimization and is easy to implement. More importantly, the dictionary can be computed offline. Therefore, the SCN representation can be quickly obtained by looking up the words in the dictionary.

13.5.2   Local Maximal Occurrence Representation

The Local Maximal Occurrence Representation [11] (LOMO) is also specially designed for person re-identification task. It is consisted of two basic features, namely the joint HSV histograms and the Scale Invariant Local Ternary Pattern [35] (SILTP) descriptor. The former is used to capture the color information, while the latter can capture the texture appearance. The computation of LOMO is very fast, and it has great robustness against the view changes in re-identification task.

Before computing the joint HSV histograms, LOMO first applies the Retinex algorithm to enhance the visual quality of person images. This also benefits to reduce the illumination variations in different cameras, thus can help to extract more discriminative feature representation. Figure 13.5 shows some example images before and after applying the Retinex, it can be found that the visual quality of processed images is clearly improved, and the brightness differences of one person's two images is depressed.

After mapping images into the HSV color space, the joint HSV histogram is computed by computing the frequency of normalized pixel values. For each channel,

(a)   **(b)**

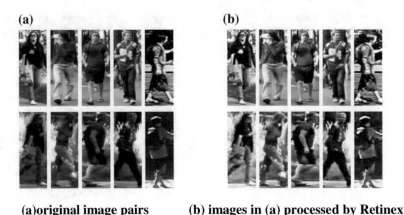

(a)original image pairs   **(b) images in (a) processed by Retinex**

Fig. 13.5 Comparison of image pairs before and after applying Retinex

we apply 8-bit quantization to the pixel values, and this leads to $8 \times 8 \times 8 = 512$ dimensional representation for each grid.

The SILTP is an extension of the Local Binary Pattern (LBP) representation by introducing the scale invariance. It should be noted here that the scale invariance indicates the pixel value scale, other than the spatial invariance. Compared to the LBP representation, SILTP is more robust to the noise pixel values. Given a pixel value at position (x_c, y_c), SILTP encode it as:

$$SILT_{N,R}^{T}(x_c, y_c) = \overset{N-1}{\underset{k-0}{\oplus}} s_\tau(I_c, I_k) \tag{13.7}$$

where I_c is the gray intensity value of the center pixel, I_k are that of its N neighborhood pixels equally spaced on a circle of radius R, $\oplus$ denotes concatenation operator of binary strings, τ is a scale factor indicating the comparing rage, and s_τ is a piecewise function defined as:

$$s_\tau(I_c, I_k) = \begin{cases} 01, & \textit{if } I_k > (1 + \tau)I_c \\ 10, & \textit{if } I_k < (1 - \tau)I_c \\ 00, & \textit{otherwise} \end{cases} \tag{13.8}$$

Since each comparison can result in one of three values, SILTP encodes it with two bits (with "11" undefined). The scale invariance of SILTP operator can be easily verified.

Figure 13.6 shows a comparison of the extraction procedure of LBP, LTP and SILTP. It can be found that LTP is more robust by introducing a small tolerative range. However, when the pixel value is multiplied by 2, the LTP is not stable enough. But after introducing a scale factor, the SILTP obtains reliable robustness against noises. Meanwhile, it is also robust to the scale variation in pixel values.

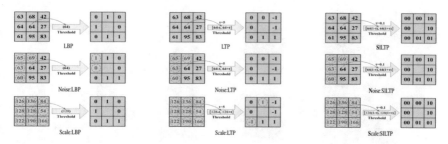

Fig. 13.6 Comparison of LBP, LTP, and SILTP operators

First row: original encodings.

Second row: encodings with noises.

Third row: encodings with scale transform (all pixel values are doubled).

The circled red pixels are changed with noises or by scale transform, and the circled red encodings are affected by those changes correspondingly.

To cope with the serious view point changes of different cameras, both joint HSV histograms and SILTP descriptor are extracted from dense grids with 50% overlapped areas along horizontal and vertical axis. The default size of each grid is 10×10 pixels and the moving step is 5 pixel along both horizontal and vertical axis. From each grid, we compute the 512-dimensional HSV histograms, and the SILTP descriptors with radius of 3 and 5. The scale factor of SILTP is set to 0.3.

To address viewpoint changes, LOMO further checks all sub-windows at the same horizontal location, and maximizes the local occurrence of each pattern (i.e. the same histogram bin) among these sub-windows. That is, only the maximal value on each bin is kept for the patterns computed from the sub-windows at the same height. The resulting histogram achieves some invariance to viewpoint changes, and at the same time captures local region characteristics of a person. Figure 13.7 shows the procedure of the proposed LOMO feature extraction.

To further consider the multi-scale information, a three-scale pyramid space is built by applying 2×2 average pooling operation to downsample the original image. By repeating the above feature extraction procedure and concatenating all the computed local maximal occurrences, the final feature representation is obtained. To suppress large bin values, the log transform is applied, and then we normalize both HSV and SILTP features to unit length.

The extraction code of LOMO implemented in MATLAB in given below. We first give the main function n which is used for reading the images from the VIPeR dataset and calling the LOMO.m function. In order to improve the efficiency of following computation, a 4-dimensional array is used to store the pedestrian images.

The code of main.m file is shown in PROGRAMME 13.1 as follows.

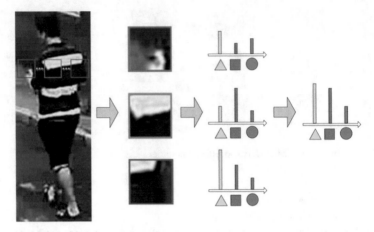

Fig. 13.7 Illustration of the LOMO feature extraction procedure

PROGRAMME 13.1: Main function of LOMO

```
% -----------------------------------------------------------------------------
--------------------
% Main function of extracting LOMO feature from VIPeR dataset.
% -----------------------------------------------------------------------------
--------------------
addpath('.\bin\');          % NOTE! the Ritenex is provided in mex file
clear; clc; close all;
images = zeros(128, 48, 3,1264, 'uint8');% The default size of VIPeR images: 128*48
dira = dir(['.\dataset\cam_a\*.bmp']);   % scan images
dirb = dir(['.\dataset\cam_b\*.bmp']);
for i = 1: 632 % 632 images in cam a
    if ~dira(i).isdir && strcmp(dira(i).name(end-2:end), 'bmp')
            images(:,:,:,i) = imread(['.\dataset\cam_a\' dirc(i).name]);
    end
end
for i = 1: 632% 632 images in cam b
    if ~dirb(i).isdir && strcmp(dirb(i).name(end-2:end), 'bmp')
        images(:,:,:,i+632) = imread(['.\dataset\cam_b\' dirc(i).name]);
    end
end
%% extract features. Run with a set of images is usually faster than that one by one, but requires
more memory.
descriptors = LOMO(images);
```

```
%% if you need to set different parameters other than the defaults, set them accordingly
%{
options.numScales = 3;
options.blockSize = 10;
options.blockStep = 5;
options.hsvBins    = [8,8,8];
options.tau        = 0.3;
options.R          = [3, 5];
options.numPoints = 4;
descriptors = LOMO(images, options);
%}
save('.\viper_lomo.mat','descriptors');
```

The LOMO.m file is shown in PROGRAMME 13.2, which calls the pyramid-MaxJointHist and pyramidMaxSILTPHist function to extract joint HSV histograms and SILTP descriptor from dense grids. Note that the input of LOMO function should be a 4-dimensional array that stores the cross-view images.

PROGRAMME 13.2: Call pyramidMaxJointHist and pyramidMaxJointHist functions to extract joint HSV and SILTP descriptor

```
function descriptors = LOMO(images, options)
% -----------------------------------------------------------------------------
% -------------------
% Function for the Local Maximal Occurrence (LOMO) feature extraction
% Input:
%     images - a set of n RGB color images. 4-dim array with size: [h, w, 3, n]
%     optioins - optional parameters. A structure containing any of the following fields:
%         - numScales: number of pyramid scales in feature extraction. Default: 3
%         - blockSize: size of the sub-window for histogram counting. Default: 10
%         - blockStep: sliding step for the sub-windows. Default: 5
%         - hsvBins: number of bins for HSV channels. Default: [8,8,8]
%         - tau: the tau parameter in SILTP. Default: 0.3
%         - R: the radius paramter in SILTP. SILTP. Default: [3, 5]
%         - numPoints: number of neiborhood points for SILTP encoding. Default: 4
%         The above default parameters are good for 128x48 and 160x60 person
%         images. You may need to adjust the numScales, blockSize,
%         and R parameters for other smaller or higher resolutions.
% Output:
%     descriptors - the extracted LOMO descriptors. Size: [d, n]
% Example:
%     I = imread('../images/000_45_a.bmp');
%     descriptor = LOMO(I);
% Version: 1.0,Date: 2015-04-29
```

```
% Author: Shengcai Liao, Email: scliao@nlpr.ia.ac.cn
% Institute: National Laboratory of Pattern Recognition, Institute of Automation,
% Chinese Academy of Sciences
% ------------------------------------------------------------------------------
-------------------
%% set parameters
numScales = 3;
blockSize = 10;
blockStep = 5;
hsvBins   = [8,8,8];
tau = 0.3;
R   = [3, 5];
numPoints = 4;
t0 = tic;
%% extract Joint HSV based LOMO descriptors
fea1 = pyramidMaxJointHist( images, numScales, blockSize, blockStep, hsvBins );
%% extract SILTP based LOMO descriptors
fea2 = [];
for i = 1 : length(R)
    fea2 = [fea2; pyramidMaxSILTPHist( images, numScales, blockSize, blockStep, tau, R(i),
numPoints )];
end
%% finishing
descriptors = [fea1; fea2];
clear Fea1 Fea2
feaTime = toc(t0);
meanTime = feaTime / size(images, 4);
fprintf('LOMO feature extraction finished. Running time: %.3f seconds in total, %.3f seconds per
image.\n', feaTime, meanTime);
end
```

PROGRAMME 13.3: Extract the joint HSV histograms from different scale spaces

```
function descriptors = pyramidMaxJointHist( oriImgs, numScales, blockSize, blockStep, hsvBins )
% ------------------------------------------------------------------------------
----------------------------------
% pyramidMaxJointHist: HSV based LOMO representation
% input:
%     oriImgs     - a set of n RGB color hsvImgs. 4-dim array with size: [h, w, 3, n]
%     numScales - number of scales to extract local max occurrence features
%     blockSize   - block size
```

```
%    blockStep - stepsize of moving block
%    hsvBins - bins of HSV histogram
% output:
%    descriptors - the extracted local max occurrence features descriptors(LOMO). size: [d, n]
% ----------------------------------------------------------------------------
----------------------------------
% set the default parameters as follows
if nargin == 1
    numScales = 3;
    blockSize = 10;
    blockStep = 5;
    hsvBins = [8,8,8];
end
totalBins = prod(hsvBins);
numImgs = size(oriImgs, 4);
hsvImgs = zeros(size(oriImgs));
hwait = waitbar(0, ' applying Retinex to images ...');
for i = 1 : numImgs
    I = oriImgs(:,:,:,i);
    I = Retinex(I);        % apply Retinex transformation
    I = rgb2hsv(I);
    I(:,:,1) = min( floor( I(:,:,1) * hsvBins(1) ), hsvBins(1)-1 );
    I(:,:,2) = min( floor( I(:,:,2) * hsvBins(2) ), hsvBins(2)-1 );
    I(:,:,3) = min( floor( I(:,:,3) * hsvBins(3) ), hsvBins(3)-1 );
    hsvImgs(:,:,:,i) = I;
    waitbar(i/numImgs, hwait);
end
close(hwait);
minRow = 1;
minCol = 1;
descriptors = [];
% scan multi-scale blocks and compute histograms
for i = 1 : numScales
    patterns = hsvImgs(:,:,3,:) * hsvBins(2) * hsvBins(1) + hsvImgs(:,:,2,:)*hsvBins(1) +···
hsvImgs(:,:,1,:);
    patterns = reshape(patterns, [], numImgs);
    height = size(hsvImgs, 1);
    width  = size(hsvImgs, 2);
    maxRow = height - blockSize + 1;
    maxCol = width  - blockSize + 1;
    [cols,rows] = meshgrid(minCol:blockStep:maxCol, minRow:blockStep:maxRow);
    cols = cols(:);
    rows = rows(:);
    numBlocks    = length(cols); % number of blocks along x axis
    numBlocksCol = length(minCol:blockStep:maxCol); % number of blocks along y axis
```

```
        if numBlocks == 0
            break;
        end
        offset = bsxfun(@plus, (0 : blockSize-1)', (0 : blockSize-1) * height);
        index = sub2ind([height, width], rows, cols);
        index = bsxfun(@plus, offset(:), index');   % (blockSize*blockSize)-by-numBlocks
    patches = patterns(index(:), :); % (blockSize * blockSize * numBlocks)-by-numImgs
    % (blockSize * blockSize)-by-(numBlocks * numChannels * numImgs)
    patches = reshape(patches, [], numBlocks * numImgs);
    fea = hist(patches, 0 : totalBins-1);   % totalBins-by-(numBlocks * numImgs)
        fea = reshape(fea, [totalBins, numBlocks / numBlocksCol, numBlocksCol, numImgs]);
        fea = max(fea, [], 3); % local maximal occurrence operation
        fea = reshape(fea, [], numImgs);
        descriptors = [descriptors; fea];
        if i < numScales % downsampling!
            hsvImgs = colorpooling(hsvImgs, 'average'); % we use 2*2 average pooling for
downsampling images.
        end
    end
end
descriptors = log(descriptors + 1);
descriptors = normc(descriptors); % normalize to unit length
end
```

Note that the Retinex function in pyramidMaxHSVHist.m is implemented in C+ +. The compiled mex file of Retinex can be downloaded from http://www.cbsr.ia.ac.c n/users/scliao/codes.html. It can be called directly in pyramidMaxHSVHist function. The pyramidMaxHSVHist also calls a colorpooling function to implement average pooling function to implement the downsampling operation. The colorpooling.m file is given below.

PROGRAMME 13.4: Downsample color images by average pooling operation

```
function outImages = colorpooling(images, method)
% ------------------------------------------------------------------------------
----------------------------------
% pooling color images using method of 'average' or 'max'
%
% input:
%     images -   a set of n hsv color imags with each plan mapped to range(0,7). 4-dim array with
%     size: [h, w, 3, n]
%     method - 'average' or 'max', pooling method
% output:
```

```
%    outImages   -   images after pooling, 4-dim array with size: [h, w, 3, n]
% -------------------------------------------------------------------------------
---------------------------------
[height, width, numChannels, numImgs] = size(images);
outImages = images;
if mod(height, 2) == 1
    outImages(end, :, :, :) = []; % prune each image's lowest pixel line to make the height even
    height = height - 1;   % adjust the height variable
end
if mod(width, 2) == 1
    outImages(:, end, :, :) = [];
    width = width - 1; % adjust the width variable
end
if height == 0 || width == 0
    error('Over scaled image: height=%d, width=%d.', height, width);
end
height = height / 2; % 2*2 pooling, downsample!
width   = width / 2;
outImages = reshape(outImages, 2, height, 2, width, numChannels, numImgs);
outImages = permute(outImages, [2, 4, 5, 6, 1, 3]);
outImages = reshape(outImages, height, width, numChannels, numImgs, 2*2);
if strcmp(method, 'average')
    outImages = floor(mean(outImages, 5)); % average pooling
else if strcmp(method, 'max')
        outImages = max(outImages, [], 5); % max pooling
    else
        error('Error pooling method: %s.', method);
    end
end
end
```

The pyramidMaxSILTPHist function extracts SILTP descriptor from dense grids in different scale spaces. Its implementation style is similar to pyramid-MaxHSVHist.m file. The max pooling operation is also applied to the patterns extracted from the grids on the same height. Note that the SILTP descriptor is computed from the gray images, so we need to transform images to gray first.

PROGRAMME 13.5: Extract SILTP descriptor from different scale spaces

```
% input: function descriptors = pyramidMaxSILTPHist( orilmgs, numScales, blockSize, blockStep,
tau, R,··· numPoints )
% ----------------------------------------------------------------------------------
-----------------------------------
% pyramidMaxSILTPHist: SILTP based LOMO representation
%    orilmgs    - a set of n RGB color hsvlmgs. 4-dim array with size: [h, w, 3, n]
%    numScales - number of scales to extract local max occurrence features
%    blockSize - block size
%    blockStep - stepsize of moving block
%    tau        - the scale parameter (>0) of SILTP. The default value is 0.03.
%    R          - radius of ltp, it should be a positive integer
%    numPoints - pixels to compare in ltp
% output:
%    descriptors - the extracted scal invariant local trinary pattern descriptor
% ----------------------------------------------------------------------------------
-----------------------------------
% set the default parameters as follows
if nargin == 1
    numScales = 3;
    blockSize = 10;
    blockStep = 5;
    tau = 0.3;
    R     = 5;
    numPoints = 4;
end
totalBins = 3^numPoints;
[imgHeight, imgWidth, ~, numImgs] = size(orilmgs); % hight, width, channels, image number
images = zeros(imgHeight,imgWidth, numImgs);
for i = 1 : numImgs
    I = orilmgs(:,:,:,i);
    I = rgb2gray(I); % convert to gray image
    images(:,:,i) = double(I) / 255; % map pixel value to 0~1
end
minRow = 1; % row start point
minCol = 1; % column start point
descriptors = [];
% Scan multi-scale blocks and compute histograms
for i = 1 : numScales
    height = size(images, 1);
    width  = size(images, 2);
    if width < R * 2 + 1
        fprintf('Skip scale R = %d, width = %d.\n', R, width);
        continue;
    end
```

```
        patterns = SILTP(images, tau, R, numPoints);% compute SILTP
        patterns = reshape(patterns, [], numImgs);
        maxRow = height - blockSize + 1;
        maxCol = width - blockSize + 1;
        [cols,rows] = meshgrid(minCol:blockStep:maxCol, minRow:blockStep:maxRow);
        cols = cols(:);
        rows = rows(:);
        numBlocks = length(cols); % number of blocks along x axis
        numBlocksCol = length(minCol:blockStep:maxCol); % number of blocks along y axis
        if numBlocks == 0
            break;
        end
offset = bsxfun(@plus, (0 : blockSize-1)', (0 : blockSize-1) * height);
index = sub2ind([height, width], rows, cols);
    index = bsxfun(@plus, offset(:), index') % (blockSize*blockSize)-by-numBlocks
patches = patterns(index(:), :); % (blockSize * blockSize * numBlocks)-by-numImgs
% (blockSize * blockSize)-by-(numBlocks * numChannels * numImgs)
patches = reshape(patches, [], numBlocks * numImgs);
fea = hist(patches, 0:totalBins-1); % totalBins-by-(numBlocks * numImgs)
    fea = reshape(fea, [totalBins, numBlocks / numBlocksCol, numBlocksCol, numImgs]);
    fea = max(fea, [], 3);
    fea = reshape(fea, [], numImgs);
    descriptors = [descriptors; fea];
    if i < numScales
        images = pooling(images, 'average'); % we use 2*2 average pooling.
    end
end
descriptors = log(descriptors + 1);
descriptors = normc(descriptors);
end
```

The most important part in the pyramidMaxSILTPHist.m file is calling SILTP function to extract SILTP descriptor from each grid. The code of SILTP.m is shown below.

PROGRAMME 13.6: Extract SILTP from grids

```
function J = SILTP(I, tau, R, numPoints, encoder)
% ---------------------------------------------------------------------------------
% ----------------------------------
%   Returns scale invariant local ternary patterns of image I (or multi images).
% input:
%   I    - input image, or multi images arranged in 3-d arrays. The size of each
%          image must be at least (2R+1)*(2R+1) pixels.
%   tau - the scale parameter (>0) of SILTP. The default value is 0.03.
%   R    - the radius parameter of SILTP. It should be a positive integer (default is 1).
%   numPoints - the number of neighboring pixels, which can be 4 or 8 (default is 4).
%   encoder - the way of encoding. The value can be 0 or 1. Default is 0.
%           0: encoded as 0 ~ 3^numPoints-1, suitable for histogram calculation.
%           1: encoded as 0 ~ 2^(2*numPoints)
% output:
%   J    -   output of the encoded image(s), which is the same size as I.
%   Vertion 1.2
%   Written by:
%       Shengcai Liao, Center for Biometric and Security Research (CBSR) &
%       National Laboratory of Pattern Recognition (NLPR),
%       Institute of Automation, Chinese Academy of Sciences (CASIA)
% ---------------------------------------------------------------------------------
% ----------------------------------
%% set default parameters
if nargin < 5
    encoder = 0;
    if nargin < 4
        numPoints = 4;
        if nargin < 3
            R = 1;
            if nargin < 2
                tau = 0.03;
            end
        end
    end
end
%% check parameters
if tau <= 0 || floor(R) ~= R || R < 1 || ~(numPoints==4 || numPoints==8) || ~(encoder == 0···
|| encoder == 1)
    help SILTP;
    error('Error parameter values!');
end
```

```
%% check image(s) size
[h, w, n] = size(I);
if h < 2*R+1 || w < 2*R+1
    error('Too small image or too large R!');
end
%% put the image(s) in a larger container
I0 = zeros(h+2*R, w+2*R, n);
I0(R+1:end-R, R+1:end-R, :) = double(I);
%% replicate border image pixels to the outer area
I0(1:R,:,:) = repmat(I0(R+1,:,:), [R,1,1]);
I0(end-R+1:end,:,:) = repmat(I0(end-R,:,:), [R,1,1]);
I0(:,1:R,:) = repmat(I0(:,R+1,:), [1,R,1]);
I0(:,end-R+1:end,:) = repmat(I0(:,end-R,:), [1,R,1]);
%% copy image(s) in specified directions
I1 = I0(R+1:end-R, 2*R+1:end, :);
I3 = I0(1:end-2*R, R+1:end-R, :);
I5 = I0(R+1:end-R, 1:end-2*R, :);
I7 = I0(2*R+1:end, R+1:end-R, :);
if numPoints == 8
    I2 = I0(1:end-2*R, 2*R+1:end, :);
    I4 = I0(1:end-2*R, 1:end-2*R, :);
    I6 = I0(2*R+1:end, 1:end-2*R, :);
    I8 = I0(2*R+1:end, 2*R+1:end, :);
end
%% compute the upper and lower range
L = (1-tau) * I;
U = (1+tau) * I;
%% compute the scale invariant local ternary patterns
if encoder == 0
    if numPoints == 4
        J = (I1 < L) + (I1 > U) * 2 + ((I3 < L) + (I3 > U) * 2) * 3 + ((I5 < L)···
 + (I5 > U) * 2) * 9 + ((I7 < L) + (I7 > U) * 2) * 27;
    else
        J = (I1 < L) + (I1 > U) * 2 + ((I2 < L) + (I2 > U) * 2) * 3 + ((I3 < L)···
 + (I3 > U) * 2) * 3^2 + ((I4 < L) + (I4 > U) * 2) * 3^3 + ...
            + ((I5 < L) + (I5 > U) * 2) * 3^4 + ((I6 < L) + (I6 > U) * 2) * 3^5···
+((I7 < L) + (I7 > U) * 2) * 3^6 + ((I8 < L) + (I8 > U) * 2) * 3^7;
    end
else
    if numPoints == 4
        J = (I1 > U) + (I1 < L) * 2 + (I3 > U) * 2^2 + (I3 < L) * 2^3 + (I5 > U) * 2^4···
 + (I5 < L) * 2^5 + (I7 > U) * 2^6 + (I7 < L) * 2^7;
```

```
        else
            J = (l1 > U) + (l1 < L) * 2 + (l2 > U) * 2^2 + (l2 < L) * 2^3 + (l3 > U) * 2^4···
    + (l3 < L) * 2^5 + (l4 > U) * 2^6 + (l4 < L) * 2^7 + (l5 > U) * 2^8···
    + (l5 < L) * 2^9 + (l6 > U) * 2^10 + (l6 < L) * 2^11 + (l7 > U) * 2^12···
    + (l7 < L) * 2^13 + (l8 > U) * 2^14 + (l8 < L) * 2^15;
        end
end
```

There is another pooling function called in the pyramidMaxSILTPHist.m file. The pooling.m is used for downsampling gray images. It is implemented in a similar way to the colorpooling.m file. The code of pooling.m file is as follows.

PROGRAMME 13.7: The pooling operation for gray images

```
function outImages = pooling(images, method)
% ----------------------------------------------------------------------------------
-----------------------------------
% downsample gray images by 2*2 average pooling or max pooling
%
% input:
%     method     -   pooling method,'average','max'
%     images     -   4-dim array with size: [h, w, 3, n]
% output:
%     outImages-   4-dim array with size: [h, w, 3, n]
% ----------------------------------------------------------------------------------
-----------------------------------
[height, width, numImgs] = size(images);
outImages = images;
if mod(height, 2) == 1
    outImages(end, :, :) = []; % abandon the last line
    height = height - 1;
end
if mod(width, 2) == 1
    outImages(:, end, :) = []; % abandon the last column pixels
    width = width - 1;
end
if height == 0 || width == 0
    error('Over scaled image: height=%d, width=%d.', height, width);
end
height = height / 2; % downsample to 1/2 height
width   = width / 2; % downsample to 1/2 width
outImages = reshape(outImages, 2, height, 2, width, numImgs);
outImages = permute(outImages, [2, 4, 5, 1, 3]);
```

```
outImages = reshape(outImages, height, width, numImgs, 2*2);
if strcmp(method, 'average')
    outImages = mean(outImages, 4);% average pooling
else

    if strcmp(method, 'max')
        outImages = max(outImages, [], 4);% max pooling
    else
        error('Error pooling method: %s.', method);
    end
end
end
```

13.6 An Example of Metric Learning Based Person Re-identification Method-XQDA

The cross-view quadratic discriminant analysis [11] (XQDA) is an extension of keep it simple and straightforward metric learning [10] (KISSME) algorithm. It can be viewed of the combination of KISSME and the linear discriminant analysis (LDA). Due to the closed-form solution, the metric in XQDA can be computed very efficiently, thus avoiding the iterative optimization among common metric learning algorithms. Here, we first introduce the KISSME algorithm, and then the XQDA.

Consider a sample difference $\Delta = x_i - x_j$ in KISSME. If samples x_i and x_j share the same label, i.e., $y_i = y_j$, then Δ is called the intra-personal difference, otherwise we call Δ the extra-personal difference. Then we can define two classes of variations: the extra-personal variations $\Omega_E = \{\Delta = x_i - x_j | y_i \neq y_j\}$, and the intra-personal variations $\Omega_I = \{\Delta = x_i - x_j | y_i = y_j\}$. By assuming Ω_E and Ω_I follow the zero mean Gaussian distribution, the likelihoods of observing Δ in Ω_I and Ω_E are as follows:

$$P(\Delta|\Omega_I) = \frac{1}{(2\pi)^{d/2}|\Sigma_I|^{1/2}} \exp\left(-\frac{1}{2}\Delta^T \sum_I^{-1} \Delta\right) \tag{13.9}$$

$$P(\Delta|\Omega_E) = \frac{1}{(2\pi)^{d/2}|\Sigma_E|^{1/2}} \exp\left(-\frac{1}{2}\Delta^T \sum_E^{-1} \Delta\right) \tag{13.10}$$

where d is the feature dimension, Σ_I and Σ_E are the covariance matrices of Ω_I and Ω_E respectively. Since Ω_E and Ω_I both have zero means, we can obtain

$$\sum_I = \sum_{y_i=y_j} (x_i - x_j)(x_i - x_j)^T \tag{13.11}$$

$$\sum\nolimits_E = \sum_{y_i \neq y_j} (x_i - x_j)(x_i - x_j)^T \tag{13.12}$$

From a statistical inference point of view the optimal statistical decision whether a pair (x_i, x_j) is intra-personal or not can be obtained by a likelihood ratio test. By applying the log trick, we have

$$\delta(x_i, x_j) = \log \left(\frac{\frac{1}{(2\pi)^{d/2}|\Sigma_I|^{1/2}} \exp\left(-\frac{1}{2}\Delta^T \Sigma_I^{-1} \Delta\right)}{\frac{1}{(2\pi)^{d/2}|\Sigma_E|^{1/2}} \exp\left(-\frac{1}{2}\Delta^T \Sigma_E^{-1} \Delta\right)} \right) \tag{13.13}$$

By simplifying Eq. (13.14) and removing the constant terms, we can obtain the following decision function

$$f(\Delta) = \Delta^T \left(\sum_I^{-1} - \sum_E^{-1} \right) \Delta \tag{13.14}$$

And so the derived distance function between (x_i, x_j) is

$$d(x_i, x_j) = (x_i - x_j)^T \left(\sum_I^{-1} - \sum_E^{-1} \right)(x_i - x_j) \tag{13.15}$$

From above derivation we can find the metric in KISSME is just $M = \sum_I^{-1} - \sum_E^{-1}$ which can be computed efficiently due to the closed-form solution. In practice, we only need to compute two variance matrices and obtain the difference of their inverse matrices.

However, KISSME is rather sensitive to the sample dimension. Its developer Köestinger et al. suggest to reduce the sample dimension to 34 with principle component analysis (PCA) before computing $\sum_I^{-1} - \sum_E^{-1}$. Although it is a common strategy to reduce feature dimension with PCA among metric learning method, it is pointed out that such a "two-stage" processing is not optimal during learning the metric. Because the samples of different classes may be cluttered in dimension reduction. To improve this deficiency, Liao $et.at$ proposed the XQDA algorithm to learn an optimal projection subspace besides the metric. Since the metric and subspace are jointly learned, thus avoiding the "two-stage" processing. We detail the XQDA algorithm below.

Let W be the wanted discriminative projection subspace, we replace x_i, x_j in Eq. (13.15) by $W^T x_i$ and $W^T x_j$, then we obtain

$$d_w(x_i, x_j) = (x_i - x_j)^T W \left(\sum_I^{t-1} - \sum_E^{t-1} \right) W^T (x_i - x_j) \tag{13.16}$$

where $\Sigma_I^{t-1} = W^T \sum_I W$, $\Sigma_E^{t-1} = W^T \sum_E W$. Therefore, the core of XQDA is to obtain the subspace W. However, directly optimizing d_w is difficult because W is contained in two inverse matrices.

Consider $\Delta = x_i - x_j$ belongs to either Ω_E or Ω_I, we can find the optimal projection directions w using the LDA-like method. Recall that both Ω_E and Ω_I have zero mean, the projected samples of the two classes will still center at zero, but may have different variances. In this case, the traditional Fisher criterion used to derive LDA is no longer suitable. However, the variances $\sigma_I = w^T \sum_I w$ and $\sigma_E = w^T \sum_E w$ can still be used to distinguish the two classes. Therefore, we can optimize the projection direction w such that $\sigma_I(w)/\sigma_E(w)$ is maximized. Therefore, we can formulate the objective function as

$$\max_w J(w) = \frac{w^T \sum_E w}{w^T \sum_I w} \tag{13.17}$$

The maximization of $J(w)$ is equivalent to

$$\max_w w^T \sum_E w \quad s.t. \quad w^T \sum_I w = 1 \tag{13.18}$$

Similar to LDA, the above problem can be solved by the generalized eigenvalue decomposition. The obtained projection directions w are just the eigenvectors corresponding to the largest r eigenvalues. With the learned subspace $W = [w_1, \ldots w_r]$, we can compute the distance between (x_i, x_j) according to Eq. (13.16).

In numeric computation, both KISSME and XQDA have to compute the covariance matrices Σ_I and Σ_E. However, directly computing them as in Eq. (13.11) and (13.12) require $O(Nkd^2)$ and $O(nmd^2)$ multiplication operations, where n and m are the numbers of probe and gallery images, $N = \max(n, m)$, and k represents the average number of images in each class. Then we can compute the $\sum_I$ as follows

$$n_I \sum_I = \tilde{\mathbf{X}}\tilde{\mathbf{X}}^T + \tilde{\mathbf{Z}}\tilde{\mathbf{Z}}^T - \mathbf{SR}^T - \mathbf{RS}^T \tag{13.19}$$

where
$\tilde{\mathbf{X}} = (\sqrt{m_1}x_1, \sqrt{m_1}x_2, \ldots, \sqrt{m_1}x_{n_1}, \ldots, \sqrt{m_c}x_n)$,
$\tilde{\mathbf{Z}} = (\sqrt{n_1}z_1, \sqrt{n_1}z_2, \ldots, \sqrt{n_1}z_{m_1}, \ldots, \sqrt{n_c}z_m)$,
$\mathbf{S} = \left[\sum_{y_i=1} x_i, \sum_{y_i=2} x_i, \ldots, \sum_{y_i=c} x_i \right]$,
$\mathbf{R} = \left[\sum_{y_j=1} z_j, \sum_{y_j=2} z_j, \ldots, \sum_{y_j=c} z_j \right]$, y_i is the class label, n_k is the number of samples in class k of X, and m_k is the number of samples in class k of Z.

Similarly, we have the following formulation about the covariance matrix $\sum_E$:

$$n_E \Sigma_E = m X X^T + n Z Z^T - s r^T - r s^T - n_I \Sigma_I \tag{13.20}$$

where $s = \sum_{i=1}^n x_i$ and $r = \sum_{j=1}^m z_j$. It is worth noting that the above simplification reduce the computation cost of $\sum_I$ and Σ_E to $O(Nd^2)$, thus greatly benefit the acceleration of computation. The actual sample differences along with their outer product are not required.

Because the XQDA has to compute the inverse of $\sum_I$, a singular matrix may bring some numerical problems. To solve this problem, we can regularize $\sum_I$ by adding a small number to the diagonal elements of $\sum_I$. In experiment, it is found that a value of 0.001 is ok when the samples are normalized to unit length. Another issue in XQDA is the dimensionality of the subspace. In practice, it is found that having the selected eigenvalues of $\sum_I^{-1} \sum_E$ is just ok.

Using the LOMO feature detailed in Sect. 13.4, the implementation code of XQDA is shown in the following. Let us see the main function first.

PROGRAMME 13.8: Main function of XQDA algorithm for person re-identification on VIPeR

```
% ---------------------------------------------------------------------------------
-----------------
% This is a demo for the XQDA metric learning, as well as the evaluation on the VIPeR
% database. Note: this demo requires about 1.0-1.4GB of memory.
% ---------------------------------------------------------------------------------
-----------------
close all; clear; clc;
feaFile = '. /viper_lomo.mat';
numClass = 632;
numFolds = 10;
numRanks = 100;
%% load the extracted LOMO features
load(feaFile, 'descriptors');
galFea = descriptors(1 : numClass, :);
probFea = descriptors(numClass + 1 : end, :);
clear descriptors
%% set the seed of the random stream.
seed = 0;
rng(seed);
%% evaluate
cms = zeros(numFolds, numRanks);
for nf = 1 : numFolds
    p = randperm(numClass);
    galFea1 = galFea( p(1:numClass/2), : );
    probFea1 = probFea( p(1:numClass/2), : );
    t0 = tic;
```

```
[W, M] = XQDA(galFea1, probFea1, (1:numClass/2)', (1:numClass/2)');
%{
%% if you need to set different parameters, set them accordingly
options.lambda = 0.001;
options.qdaDims = -1;
options.verbose = true;
[W, M] = XQDA(galFea1, probFea1, (1:numClass/2)', (1:numClass/2)', options);
%}
clear galFea1 probFea1
trainTime = toc(t0);
galFea2 = galFea(p(numClass/2+1 : end), : );
probFea2 = probFea(p(numClass/2+1 : end), : );
t0 = tic;
dist = MahDist(M, galFea2 * W, probFea2 * W);
clear galFea2 probFea2 M W
matchTime = toc(t0);
fprintf('Fold %d: ', nf);
fprintf('Training time: %.3g seconds. ', trainTime);
fprintf('Matching time: %.3g seconds.\n', matchTime);
cms(nf,:) = EvalCMC( -dist, 1 : numClass / 2, 1 : numClass / 2, numRanks );
clear dist
fprintf(' Rank1,   Rank5, Rank10, Rank15, Rank20\n');
fprintf('%5.2f%%, %5.2f%%, %5.2f%%, %5.2f%%, %5.2f%%\n\n', ···
cms(nf,[1,5,10,15,20]) * 100);
end
%% show result
meanCms = mean(cms);
plot(1 : numRanks, meanCms);
fprintf('The average performance:\n');
fprintf(' Rank1,   Rank5, Rank10, Rank15, Rank20\n');
fprintf('%5.2f%%, %5.2f%%, %5.2f%%, %5.2f%%, %5.2f%%\n\n', ···
meanCms([1,5,10,15,20]) * 100);
```

The most important part in above codes is calling XQDA function to learn the subspace W and the metric $\sum_I^{-1} \sum_E^{-1}$. The implementation code of XQDA function is as follows.

PROGRAMME 13.9: Learning the subspace and metric of XQDA

```
function [W, M, inCov, exCov] = XQDA(galX, probX, galLabels, probLabels, options)
% ------------------------------------------------------------------------------
-------------------
% Cross-view Quadratic Discriminant Analysis for subspace and metric learning
%
% input:
%     galX - features of gallery samples. Size: [n, d]
%     probX - features of probe samples. Size: [m, d]
%     galLabels - class labels of the gallery samples
%     probLabels - class labels of the probe samples
%     [options]- optional parameters. A structure containing any of the following fields:
%             lambda    - the regularizer. Default: 0.001
%             qdaDims – the dimensions to be preserved in the learned subspace.
%             Negative values indicate automatic dimension selection by perserving
%             latent values larger than 1. Default: -1.
%             verbose- whether to print the learning details. Default: false
%
% output:
%     W - the subspace projection matrix. Size: [d, r], r is the subspace dimension.
%     M - the learned metric kernel. Size: [r,r]
%     inCov - covariance matrix of the intra-personal difference class. Size: [r,r]
%     exCov - covriance matrix of the extra-personal difference class. Size: [r,r]
% ------------------------------------------------------------------------------
-------------------
lambda = 0.001;
qdaDims = -1;
verbose = false;
% set default parameters
if nargin >= 5 && ~isempty(options)
if isfield(options,'lambda') && ~isempty(options.lambda)···
&& isscalar(options.lambda) && isnumeric(options.lambda) ...
        lambda = options.lambda;
    end
if isfield(options,'qdaDims') && ~isempty(options.qdaDims)···
 && isscalar(options.qdaDims) && isnumeric(options.qdaDims)···
 && options.qdaDims > 0
        qdaDims = options.qdaDims;
    end
if isfield(options,'verbose') && ~isempty(options.verbose)···
 && isscalar(options.verbose) && islogical(options.verbose)
        verbose = options.verbose;
    end
end
```

```
if verbose == true
    fprintf('options.lambda = %g.\n', lambda);
    fprintf('options.qdaDims = %d.\n', qdaDims);
    fprintf('options.verbose = %d.\n', verbose);
end
[numGals, d] = size(galX); % number of gallery images
numProbs = size(probX, 1); % number of probe images
% If d > numGals + numProbs, it is not necessary to apply XQDA on the high
% space. In this case we can apply XQDA on QR decomposed space, achieving the
% same performance but much faster.
if d > numGals + numProbs
    if verbose == true
        fprintf('\nStart to apply QR decomposition.\n');
    end
    t0 = tic;
    [W, X] = qr([galX', probX'], 0); % [d, n]
    galX = X(:, 1:numGals)';
    probX = X(:, numGals+1:end)';
    d = size(X,1);
    clear X;
    if verbose == true
        fprintf('QR decomposition time: %.3g seconds.\n', toc(t0));
    end
end
labels = unique([galLabels; probLabels]);
c = length(labels);
if verbose == true
    fprintf('#Classes: %d\n', c);
    fprintf('Compute intra/extra-class covariance matrix...');
end
t0 = tic;
galW = zeros(numGals, 1);
galClassSum = zeros(c, d);
probW = zeros(numProbs, 1);
probClassSum = zeros(c, d);
ni = 0;
for k = 1 : c
    galIndex = find(galLabels == labels(k));
    nk = length(galIndex);
    galClassSum(k, :) = sum( galX(galIndex, :), 1 );
    probIndex = find(probLabels == labels(k));
    mk = length(probIndex);
```

```
        probClassSum(k, :) = sum( probX(probIndex, :), 1 );
        ni = ni + nk * mk;
        galW(galIndex) = sqrt(mk);
        probW(probIndex) = sqrt(nk);
end
galSum = sum(galClassSum, 1);
probSum = sum(probClassSum, 1);
galCov = galX' * galX;
probCov = probX' * probX;
galX = bsxfun( @times, galW, galX );
probX = bsxfun( @times, probW, probX );
inCov = galX' * galX + probX' * probX - galClassSum' * probClassSum - probClassSum' *
galClassSum;
exCov = numProbs * galCov + numGals * probCov - galSum' * probSum - probSum' * galSum -
inCov;
ne = numGals * numProbs - ni;
inCov = inCov / ni;
exCov = exCov / ne;
inCov = inCov + lambda * eye(d);
if verbose == true
        fprintf(' %.3g seconds.\n', toc(t0));
        fprintf('#Intra: %d, #Extra: %d\n', ni, ne);
        fprintf('Compute eigen vectors...');
end
t0 = tic;
[V, S] = svd(inCov \ exCov);
if verbose == true
        fprintf(' %.3g seconds.\n', toc(t0));
end
latent = diag(S);
[latent, index] = sort(latent, 'descend');
energy = sum(latent);
minv = latent(end);
r = sum(latent > 1);
energy = sum(latent(1:r)) / energy;
if qdaDims > r
        qdaDims = r;
end
if qdaDims <= 0
        qdaDims = max(1,r);
end
if verbose == true
        fprintf('Energy remained: %f, max: %f, min: %f, all min: %f, #opt-dim: %d, qda-dim: %d.\n',...
          energy, latent(1), latent(max(1,r)), minv, r, qdaDims);
end
```

```
V = V(:, index(1:qdaDims));
if ~exist('W', 'var');
    W = V;
else
    W = W * V;
end
if verbose == true
    fprintf('Compute kernel matrix...');
end
t0 = tic;
inCov = V' * inCov * V;
exCov = V' * exCov * V;
M = inv(inCov) - inv(exCov);
if verbose == true
    fprintf(' %.3g seconds.\n\n', toc(t0));
end
end
```

It is worth noting that QR decomposition is applied if the feature dimension is higher than the sample number in the XQDA.m file. The advantage is that the computation cost can be greatly reduced in this way. To obtain the eigenvectors of $\Sigma_I^{-1} \Sigma_E$, the singular value decomposition (SVD) is used instead of eigenvalue decomposition to achieve numeric stability.

There are two other functions of MahDist and EvalCMC in the main function of XQDA. The MahDist function implement the computation of Mahalanobis distance between every pair in two feature matrices. And the EvalCMC function computes the cumulative matching accuracies on each rank. The code of MahDist function is given below.

PROGRAMME 13.10: Compute the Mahalanobis distance between every sample pair in two feature matrices

```
function dist = MahDist(M, Xg, Xp)
% ------------------------------------------------------------------------
% --------------------
% Mahalanobis distance, Note: MahDist(M, Xg) is the same as MahDist(M, Xg, Xg).
%
% Input:
%    M    - the metric kernel
%    Xg   - features of the gallery samples. Size: [n, d]. Each row represents a sample.
%    Xp   - features of the probe samples. Optional. Size: [m, d]
%
```

```
% Output:
%     dist: the computed distance matrix between Xg and Xp
% -------------------------------------------------------------------------------
--------------------
if nargin == 2 % compute the pairwise Mahalanobis distances between each pair in feature
matrix Xg
    D = Xg * M * Xg';
    u = diag(D);
    dist = bsxfun(@plus, u, u') - 2 * D;
else % compute the pairwise Mahalanobis distances between each pair in two feature matrices
    u = sum( (Xg * M) .* Xg, 2);
    v = sum( (Xp * M) .* Xp, 2);
    dist = bsxfun(@plus, u, v') - 2 * Xg * M * Xp';
end
```

Based on the distance matrix of the probe images and the gallery images, we can rank the gallery images according their distances with each probe image. Then we can obtain the matching accuracies on each rank. By accumulating the accuracies and plotting them, a CMC curve is obtained. To obtain a robust performance, the experiment is usually repeated 10 times to average the CMC curve. This can be found in the main function of XQDA. The code of EvalCMC function has been given in Sect. 6.5.6, so we omit it here.

On VIPeR dataset, we can obtain a 40% rank-1 matching accuracy at rank-1 by running the main function of XQDA with the LOMO feature. Due to the closed-form solution, it only takes a general 1.5 s to learn the metric and subspace. So the XQDA is a very efficient and powerful metric learning algorithm whose re-identification result is rather impressive. The plotted CMC curve is shown in Fig. 13.8.

Fig. 13.8 The CMC curve of XQDA+LOMO on the VIPeR dataset

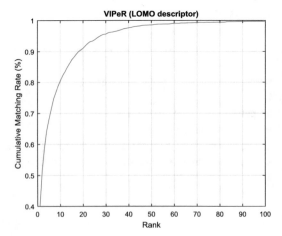

References

1. Wang T, Gong S, Zhu X, Wang S (2016) Person re-identification by discriminative selection in video ranking. IEEE Trans Pattern Anal Mach Intelligen pp 1–1
2. You J, Wu A, Li X, Zheng WS (2016) Top-push video-based person re-identification, pp 1345–1353
3. Hirzer M, Beleznai C, Roth PM, Bischof H (2011) Person re-identification by descriptive and discriminative classification. Image Anal 91–102. Springer
4. Loy CC, Xiang T, Gong S (2009) Multi-camera activity correlation analysis. In: IEEE conference on computer vision and pattern recognition, CVPR 2009, pp 1988–1995. IEEE
5. Cheng DS, Cristani M, Michele S, Loris B, Vittorio M (2011) Custom pictorial structures for re-identification. In BMVC, vol 2, p 6. Citeseer
6. Farenzena M, Bazzani L, Perina A, Murino V, Cristani M (2010) Person re-identification by symmetry-driven accumulation of local features. In: 2010 IEEE conference on computer vision and pattern recognition, CVPR, pp 2360–2367. IEEE
7. Kviatkovsky Igor, Adam Amit, Rivlin Ehud (2013) Color invariants for person reidentification. IEEE Trans Pattern Anal Mach Intelligen 35(7):1622–1634
8. Pedagadi S, Orwell J, Velastin S, Boghossian B (2013) Hierarchical Gaussian descriptor for :Person re-identification. In: Proceedings of the IEEE conference on computer vision and pattern recognition, pp 1363–1372
9. Yang Y, Yang J, Yan J, Liao S, Yi D, Li SZ (2014) Salient color names for person re-identification. In: ECCV, pp 536–551
10. Koestinger M, Hirzer M, Wohlhart P, Roth PM, Bischof H (2012) Large scale metric learning from equivalence constraints. In: 2012 IEEE conference on computer vision and pattern recognition (CVPR), pp 2288–2295. IEEE
11. Liao S, Hu Y, Zhu X, Li SZ (2015) Person re-identification by local maximal occurrence representation and metric learning. In: Proceedings of the IEEE conference on computer vision and pattern recognition, pp 2197–2206
12. Weinberger KQ, Saul LK (2009) Distance metric learning for large margin nearest neighbor classification. J Mach Learn Res 10:207–244
13. Zheng Wei-Shi, Gong Shaogang, Xiang Tao (2013) Reidentification by relative distance comparison. IEEE Trans Pattern Anal Mach Intelligen 35(3):653–668
14. Dalal N, Triggs B (2005) Histograms of oriented gradients for human detection. In: IEEE computer society conference on computer vision & pattern recognition, pp 886–893, 2005
15. Gray D, Brennan S, Tao H (2007) Evaluating appearance models for recognition, reacquisition, and tracking. In: Proceedings of IEEE international workshop on performance evaluation for tracking and surveillance (PETS), vol. 3. Citeseer
16. Bingpeng Ma YuSu, Jurie Frederic (2014) Covariance descriptor based on bio-inspired features for person re-identification and face verification. Image Vis Comput 32(6):379–390
17. Das A, Chakraborty A, Roy-Chowdhury AK (2014) Consistent re-identification in a camera network. In: European conference on computer vision, vol 8690. Lecture Notes in Computer Science, pp 330–345. Springer
18. Ahonen Timo, Hadid Abdenour, Pietikainen Matti (2006) Face description with local binary patterns: application to face recognition. IEEE Trans Pattern Anal Mach Intelligen 28(12):2037–2041
19. Zhao R, Ouyang W, Wang X (2014) Learning mid-level filters for person re-identification. In: Proceedings of the IEEE conference on computer vision and pattern recognition, pp 144–151
20. Bengio Yoshua (2009) Learning deep architectures for AI. Foundat Trends Machine Learn 2(1):1–127

21. Xiao T, Li H, Ouyang W, Wang X (2016) Learning deep feature representations with domain guided dropout for person re-identification. In: IEEE conference on computer vision and pattern recognition

22. Zhao H, Tian M, Sun S, Shao J, Yan J, Yi S, Wang X, Tang X (2017) Spindle Net: person re-identification with human body region guided feature decomposition and fusion. In: IEEE Conference on computer vision and pattern recognition (2017)

23. Zhao L, Li X, Wang J, Zhuang Y (2017) Deeply-learned part-aligned representations for person re-identification. In: IEEE international conference on computer vision (2017)

24. Roth PM, Hirzer M, Köstinger M, Beleznai C, Bischof H (2014) Mahalanobis distance learning for Person Re-identification

25. Baltieri D, Vezzani R, Cucchiara R (2011) 3dpes: 3d people dataset for surveillance and forensics. In Proceedings of the 1st international ACM workshop on multimedia access to 3D human objects, pp 59–64. Scottsdale, Arizona, USA

26. Li W, Zhao R, Xiao T, Wang X (2012) Human reidentification with transferred metric learning. In Computer Vision–ACCV 2012, pp 31–44. Springer

27. Li W, Wang X (2013) Locally aligned feature transforms across views. In: IEEE conference on computer vision & pattern recognition, pp 3594–3601

28. Li W, Zhao R, Xiao T, Wang X (2014) Deepreid: deep filter pairing neural network for person re-identification. In: 2014 IEEE conference on computer vision and pattern recognition (CVPR), pp 152–159. IEEE

29. Bedagkar-Gala A, Shishir K Shah. A survey of approaches and trends in person re-identification. Image and Vision Computing, 32(4):270–286, 2014

30. William Robson Schwartz and Larry S. Davis

31. Bialkowski A, Denman S, Sridharan S, Fookes C, Lucey P (2013) A database for person re-identification in multi-camera surveillance networks. In International conference on digital image computing techniques and applications, pp 1–8

32. Wang T, Gong S, Zhu X, Wang S (2014) Person re-identification by video ranking. In: Computer vision–ECCV 2014, pp 688–703. Springer

33. Deng J, Dong W, Socher R, Li LJ, Li K, Fei-Fei L (2009) Imagenet: a large-scale hierarchical image database. In: IEEE conference on computer vision and pattern recognition, 2009. CVPR 2009, pp 248–255. IEEE

34. Pedagadi S, Orwell J, Velastin S, Boghossian B (2013) Local fisher discriminant analysis for pedestrian re-identification. In: Proceedings of the IEEE conference on computer vision and pattern recognition, pp 3318–3325

35. Liao S, Zhao G, Kellokumpu V, Pietikäinen M, Li SZ (2010) Modeling pixel process with scale invariant local patterns for background subtraction in complex scenes. In: Computer vision and pattern recognition, pp 1301–1306

Chapter 14
Image and Video Understanding Based on Deep Learning

Abstract In this chapter we firstly introduce the development and the main reasons of the success of deep learning, then the structure and principle of the deep CNN are explored, and several classical convolution network models are analyzed, finally two instances based on CNN architecture are given.

14.1 Introduction

Rumelhart et al. proposed Back Propagation (BP) algorithm of artificial neural network in 1986 [1], which inspired the enthusiasm for the research of neural network in machine learning. But because the BP neural network is easy to meet over fitting, long training time and other problems, in the 90s, support vector machines (SVM) based on statistical learning theory became more popular [2]. SVM has a strong learning ability of small sample, and its learning effect is also superior to BP neural networks, which led to the study of neural networks falling into a ditch again.

Hinton et al. proposed deep learning in Science in 2006 [3], in which two main ideas were given: (a) neural network with multi-layer artificial has excellent feature learning ability, and the learned data can better reflect the essential characteristics of the data, which is conducive to the visualization or classification; (b) The training difficulty of deep neural network can be effectively overcome by layer-wise unsupervised training.

Theoretical research shows that in order to learn complex functions that can represent high-level abstract features, a deep network is needed. The deep network is composed of multilayer nonlinear operators, and the typical design is a neural network with multi-layer hidden nodes. However, as network layer increases, how to search the parameter space of the deep architecture becomes a challenging task.

© Springer International Publishing AG, part of Springer Nature 2019   513
S. Gong et al., *Advanced Image and Video Processing Using MATLAB*,
Modeling and Optimization in Science and Technologies 12,
https://doi.org/10.1007/978-3-319-77223-3_14

In recent years, the main reasons of the success of deep learning includes:

(a) On training data, the emergence of large-scale training data (such as ImageNet) provides good training resources for deep learning.
(b) The rapid development of computer hardware (especially the advent of GPU) has made it possible to train large-scale neural networks.

Convolutional neural networks (CNN) is a kind of neural networks with convolution structure, which reduces the memory using the method of weighting sharing in the deep network, also reduces the number of network parameters, relieves overfitting problem.

In order to guarantee a certain amount of translation, scaling, distortion invariance, local receptive field, shared weight and space or time downsampling are designed in CNN. Convolutional neural network LeNet-5 is put forward for character recognition [4], which is composed of convolutional layers, downsampling layers and a whole connecting layer. It achieves better results in the small handwritten digital recognition. In 2012, Krizhevsky et al. designed a convolutional neural network, named AlexNet [5], won the first place in Image classification task of ImageNet challenge, which proclaimed huge success of CNN in large-scale image classification. AlexNet possesses deeper architecture with ReLU (Rectified linear unit) as nonlinear activation function and dropout to avoid overfitting.

After AlexNet, the researchers proposed deeper neural networks, such as the Google's GoogLeNet [6] and the 152-layer residual network designed by MSRA [7]. Table 14.1 is the leading result of ImageNet's image classification task over the years, and it can be seen that the network with the deeper layers often gains better classification results.

The rest of the chapter is organized as follows. The structure and principle of the deep CNN are dissected in following section, then several classical convolution network models are analyzed, and finally two instances based on CNN are given.

Table 14.1 The results of the image classification task on ImageNet

Time	Organization	Top-5 error rate (%)	Net name	Depth
2015.12.10	MSRA	3.57	ResNet [7]	152
2014.8.18	Google	6.67	GoogLeNet [6]	22
2013.11.14	NYU	11.7	Clarifai [8]	10
2012.10.13	U.Toronto	15.0	Alexnet [5]	8

14.2 Model Analysis of CNN

14.2.1 Basic Modules of CNN

The basic modules of CNN can be divided into four parts: input layer, convolutional layer, fully-connected layer and output layer.

Input layer. The convolutional input layer can directly affect the raw input data. If input is image, the input data are the pixel values of the image.

Convolutional layer. The convolution layer of the CNN, also known as the feature extraction layer, consists of two parts. The first part is the real convolutional layer. The main role is to extract the features of input data. Different convolution kernel extracts different characteristics of input data. The more convolution kernels in the convolutional layer, the more the features of the input data can be extracted. The second part is the pooling layer, also called subsampling layer. The main purpose is to reduce the amount of data processing on the basis of retaining useful information and speed up the training process. In general, CNN contains at least two convolutional layers, namely convolutional layer—pooling layer—convolutional layer—pooling layer.

Fully-connected layer. Fully-connected layers are actually the hidden layers of the Multilayer Perceptrons. In general, the neurons in the following layers are connected to each neuron in the previous layer, and there is no connection between the neurons in the same layer.

Output layer. The number of neural nodes in the output layer is set according to specific application tasks. If it is a classification task, the CNN output layer is usually a classifier, such as a Softmax classifier.

14.2.2 Convolution and Pooling

(1) Convolution

Convolution is often used for image feature extraction, the most important of which is the convolution kernel. The key design points generally involve the size, the number and the stride of the convolution kernel. Theoretically, the number means the number of feature maps obtained from the upper layer through convolution filter. The more features you extract, the more feature space the network represents, and the final recognition result will be more accurate.

But if the number of convolution kernels is too large, the complexity of the network and the number of parameters will increase, which leads to the increment of calculation complexity and overfitting phenomenon. So the number of convolution kernels shall be determined according to the size of the specific image datasets.

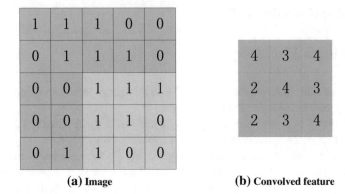

(a) Image   **(b)** Convolved feature

Fig. 14.1 Convolution diagram of image

Image convolution feature extraction will be realized on a $(n_h \times n_w)$ image by setting a convolution kernel filter with size of $w \times w$ and stride k, then a feature map will be generated with size of $\frac{n_h - w + k}{k} \times \frac{n_w - w + k}{k}$, shown in Fig. 14.1. In general, the smaller size of the convolution kernel, the higher quality of the feature. Nonetheless, the size should be determined according to the size of the input image.

(2)   Pooling

Feature map of image is obtained by convolution of the input image, and then new features will be produced in small neighborhood of the feature map by using the pooling technique. By means of pooling to upper layers, parameters (the feature dimension) can reduce, and the enhanced features make the final expression keep invariance (rotation, translation, scaling, etc.). So the essence of pooling is a dimension reduction process. The common pooling methods include mean- ooling, max-pooling, and so on.

According to relevant theories, the error of extracted feature mainly comes from two aspects:

(a)   the increase of estimation variance for limited neighborhood size;
(b)   the offset of estimated mean caused by convolution layer parameter error. Generally speaking, mean-pooling reduces the first error, and more retention of the background information of the image. Max-pooling reduces the second error, and retains more texture information.

14.2.3   Activation Function

Activation functions often used in the neural network include Sigmoid function, Tanh function and ReLU function, etc. The first two activation function in traditional BP neural network are used more. ReLU function is used more in deep learning. ReLU

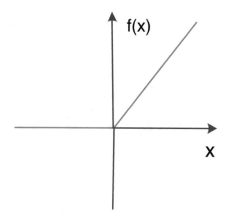

Fig. 14.2 ReLU function

function is a rectified linear unit proposed by Hinton et al. [5], shown as Fig. 14.2. Training on CNN using ReLU will be faster than sigmoid and tanh function.

Assuming that the activation function of a neural node is $h^{(i)}$, the expression of the ReLU function is:

$$h^{(i)} = \max((w^{(i)})^T x, 0) = \begin{cases} (w^{(i)})^T x & (w^{(i)})^T x > 0 \\ 0 & else \end{cases} \tag{14.1}$$

where i represents the number of hidden layer nodes, $w^{(i)}$ indicates the weight of hidden layer nodes.

Because ReLU function has the form of linear, unsaturated, unilateral suppression and sparse activation, its use in convolution neural network is more common than sigmoid and tanh function.

14.2.4 Softmax Classifier and Cost Function

When CNN is applied to image classification task, a softmax classifier is often attached to the last fully-connected layer of the neural network to predict the image label.

In softmax regression, our goal is to solve multiple classification problems. Label y may have k different values (rather than 2). Therefore, for the training dataset $\{(x^{(1)}, y^{(1)}), \ldots, (x^{(m)}, y^{(m)})\}$, we have $y^{(i)} \in \{1, 2, \ldots, k\}$.

For a given test input x, we want to estimate the probability value $p(y = j|x)$ for each category j by the hypothesis function. That is to say, we want to estimate the probability of each category of x. Consequently, our hypothetical function will output a vector with k dimension (the sum of the vector element is 1) to represent the probability value of this estimate. Specifically, our hypothetical function $h_\theta(x)$ is as follows:

$$h_\theta(x^{(i)}) = \begin{bmatrix} p(y^{(i)} = 1 | x^{(i)}; \theta) \\ p(y^{(i)} = 2 | x^{(i)}; \theta) \\ \cdots \\ p(y^{(i)} = k | x^{(i)}; \theta) \end{bmatrix} = \frac{1}{\sum_{j=1}^{k} e^{\theta_j^T x^{(i)}}} \begin{bmatrix} e^{\theta_1^T x^{(i)}} \\ e^{\theta_2^T x^{(i)}} \\ \cdots \\ e^{\theta_k^T x^{(i)}} \end{bmatrix} \qquad (14.2)$$

For convenience, we also use symbol θ to represent all model parameters. The probability of $x^{(i)}$ belongs to j is:

$$p(y = y^{(i)} | x^{(i)}; \theta) = \frac{e^{\theta_j^T x^{(i)}}}{\sum_{l=1}^{k} e^{\theta_l^T x^{(i)}}} \qquad (14.3)$$

When conditional probability $p(y = y^{(i)} | x^{(i)}; \theta)$ of each sample is the largest, recognition rate of classifier is the highest, which is equivalent to maximizing the likelihood function as follows:

$$L(\theta | x) = \prod_{i=1}^{m} p(y = y^{(i)} | x^{(i)}; \theta) \qquad (14.4)$$

To reduce the amount of computation and prevent overflow, after taking the logarithm of likelihood function, the appropriate deformation is:

$$J(\theta) = -\frac{1}{m} \left[\sum_{i=1}^{m} \sum_{j=1}^{k} 1\{y^{(i)} = j\} \log \frac{e^{\theta_j^T x^{(i)}}}{\sum_{l=1}^{k} e^{\theta_l^T x^{(i)}}} \right] \qquad (14.5)$$

where $1\{.\}$ indicates indicator function, $1\{\text{true}\} = 1$, $1\{\text{false}\} = 0$. At this point, maximizing likelihood function is equivalent to minimizing cost function, so gradient descent method is used to solve the minimum value of $J(\theta)$ and determine the parameter θ. The gradient of cost function $J(\theta)$ is:

$$\nabla_{\theta_j} J(\theta) = -\frac{1}{m} \sum_{i=1}^{m} \left[x^{(i)} (1\{y^{(i)} = j\} - p(y^{(i)} = j | x^{(i)}; \theta)) \right] \qquad (14.6)$$

In practical use, we usually add regularization $\frac{\lambda}{2} \sum_{i=1}^{m} \sum_{j=0}^{n} \theta_{ij}^2$ (L2 norm) to the cost function to prevent the overfitting problem, thus the cost function can be transformed into:

$$J(\theta) = -\frac{1}{m} \left[\sum_{i=1}^{m} \sum_{j=1}^{k} 1\{y^{(i)} = j\} \log \frac{e^{\theta_j^T x^{(i)}}}{\sum_{l=1}^{k} e^{\theta_l^T x^{(i)}}} \right] + \frac{\lambda}{2} \sum_{i=1}^{m} \sum_{j=0}^{n} \theta_{ij}^2 \qquad (14.7)$$

The second item in the upper equation will punish the larger parameter value, also known as the weight attenuation term. The proper λ can reduce the order of magnitudes of weight, so that the value of network parameters can be controlled to prevent overfitting.

14.2.5   Learning Algorithm

Neural networks mainly utilize back propagation algorithm to implement the gradient calculation and update the parameters using the gradient. The two main methods are Stochastic Gradient Decent (SGD), Adaptive Moment Estimation (Adam).

Usually, training datasets are very large. If loading all the training samples in one time, there will be memory overflow problems. So we actually use a mini-batch of datasets, the number of mini-batch $N \ll |D|$, thus the cost function will be:

$$J(\theta) = \frac{1}{|N|} \sum_i^{|N|} f_\theta(x^{(i)}) + \lambda r(\theta) \tag{14.8}$$

(1)   Stochastic Gradient Descent

Network loads a mini-batch each time for training in SGD method. Since each mini-batch is selected randomly, cost function in each iteration is different. The gradient of current batch has a far greater impact on the update of network parameters. To reduce this effect, momentum coefficient will be usually introduced to improve the traditional stochastic gradient descent method.

Momentum simulates the inertia of a moving object, that is, when the update operation is performed, to some degree, keep the updated direction. At the same time, the current batch gradient is used to fine-tune the final update direction, which can enhance its stability to a certain extent. Thus the network will learn faster, and there is a certain ability to get rid of local optimality. Iterative equation of SGD with momentum is shown as follows:

$$V_{t+1} = \mu V_t - \eta \nabla J(\theta_t) \tag{14.9}$$

$$\theta_{t+1} = \theta_t + V_{t+1} \tag{14.10}$$

where V_t is the last weight update amount, μ is momentum coefficient between 0 and 1, which indicates to what extent to keep the original direction. η is learning rate.

Characteristic of SGD can be summarized as follows:

(a) At the beginning of the descent, the last parameter is used to update. Since descent direction is consistent, network, multiplied by a larger number μ, is able to accelerate well.
(b) In the middle and later stages, when the local minimum is oscillating back and forth, $gradient \rightarrow 0$, μ makes update amplitude increase, which can beat local minimum trap.
(c) When the gradient changes direction, μ can reduce updates. In general, the momentum term can accelerate the SGD in the relevant direction and suppress the oscillation, thus accelerate the convergence.

(2) Adaptive Moment Estimation

Adam is a RMSprop with a momentum term in essence. It uses the first order reliability estimation and the second order reliability estimation of the gradient to dynamically adjust the learning rate of each parameter. The main advantage of Adam is that learning rates have a defined scope in each iteration after bias correction, which makes the parameter stable. Iteration equations are as follows:

$$m_t = \mu m_{t-1} + (1 - \mu)\nabla J(\theta_t) \tag{14.11}$$

$$n_t = v n_{t-1} + (1 - v)\nabla^2 J(\theta_t) \tag{14.12}$$

$$\hat{m}_t = \frac{m_t}{1 - \mu^t} \tag{14.13}$$

$$\hat{n}_t = \frac{n_t}{1 - v^t} \tag{14.14}$$

$$\Delta\theta_t = -\frac{\hat{m}_t}{\sqrt{\hat{n}_t} + \varepsilon}\eta \tag{14.15}$$

where m_t, n_t are the first order reliability estimation and the second order reliability estimation of the gradient, which can be seen as estimates of expectations $E|\nabla J(\theta_t)|$ and $E|\nabla^2 J(\theta_t)|$. $\hat{m}_t$, $\hat{n}_t$ are the correction to m_t, n_t, which can be approximated to an unbiased estimate of expectations. As can be seen, the moment estimation of the gradient has no additional requirement for memory, and can be dynamically adjusted according to the gradient, while $-\frac{\hat{m}_t}{\sqrt{\hat{n}_t}+\varepsilon}$ is a dynamic constraint to learning rate with a clear range.

Characteristic of Adam can be summarized as follows:

(a) Being good at handling sparse gradient and non-stationary target.
(b) Small memory requirements.
(c) Different adaptive learning rates are calculated for different parameters.

(d) Most non-convex optimization problems are applicable, so do big data sets and high dimensional space.
(e) Usually, its iteration speed is faster than SGD, but its convergence accuracy is generally inferior to SGD.

14.2.6   Dropout

Weight decay (L2 regularization) is implemented by modifying the cost function, while dropout is realized by modifying the architecture of neural network, which is an optimization method used in training neural networks. Dropout can make some units of the hidden layer of the network don't work randomly during model training phase. Those unworking units won't be calculated but their weights will be kept (temporarily not updated), since it might work again along with the input of the next sample. During the training phase, dropout sets the output of the hidden layer node to 0 at a certain probability $1 - p$. The neural network structures without dropout and with dropout are compared in as Fig. 14.3.

One advantage of dropout is that it is computationally cheap. Using dropout during training, it requires only O(n) computation per example per update, to generate n random binary numbers and multiply them by the state. Another significant advantage of dropout is that it does not significantly limit the type of model or training procedure that can be used. It works well with nearly any model that uses a distributed representation and can be trained with stochastic gradient descent.

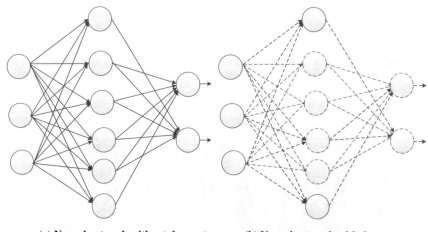

(a) Neural network without dropout   (b) Neural network with dropout

Fig. 14.3 Dropout schematic diagram

14.2.7   Batch Normalization

In the process of training deep neural networks, there are usually "gradient diffusion" problems. That is to say, when the back propagation method is used to compute the gradient derivative, with the increase of network depth, the amplitude of gradient of back propagation (from the output layer to the first layer of the network) will decrease dramatically. To solve the gradient diffusion problem, Google proposed the Batch Normalization method in ICML conference in 2015 [9]. Batch Normalization, that is, when stochastic gradient descent computed, the corresponding activation output will be normalized by mini-batch, so the mean of the results is 0 and the variance is 1. By this mean, the output that is going to decrease gets bigger. Therefore, the problem of gradient diffusion is solved in a large part, and the training of deep neural network will be accelerated.

14.3   Typical CNN Models

14.3.1   LeNet

LeNet is a classical convolution neural network model for handwritten character recognition by Yan Lecun in 1998 [4]. The architecture of which is shown as Fig. 14.4.

The architecture of LeNet contains 7 layers, including 3 convolutional layers. The first convolutional layer C1 consists of 6 feature maps, 156 trainable parameters, and 122,304 (156 × (28 × 28) − 122,304) connections. Each unit in each feature map is connected to a 5 × 5 neighborhood in the input. The size of the feature maps is 28 × 28 which prevents connection from the input from falling off the boundary. Convolutional layer C3 has 1,516 trainable parameters and 151,600 connections. The connection between S2 and C3 is shown as Fig. 14.4. Each unit in each feature map is connected to several 5 × 5 neighborhoods at identical locations in a subset of

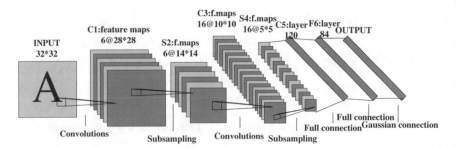

Fig. 14.4   Architecture of LeNet-5

S2's feature maps. Layer C5 is a convolutional layer with 120 feature maps, the size of C5's feature maps is 1 × 1: this amounts to a full connection between S4 and C5.

The architecture of LeNet contains 2 sub-sampling layers, too. Layer S2 has 6 feature maps while S2 has 16 feature maps. Layer S2 has 12 trainable parameters and 5,880 connections. Likewise, Layer S4 has 32 trainable parameters and 156,000 connections.

Layer F6, contains 84 units and is fully connected to C5. It has 10,164 trainable parameters.

14.3.2 AlexNet

AlexNet is the convolutional neural network model used in the ImageNet Large Scale Visual Recognition Challenge (ILSVRC) 2012 by Hinton team [5], and won the first prize. They achieved a winning top-5 test error rate of 15.3%, which is more 10% than the second-best entry. AlexNet has five convolutional layers and three fully-connected layers (Fig. 14.5).

The network's input is 150,528-dimensional, output is 1000-dimensional.

AlexNet contains five convolutional layers and three fully-connected layers, outputs one thousand categories classified by softmax classifier. AlexNet applied a variety of new technologies in the network, including:

(1) Data augmentation

The most common method to reduce overfitting on image data is to artificially enlarge the dataset. AlexNet employed several distinct forms of data augmentation, including:

(a) The first form of data augmentation consists of generating image translations and horizontal reflections by extracting random 224 × 224 patches from the 256 × 256 images.

(b) At test time, the network makes a prediction by extracting five 224 × 224 patches (the four corner patches and the center patch) as well as their horizontal reflections (hence ten patches in all), and averaging the predictions.

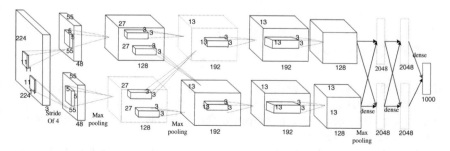

Fig. 14.5 The architecture of AlexNet

(c) The network performs PCA on the set of RGB pixel. To each training image, multiples of the found principal components are added, with magnitudes proportional to the corresponding eigenvalues times a random variable drawn from a Gaussian with mean zero and standard deviation 0.1.

(2) ReLU activation function

The standard activation function, such as tanh or sigmoid, in terms of training time with gradient descent, these saturating nonlinearities are much slower than the non-saturating nonlinearity. Nonlinearity as Rectified Linear Units (ReLUs) can void this problem. Deep convolutional neural networks with ReLUs train several times faster than their equivalents with tanh units.

(3) Dropout

The neurons which are "dropped out" in this way do not contribute to the forward pass and do not participate in back-propagation. So every time an input is presented, the neural network samples a different architecture, but all these architectures share weights. This technique reduces complex co-adaptations of neurons, since a neuron cannot rely on the presence of particular other neurons. It is, therefore, forced to learn more robust features that are useful in conjunction with many different random subsets of the other neurons.

(4) Training on Multiple GPUs

The memory of a single GPU is too small, which limits the maximum size of the networks. AlexNet spreads the net across two GPUs. The parallelization scheme essentially puts half of the kernels (or neurons) on each GPU, with one additional trick: the GPUs communicate only in certain layers.

(5) Local Response Normalization

ReLUs have the desirable property that they do not require input normalization to prevent them from saturating. If at least some training examples produce a positive input to a ReLU, learning will happen in that neuron. However, local normalization scheme aids generalization. This sort of response normalization implements a form of lateral inhibition inspired by the type found in real neurons, creating competition for big activities amongst neuron outputs computed using different kernels.

14.3.3 GoogLeNet

GoogLeNet is the championship of ILSVRC 2014 [6], which is a 22 layers deep network and obtains a top-5 error of 6.67% on both the validation and testing data, ranking the first among other participants. The main contribution of GoogLeNet is their Inception architecture.

The most straightforward way of improving the performance of deep neural networks is by increasing their size. This includes both increasing the depth: the number of network levels, and its width: the number of units at each level. Bigger size typically means a larger number of parameters, which makes the enlarged network more prone to overfitting, especially if the number of labeled examples in the training set is limited. The other drawback of uniformly increased network size is the dramatically increased use of computational resources.

A fundamental way of solving both of these issues would be to introduce sparsity and replace the fully connected layers by the sparse ones, even inside the convolutions. The main idea of the Inception architecture is to consider how an optimal local sparse structure of a convolutional vision network can be approximated and covered by readily available dense components. The naïve version of the Inception architecture are restricted to filter sizes 1×1, 3×3 and 5×5. Additionally, since pooling operations have been essential for the success of current convolutional networks, it suggests that adding an alternative parallel pooling path in each such stage should have additional beneficial effect, shown as Fig. 14.6.

One big problem with the above modules, at least in this naïve form, is that even a modest number of 5×5 convolutions can be prohibitively expensive on top of a convolutional layer with a large number of filters.

This leads to the second idea of the Inception architecture: judiciously reducing dimension wherever the computational requirements would increase too much otherwise. That is, 1×1 convolutions are used to compute reductions before the expensive 3×3 and 5×5 convolutions. Besides being used as reductions, they also include the use of rectified linear activation making them dual-purpose. The final result is depicted in Fig. 14.7.

The hierarchical structure of GoogLeNet is as follows:

Input dimensionality of initial data is $224 \times 224 \times 3$.

The first convolutional layer conv1, has 64 features with pad 3, 7×7 filter size, and stride 2, resulting in a $112 \times 112 \times 64$ output. After ReLU calculation, a pooling layer pool1 is added for dimension reduction, with 3×3 patch size, and stride 2. $[(112 - 3 + 1)/2] + 1 = 56$, output dimensionality is $56 \times 56 \times 64$.

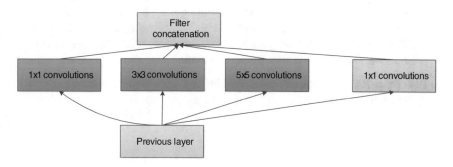

Fig. 14.6 Inception module in naïve version

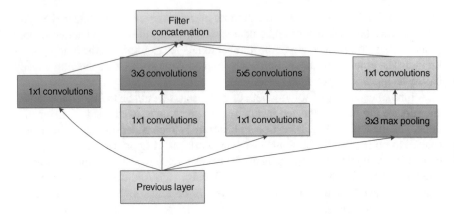

Fig. 14.7 Inception module with dimensionality reduction

The second convolutional layer conv2, has 192 features with pad 1, 3 × 3 filter size, and stride 1, resulting in a 56 × 56 × 192 output. After ReLU calculation, a pooling layer pool2 is added for dimension reduction, with 3 × 3 patch size and stride 2, resulting in a 28 × 28 × 192 output.

The third convolutional layer is an inception layer named (3a), which is composed of inception module using different scale convolution kernel. (3a) contains four branches:

(1) A 1 × 1 convolution with 64 filters (And then executing the ReLU calculation), resulting in a 28 × 28 × 64 output.

(2) A 1 × 1 convolution with 96 filters for dimension reduction, leading to a 28 × 28 × 96 intermediate result, after ReLU calculation, 3 × 3 convolution is conducted with 128 filters with pad 1, resulting in a 28 × 28 × 128 output.

(3) A 1 × 1 convolution with 16 filters for dimension reduction, leading to a 28 × 28 × 16 intermediate result, after ReLU calculation, 5 × 5 convolution is conducted with 32 filters with pad 2, resulting in a 28 × 28 × 32 output.

(4) Pooling layer, 3 × 3 convolution with pad 1, bringing about a 28 × 28 × 192 intermediate result, then a 1 × 1 convolution with 32 filters for dimension reduction, resulting in a 28 × 28 × 32 output.

Four outputs can be concatenated, resulting in a 28 × 28 × 256 output.
In the same way, data evolves like Table 14.2.

Table 14.2 GoogLeNet incarnation of the inception architecture

Type	Patch size/stride	Outputsize	Depth	#1 × 1	#3×3 reduce	#3 × 3	#5 × 5r educe	#5 × 5	Pool proj	Params	Ops
Convolution	7 × 7/2	112 × 112 × 64	1							2.7K	34M
Max pool	3 × 3/2	56 × 56 × 64	0								
Convolution	3 × 3/1	56 × 56 × 192	2		64	192				112K	360M
Max pool	3 × 3/2	28 × 28 × 192	0								
Inception(3a)		28 × 28 × 256	2	64	96	128	16	32	32	159K	128M
Inception(3b)		28 × 28 × 480	2	128	128	192	32	96	64	380K	304M
Max pool	3 × 3/2	14 × 14 × 480	0								
Inception(4a)		14 × 14 × 512	2	192	96	208	16	48	64	364K	73M
Inception(4b)		14 × 14 × 512	2	160	112	224	24	64	64	437k	88M
Inception(4c)		14 × 14 × 512	2	128	128	256	24	64	64	463K	100M
Inception(4d)		14 × 14 × 528	2	112	144	288	32	64	64	580K	119M
Inception(4e)		14 × 14 × 832	2	256	160	320	32	128	128	840K	170M
Max pool	3 × 3/2	7 × 7 × 832	0								
Inception(5a)		7 × 7 × 832	2	256	160	320	32	128	128	1072K	54M
Inception(5b)		7 × 7 × 1024	2	384	192	384	48	128	128	1388K	71M
Avg pool	7 × 7/1	1 × 1 × 1024	0								
Dropout (40%)		1 × 1 × 1024	0								
Linear		1 × 1 × 1000	1							1000K	1M
Softmax		1 × 1 × 1000	0								

14.3.4   VGGNet

VGGNet was proposed by Visual Geometry Group won of Oxford, and won the first place in localization task and the second place in classification of ILSVRC 2014 [10]. The main contributions of VGGNet including: a very small convolution (3 × 3) and a deeper network can effectively improve the effect of the model. VGGNet has good generalization ability on other dataset.

(1)   Network architecture

The input to a ConvNet is a fixed-size 224 × 224 RGB image. The only pre-processing is subtracting the mean RGB value, computed on the training set, from each pixel. The image is passed through a stack of convolutional (conv.) layers, where filters are used with a very small receptive field: 3 × 3 (which is the smallest size to capture the notion of left/right, up/down, center). Experiments with 1 × 1 convolution filters are conducted, which can be seen as a linear transformation of the input channels (followed by non-linearity). The convolution stride is fixed to 1 pixel; the spatial padding of conv. layer input is such that the spatial resolution is preserved after convolution, i.e. the padding is 1 pixel for 3 × 3 conv. layers. Spatial pooling is carried out by 5 max-pooling layers, which follow some of the conv. layers (not all the conv. layers are followed by max-pooling). Max-pooling is carried out over a 2 × 2 pixel window, with stride 2.

A stack of convolutional layers is followed by three Fully-Connected (FC) layers: the first two have 4096 channels each, the third performs 1000-way classification and thus contains 1000 channels (one for each class). The final layer is the soft-max transform layer.

Table 14.3 refers to the nets by their names (A–E). All configurations differ only in the depth: from 11 weight layers in the network A(8 conv. and 3 FC layers) to 19 weight layers in the network E (16 conv. and 3 FC layers). The width of conv. layers (the number of channels) is rather small, starting from 64 in the first layer and then increasing by a factor of 2 after each max-pooling layer, until it reaches 512.

(2)   Training

The training is carried out using mini-batch gradient descent with momentum. The batch size was set to 256, momentum—to 0.9. The training weight decay was set to 0.0005 and dropout regularization for the first two fully-connected layers (dropout ratio set to 0.5). The learning rate was initially set to 0.01, and then decreased by a factor of 10 when the validation set accuracy stopped improving. For random initialization, we sampled the weights from a normal distribution with the zero mean and 0.01 variance. The biases were initialized with 0.

To obtain 224 × 224 input images, they were randomly cropped from the full-size (non-cropped) training images, isotropically rescaled so that the smallest side equals $S \geq 224$. To further augment the training set, the crops underwent random horizontal flipping and random RGB color shift.

Table 14.3 Network architecture of VGGNet

ConvNet configuration					
A	A-LRN	B	C	D	E
11 weight layers	11 weight layers	13 Weight layers	16 weight layers	16 weights layers	19 weights layers
Input (224 × 224 RGB image)					
conv3-64	conv3-64 **LRN**	conv3-64 **conv3-64**	conv3-64 conv3-64	conv3-64 conv3-64	conv3-64 conv3-64
Maxpool					
conv3-128	conv3-128	conv3-128 **conv3-128**	conv3-128 conv3-128	conv3-128 conv3-128	conv3-128 conv3-128
Maxpool					
conv3-256 conv3-256	conv3-256 conv3-256	conv3-256 conv3-256	conv3-256 conv3-256 **conv1-256**	conv3-256 conv3-256 **conv3-256**	conv3-256 conv3-256 conv3-256 **conv3-256**
Maxpool					
conv3-512 conv3-512	conv3-512 conv3-512	conv3-512 conv3-512	conv3-512 conv3-512 **conv1-512**	conv3-512 conv3-512 **conv3-512**	conv3-512 conv3-512 conv3-512 **conv3-512**
Maxpool					
conv3-512 conv3-512	conv3-512 conv3-512	conv3-512 conv3-512	conv3-512 conv3-512 **conv1-512**	conv3-512 conv3-512 **conv3-512**	conv3-512 conv3-512 conv3-512 **conv3-512**
Maxpool					
FC-4096					
FC-4096					
FC-1000					
Soft-max					

(3) Testing

At test time, an input image is not necessarily equal to the image size in training phase. Namely, the fully-connected layers are first converted to the convolutional layers (the first FC layer—to a 7 × 7 convolutional layer, the last two FC layers—to 1 × 1 convolutional layers). The resulting net, which now contains only convolutional layers, is applied to the whole (uncropped) image by convolving the filters in each layer with the full-size input. The resulting output feature map is a class score map with the number of channels equal to the number of classes, and the variable spatial resolution, dependent on the input image size. Finally, to obtain a fixed-size vector of class scores for the image, the class score map is spatially averaged (sum-pooled).

14.3.5 ResNet

ResNet is the Residual Networks of Kaiming He [7]. ResNet achieved overwhelming success in ILSVRC2015, won the first place in classification, detection, localization task on ImageNet Dataset and detection, segmentation task on COCO Dataset. What's more, Deep Residual Learning for Image Recognition was awarded the best paper of CVPR2016.

The essential motivation of ResNet is to resolve degradation problem: with the network depth increasing, accuracy gets saturated and then degrades rapidly. However, with the depth increase of the model, the learning ability of the network strengthens, deeper model shouldn't get higher error rate. The reason of degradation problem is the difficulty in optimizing the network. For this reason, a residual structure is proposed, shown as Fig. 14.8.

Instead of hoping each few stacked layers directly fit a desired underlying mapping, these layers may fit a residual mapping. Formally, denoting the desired underlying mapping as $H(x)$, it will be reasonable to let the stacked nonlinear layers fit another mapping of $F(x) = H(x) - x$. The original mapping is recast into $F(x) + x$. The formulation of $F(x) + x$ can be realized by feedforward neural networks with "shortcut connections", shown as Fig. 14.8.

Its main advantages embodied in: Deep residual nets are easy to optimize, while the counterpart simply stacked nets exhibit higher training error when the depth increases; deep residual nets can easily enjoy accuracy gains from greatly increased depth, then degradation problem can be well solved.

Figure 14.9 constructs a 34-layer deep residual network. It is worth noticing that residual model has fewer filters and lower complexity than VGG nets. A 34-layer deep residual network has 3.6 billion FLOPs (multiply-adds), which is only 18% of VGG-19 (19.6 billion FLOPs).

Experimental results indicate that the 34-layer ResNet is better than the shallower ResNet. More importantly, the 34-layer ResNet exhibits considerably lower training error and is generalizable to the validation data. This indicates that the degradation problem is well addressed in this setting and accuracy gains can be obtained from increased depth.

Fig. 14.8 Residual structure

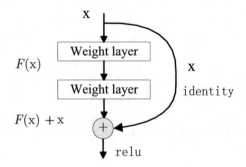

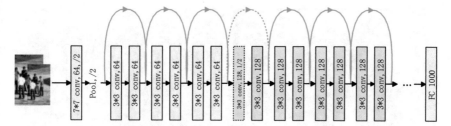

Fig. 14.9 A residual network with 34 parameter layers

ResNet constructed 50-layer, 101-layer and 152-layer ResNets by using more 3-layer blocks. The 50/101/152-layer ResNets are more accurate than the 34-layer ones by considerable margins. Above all, degradation problem don't occur while significant accuracy improvement is achieved from considerably increased depth. Their final result is 3.57% top-5 error on the test set, which won the 1st place in ILSVRC 2015.

14.4 Deep Learning Model for Lip Recognition Instance

Lip as a kind of biological characteristic can be used to recognize person. Lip recognition method locates the lip region firstly from image or video, then the features of lip region will be extracted, which is exploited to match the lip feature with standard lip model in the library. In this section, we design a lip recognition instance based on deep learning model, in which a VGG architecture will be utilized to train a deep model, and the training process will be detailed and explained.

14.4.1 Testing Dataset

Plip Dataset is a lip dataset for internal use in research institutes. The Dataset collects lip information from 26 persons, refers to different condition, such as facial expression, illumination, and so on, shown as Fig. 14.10.

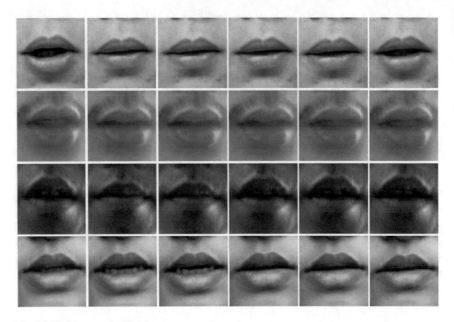

Fig. 14.10 Lip cases in plip dataset

14.4.2   Deep Network Training

VGG-FACE deep network fine-tuned on VGG architecture is shown as Fig. 14.11.

The next programmers describe the process of training a deep network. The following code is the shell script of creating lmdb database.

Layer	Conv 1_1	Conv 1_2	conv 2_1	conv 2_2	conv 3_1	conv 3_2	conv 3_3	conv 4_1	conv 4_2	conv 4_3	conv 5_1	conv 5_2	conv 5_3	FC-6	FC-7	FC-8
channel	64	64	128	128	256	256	256	512	512	512	512	512	512	4096	4096	26
map size	224×224	224×224	112×112	112×112	56*56	56*56	56*56	28*28	28*28	28*28	14*14	14*14	14*14			

Fig. 14.11 VGG-FACE deep model fine-tuned on VGG architechture

PROGRAMME 14.1: create_lip_net.sh

```
#!/usr/bin/envsh
# Create the lmdb inputs
EXAMPLE=vgglip
DATA=vgglip
TOOLS=./build/tools
TRAIN_DATA_ROOT=vgglip/train/
VAL_DATA_ROOT=vgglip/val/
# Set RESIZE=true to resize the images to 256x256. Leave as false if images have
# already been resized using another tool.
RESIZE=true
if $RESIZE; then
  RESIZE_HEIGHT=224
  RESIZE_WIDTH=224
else
  RESIZE_HEIGHT=0
  RESIZE_WIDTH=0
fi
if [ ! -d "$TRAIN_DATA_ROOT" ]; then
echo "Error: TRAIN_DATA_ROOT is not a path to a directory: $TRAIN_DATA_ROOT"
echo "Set the TRAIN_DATA_ROOT variable in create_imagenet.sh to the path" \
        "where the ImageNet training data is stored."
exit 1
fi
if [ ! -d "$VAL_DATA_ROOT" ]; then
echo "Error: VAL_DATA_ROOT is not a path to a directory: $VAL_DATA_ROOT"
echo "Set the VAL_DATA_ROOT variable in create_imagenet.sh to the path" \
        "where the ImageNet validation data is stored."
exit 1
fi
echo "Creating train lmdb..."
GLOG_logtostderr=1 $TOOLS/convert_imageset.bin \
    --resize_height=$RESIZE_HEIGHT \
    --resize_width=$RESIZE_WIDTH \
```

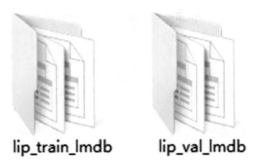

Fig. 14.12 Lmdb database creation

```
    $TRAIN_DATA_ROOT \
    $DATA/train.txt \
    $EXAMPLE/lip_train_lmdb
echo "Creating vallmdb..."
GLOG_logtostderr=1 $TOOLS/convert_imageset.bin \
    --resize_height=$RESIZE_HEIGHT \
    --resize_width=$RESIZE_WIDTH \
    $VAL_DATA_ROOT \
    $DATA/val.txt \
    $EXAMPLE/lip_val_lmdb
echo "Done."
```

Under Linux OS, the shell script can be executed as:
sh create_lip_net.sh
Then two files will be generated, as the following (Fig. 14.12).

(2)   The following code is the shell script of creating mean file.

PROGRAMME 14.2: make_lip_mean.sh

```
#!/usr/bin/envsh
# Compute the mean image from the imagenet training lmdb
# N.B. this is available in data/ilsvrc12
EXAMPLE=vgglip
DATA=vgglip
TOOLS=./build/tools
$TOOLS/compute_image_mean $EXAMPLE/VIRAT10_train_lmdb \
    $DATA/lip_mean.binaryproto
echo "Done."
```

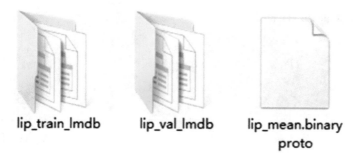

lip_train_lmdb   lip_val_lmdb   lip_mean.binary
                                               proto

Fig. 14.13 Mean file creation

The shell script can be executed as:
sh make_lip_mean.sh.sh
Another file lip_mean.binaryproto will be produced (Fig. 14.13).

(3) The following code is the shell script of creating solver file.

PROGRAMME 14.3: lip_solver.sh

```
net: "vgglip/vgglip_train_test.prototxt"
test_iter: 50
test_interval: 380
test_initialization: false
display: 40
average_loss: 40
base_lr: 0.0001
lr_policy: "step"
stepsize: 10000
gamma: 0.96
max_iter: 10000
momentum: 0.9
weight_decay: 0.0002
snapshot: 10000
snapshot_prefix: "vgglip/snapshots/vgg"
solver_mode: GPU
```

Fig. 14.14 Caffe model file
creation

**vgg_iter_10000.
caffemodel**

(4)   The following code is the shell script of training the model.

PROGRAMME 14.4: vgg_lip_training.sh

```
#!/usr/bin/envsh
./build/tools/caffe train \
    -solver=vgglip/lip_solver.prototxt \
-weights=vgglip/VGG_FACE.caffemodel -gpu=1
```

The shell script can be executed as:
sh vgg_lip_training.sh
And then a caffe model file will be created, shown as Fig. 14.14, which will be
used in the next section.

14.4.3   Code Analysis

Taking Plip Dataset as instance, we display the code of lip recognition based on deep
learning model, main function shown as PROGRAMME 14.5. The images in folder
will be scanned firstly. The features of each image are extracted and inputted into
SVM classifier to obtain the final recognition results.

PROGRAMME 14.5: Main function of lip recognition based on deep learning model

```
clear; close all;
addpath /home/workspace/caffe-master/matlab
saveDir = '/home/workspace/caffe-master/vgglip/_temp/feature4/';
if ~exist(saveDir, 'dir')
    mkdir(saveDir);
end
model = '/home/workspace/caffe-master/vgglip/VGG_lip_deploy.prototxt';
weights = '/home/workspace/caffe-master/vgglip/snapshots/vgg_iter_10000.caffemodel';
[train_vid, train_label] = textread('plipTrainTestlist\trainlist_plip.txt', '%s%d');
[test_vid, test_label] = textread('plipTrainTestlist\testlist_plip.txt', '%s%d');
caffe.set_mode_gpu();
caffe.set_device(0);
net = caffe.Net(model, weights, 'test'); % create net and load weights
isize = net.blobs('data').shape;
net.blobs('data').reshape([isize(1) isize(2) 3 1]); % reshape blob 'data'
net.reshape();
mean=load('/home/workspace/caffe-master/matlab/+caffe/imagenet/ilsvrc_2012_mean.mat');
image_mean=mean.mean_data;
%% extract feature for train
if 1
    fid = fopen('/home/workspace/caffe-master/vgglip/dataset/gtrain10.txt', 'r');
    trainList = textscan(fid, '%s%s');
    trainList = trainList{1};
    fclose(fid);
gtrain10= zeros(length(trainList),4096);
for f=1:length(trainList)
    name = trainList{f};
    fprintf('train-image %s\n',name);
    infile = ['/home/workspace/caffe-master/vgglip/dataset/gtrain10/' name];
    im = imread(infile);
    if size(im,3)<=1
        im = cat(3,im,im,im);
    end
    im_data = im(:, :, [3, 2, 1]);   % permute channels from RGB to BGR
    im_data = permute(im_data, [2, 1, 3]);   % flip width and height
    im_data = single(im_data);   % convert from uint8 to single
```

```
            im_data = imresize(im_data, [256, 256], 'bilinear');    % resize im_data
            im_data = im_data - image_mean;    % subtract mean_data
            im_data = imresize(im_data, [224 224], 'bilinear');
            res = net.forward({im_data});
            feat = net.blobs('fc7').get_data();
            gtrain10(f,:) = feat;
    end
    save('/home/workspace/caffe-master/vgglip/_temp/gtrain10.mat','gtrain10');
end
%% extract feature for test
if 1
    fid = fopen('/home/workspace/caffe-master/vgglip/dataset/gtest10.txt', 'r');
    testList = textscan(fid, '%s%s');
    testList = testList{1};
    fclose(fid);
    gtest10 = zeros(length(testList),4096);
    for f=1:length(testList)
        name = testList{f};
        fprintf('test-image %s\n',name);
        infile = ['/home/workspace/caffe-master/vgglip/dataset/gtest10/' name];
        im = imread(infile);
        im_data = im(:, :, [3, 2, 1]);    % permute channels from RGB to BGR
        im_data = permute(im_data, [2, 1, 3]);    % flip width and height
        im_data = single(im_data);    % convert from uint8 to single
        im_data = imresize(im_data, [256, 256], 'bilinear');    % resize im_data
        im_data = im_data - image_mean;    % subtract mean_data
        im_data = imresize(im_data, [224 224], 'bilinear');
        res = net.forward({im_data});
        feat = net.blobs('fc7').get_data();
        gtest10(f,:) = feat;
    end
    save('/home/workspace/caffe-master/vgglip/_temp/gtest10.mat','gtest10');
end


disp('svm...');
model = svmtrain(double(train_label),double(gtrain10)',svm_option);
[pred_labels, accuracy, prob_e] = svmpredict(double(test_label), double(gtest10)', model, '-b 1');
fprintf('Final accuracy: %f.\n', accuracy);
fprintf('\n Successfully finished... \n');
```

The main function can be executed to obtain the final recognition result. The accuracy of lip recognition based on deep learning method on Plip Dataset is about 90%.

14.5 Deep CNN Architecture for Event Recognition Instance

Event recognition indicates the process of recognizing spatial-temporal visual model from video. Along with the widely use of video monitoring system in real life, surveillance video event recognition has been widely utilized [11]. In this section, we introduce an event recognition instance based on two-stream CNNs fusion architecture [12], in which a deep CNN architecture is introduced firstly, then a spatial and temporal convolutional layer feature fusion method is designed, and a Fisher vector (FV) method is given to encode the feature. At last, the encoded features are input into SVM classifier to obtain the final recognition result.

14.5.1 Testing Dataset

VIRAT 2.0 Dataset includes about 8 h of surveillance videos recorded from total 11 scenes, captured by stationary HD cameras ($1280 \times 720p$ or $1920 \times 1080p$) installed at different school parking lots, shop entrances, and construction sites [13]. 11 categories of person-vehicle interaction events and other interaction events including: (1) loading an object to a vehicle (LAV), (2) unloading an object from a vehicle (UAV), (3) opening a vehicle trunk (OAT), (4) closing a vehicle trunk (CAT), (5) getting into a vehicle (GIV), and (6) getting out of a vehicle (GOV), (7) gesturing (GES), (8) carrying an Object (CAO), (9) running (RUN), (10) entering a facility (EAF), (11) exiting a facility (XAF) (Table 14.4).

Table 14.4 Event cases in VIRAT 2.0Dataset

| A. EAF | B. XAF | C.LAV | D.UAV | E.GIV | F.GOV |

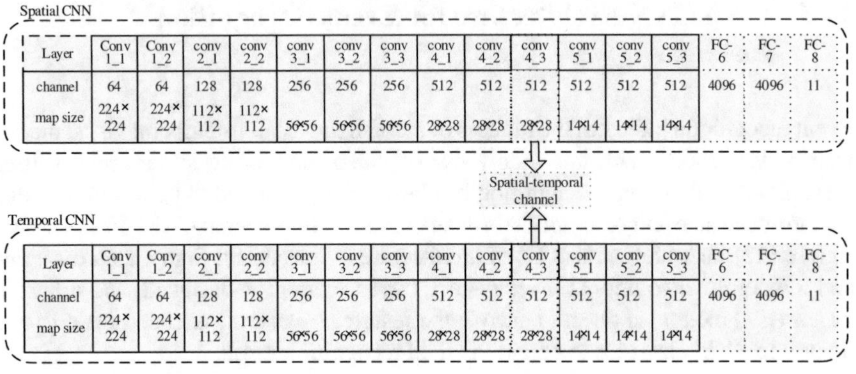

Spatial CNN

Layer	Conv 1_1	Conv 1_2	conv 2_1	conv 2_2	conv 3_1	conv 3_2	conv 3_3	conv 4_1	conv 4_2	conv 4_3	conv 5_1	conv 5_2	conv 5_3	FC-6	FC-7	FC-8
channel	64	64	128	128	256	256	256	512	512	512	512	512	512	4096	4096	11
map size	224×224	224×224	112×112	112×112	56×56	56×56	56×56	28×28	28×28	28×28	14×14	14×14	14×14			

Spatial-temporal channel

Temporal CNN

Layer	Conv 1_1	Conv 1_2	conv 2_1	conv 2_2	conv 3_1	conv 3_2	conv 3_3	conv 4_1	conv 4_2	conv 4_3	conv 5_1	conv 5_2	conv 5_3	FC-6	FC-7	FC-8
channel	64	64	128	128	256	256	256	512	512	512	512	512	512	4096	4096	11
map size	224×224	224×224	112×112	112×112	56×56	56×56	56×56	28×28	28×28	28×28	14×14	14×14	14×14			

Fig. 14.15 CNN architecture

14.5.2   Deep Feature Extraction

The architecture of CNNs is fine-tuned on the very deep two-stream models [14], which combines the merit of two-stream CNNs and VGG model. Fine-tuning has been verified as an effective way to initialize the CNNs [15]. For spatial network, we first extract frames of the videos, and set the input channel number as 3. For temporal net, we first extract optical flows of the videos, then set the input channel number as 20 (10 pairs of flow-x and flow-y), which is different with spatial network (20 vs. 3). We use convolutional layers as the output, and only extract convolutional features, the later layers will be removed, as shown in Fig. 14.15. At the same time, some convolutional layers are used for spatial-temporal fusion, which will be introduced in the next section.

14.5.3   Spatial-Temporal Feature Fusion

In general, the event factors including: two or more objects and the interaction between objects. Take the event of loading an object to a vehicle (LAV) as an example, this motion will be recognized by temporal CNNs. At the same time, spatial CNNs can recognize the appearance, and their combination can discriminates the activity successfully. The spatial consistency is easily achieved when the two network feature maps have the same resolution at the layers to be fused. Hence, we conduct spatial and temporal networks co-evolve at their feature maps.

Fig. 14.16 Spatial and temporal convolutional layer feature maps fuse by a 2D pooling

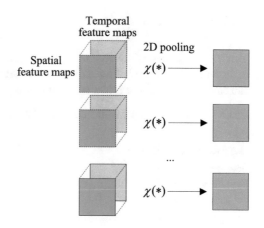

Fusion function $f : \mathbf{x}^a, \mathbf{x}^b \rightarrow y$ fuses two feature maps, here $\mathbf{x}^a \in \mathbb{R}^{H \times W \times D}$ and $\mathbf{x}^b \in \mathbb{R}^{H' \times W' \times D'}$, and produces an output $y \in \mathbb{R}^{H'' \times W'' \times D''}$, where W, H and D are the width, height and feature maps number. We use a 2D pooling method between the spatial feature maps and temporal feature maps in an appropriate convolutional layer, as shown in Fig. 14.16.

First, we concatenate the spatial feature maps and temporal feature maps by:

$$\mathbf{y}^{cat}_{i,j,2d} = \mathbf{x}^a_{i,j,d} \tag{14.16}$$

$$\mathbf{y}^{cat}_{i,j,2d-1} = \mathbf{x}^b_{i,j,d} \tag{14.17}$$

$$\mathbf{y}^{cat} = f^{cat}(\mathbf{x}^a, \mathbf{x}^b) \tag{14.18}$$

where $\mathbf{y} \in R^{H \times W \times 2D}$.

On the feature maps in Eq. 14.18, we carry out 2D pooling fusion, and get a $H \times W \times D$ output as spatial-temporal layer, depicted as follows:

$$\mathbf{y}^{fus} = \chi(f^{cat}(\mathbf{x}^a, \mathbf{x}^b)) \tag{14.19}$$

14.5.4 Fisher Vector Encoding

Fisher vector is verified as an effective high dimensional feature representation method for action recognition and so on [16]. We choose Fisher Vector to encode our video feature. Firstly, we reduce its dimension to D. Secondly, we train a GMM with K mixtures, and obtain a 2KD-dimensional vector.

14.5.5   Code Analysis

This case is a deep CNN architecture for event recognition. The following code is main function of deep feature extraction implemented on VIRAT2.0 Dataset, which can scan the videos in a folder and extract deep features.

PROGRAMME 14.6: Main function of deep feature extraction

```
clear;   clc;
dataPath = '/disk5/ygli/VIRAT2.0/';
directory = dir(dataPath);
for i = 3 : size(directory)
    if directory(i).isdir
        videos = dir([dataPath directory(i).name    '/*.avi'])
        for j = 1 : size(videos)
            VIRAT20_extractFeatures_fusion([dataPath directory(i).name '/' videos(j).name])
        end
    end
end
```

The following code is two-stream CNNs feature extraction and fusion function. Spatial convolution feature is extracted from video frames, while temporal convolutional feature is extracted from flow groups. Therefore, horizontal and vertical flow frames should be drawn as files from video before temporal convolutional feature extracted. Deep model adopted is the VGG16 architecture fine-tuned as before. Feature of each video is saved as a file with v7.3 format.

PROGRAMME 14.7: Two-stream CNNs feature extraction and fusion function

```
function [ features ] = VIRAT20_extractFeatures_fusion( vid_path )
    splPath = regexp(vid_path, '/', 'split');
    folder_name = splPath{end-1};
    vid_name = splPath{end};
    splPath2 = regexp(vid_name, '\.', 'split');
    vid = splPath2{1};
    result = cell(2, 1);
    result_folder = ['result_layer_cnn/VIRAT20_layerfusion/', folder_name, '/'];
    result_name = [result_folder, vid, '.mat'];
    if (~exist(result_folder, 'dir'))
        mkdir(result_folder);
    end
    if (~exist(result_name, 'file'))
        % flow extraction
        display('Extract TVL1 optical flow field...');
        flow_folder = ['flow_VIRAT20/', folder_name, '/', vid];
        if (~exist(flow_folder, 'dir'))
            mkdir(flow_folder)
            system(['export LD_LIBRARY_PATH=/usr/local/lib: $LD_LIBRARY_ PATH; ./denseFlow_gpu224
    -f ', vid_path, ' -x ', flow_folder, '/flow_x -y ', flow_folder, '/flow_y -b 20 -t 1 -d 1']);
        end
        sizes = [480,640; 340,454; 224,224; 240,320; 120,160];
% Extract spatial feature
    display('Extract spatial feature...');
    scale = 3;
    gpu_id = 0;
    model_def_file = [ 'model_proto/spatial_vgg_16.prototxt'];
    model_file = 'weights/vgg_16_rgb.caffemodel';
    caffe.reset_all();
    caffe.set_mode_gpu();
    caffe.set_device(gpu_id);
    net = caffe.Net(model_def_file, model_file, 'test');
    [sfeature_c4_3] = SpatialCNNFeature(vid_path, net, sizes(scale,1), sizes(scale,2));
    [sfeature_conv4_3n] = FeatureMapNormalization(sfeature_c4_3);
```

```
% Extract temporal feature
    display('Extract temporal feature...');
    model_def_file = ['model_proto/temporal_vgg_16_flow.prototxt'];
    model_file = 'weights/vgg_16_flow.caffemodel';
    caffe.set_mode_gpu();
    caffe.set_device(gpu_id);
    net = caffe.Net(model_def_file, model_file, 'test');
    [tfeature_conv4_3] = TemporalCNNFeature([flow_folder,'/'], net, sizes(3,1), sizes(3,2));
    [tfeature_conv4_3n] = FeatureMapNormalization(tfeature_conv4_3);
    [STfusion_conv4_3] = STfusion_feature(sfeature_conv4_3n,tfeature_conv4_3n);
    result = STfusion_conv4_3;
    save(result_name, 'result', '-v7.3');
    end
end


function [FCNNFeature_c4_3] = SpatialCNNFeature(vid_name, net, NUM_HEIGHT, NUM_WIDTH)
    % Input video
    vidObj = VideoReader(vid_name);
    duration = vidObj.NumberOfFrame;
    video = zeros(NUM_HEIGHT, NUM_WIDTH, 3, duration,'single');
    for i = 1 : duration
        tmp = read(vidObj,i);
        video(:,:,:,i) = imresize(tmp, [NUM_HEIGHT, NUM_WIDTH], 'bilinear');
    end
    d = load('VGG_mean');
    IMAGE_MEAN = d.image_mean;
    IMAGE_MEAN = imresize(IMAGE_MEAN,[NUM_HEIGHT,NUM_WIDTH]);
    video = video(:,:,[3,2,1],:);
    video = bsxfun(@minus,video,IMAGE_MEAN);
    video = permute(video,[2,1,3,4]);
    batch_size = 50;
```

The following code is spatial CNN convolutional feature extraction function, which draws frames from the video and extracts con4_3 convolutional feature from the video frames.

PROGRAMME 14.8: Spatial CNN convolutional feature extraction function

```
num_images = size(video,4);
num_batches = ceil(num_images/batch_size);
FCNNFeature_c4_3 = [];
images = zeros(NUM_WIDTH, NUM_HEIGHT, 3, batch_size, 'single');
for bb = 1 : num_batches
    range = 1 + batch_size*(bb-1): min(num_images,batch_size*bb);
    tmp = video(:,:,:,range);
    images(:,:,:,1:size(tmp,4)) = tmp;
    net.blobs('data').set_data(images);
    net.forward_prefilled();
    feature_c4_3 = permute(net.blobs('conv4_3').get_data(),[2,1,3,4]);
    if isempty(FCNNFeature_c4_3)
        FCNNFeature_c4_3  =  zeros(size(feature_c4_3,1),  size(feature_c4_3,2),  size(feature_c4_3,3),
num_images, 'single');
    end
    FCNNFeature_c4_3(:,:,:,range) = feature_c4_3(:,:,:,mod(range-1,batch_size)+1);
    end
end
    function [FCNNFeature_conv4_3] = TemporalCNNFeature(vid_name, net, NUM_HEIGHT, NUM_WIDTH)
    L = 10;
    % Input video
    filelist =dir([vid_name,'*_x*.jpg']);
    video = zeros(NUM_HEIGHT,NUM_WIDTH,L*2,length(filelist));
    for i = 1: length(filelist)
        flow_x = imread(sprintf('%s_%04d.jpg',[vid_name,'flow_x'],i));
        flow_y = imread(sprintf('%s_%04d.jpg',[vid_name,'flow_y'],i));
        video(:,:,1,i) = imresize(flow_x,[NUM_HEIGHT,NUM_WIDTH],'bilinear');
        video(:,:,2,i) = imresize(flow_y,[NUM_HEIGHT,NUM_WIDTH],'bilinear');
    end

for i = 1:L-1
    tmp = cat(4, video(:,:,(i-1)*2+1:i*2,2:end),video(:,:,(i-1)*2+1:i*2,end));
    video(:,:,i*2+1:i*2+2,:)   = tmp;
end
d   = load('flow_mean');
FLOW_MEAN = d.image_mean;
FLOW_MEAN = imresize(FLOW_MEAN,[NUM_HEIGHT,NUM_WIDTH]);
batch_size = 50;
num_images = size(video,4);
num_batches = ceil(num_images/batch_size);
FCNNFeature_conv4_3 = [];
for bb = 1 : num_batches
    range = 1 + batch_size*(bb-1): min(num_images,batch_size*bb);
```

The following code is temporal CNN convolutional feature extraction function, which loads a flow group composed of 10 pairs of flow frames (flow-x and flow-y) extracted from the video.

PROGRAMME 14.9: Temporal CNN convolutional feature extraction function

```
images = zeros(NUM_WIDTH, NUM_HEIGHT, L*2, batch_size, 'single');
tmp = video(:,:,:,range);
for ii = 1 : size(tmp,4)
    img = single(tmp(:,:,:,ii));
    images(:,:,:,ii) = permute(img -FLOW_MEAN,[2,1,3]);
end
net.blobs('data').set_data(images);
net.forward_prefilled();
feature_conv4_3 = permute(net.blobs('conv4_3').get_data(),[2,1,3,4]);
if isempty(FCNNFeature_conv4_3)
    FCNNFeature_conv4_3       =       zeros(size(feature_conv4_3,1),       size(feature_conv4_3,2),
size(feature_conv4_3,3), num_images, 'single');
end
FCNNFeature_conv4_3(:,:,:,range) = feature_conv4_3(:,:,:,mod(range-1,batch_size)+1);
end
end


function [cnn_feature1] = FeatureMapNormalization(cnn_feature)
r = size(cnn_feature,1);
c = size(cnn_feature,2);
f = size(cnn_feature,3);
t = size(cnn_feature,4);
cnn_feature1 = permute(cnn_feature,[1,2,4,3]);
cnn_feature1 = reshape(cnn_feature1,r*c*t,[]);
cnn_feature1 = bsxfun(@rdivide,cnn_feature1,max(cnn_feature1,[],1)+eps);
cnn_feature1 = reshape(cnn_feature1,r,c,t,f);
cnn_feature1 = permute(cnn_feature1,[1,2,4,3]);
end
```

The following code is convolutional feature normalization function, which can improve the generalization ability of the feature.

PROGRAMME 14.10: Convolutional feature normalization function

```
function [STfusion_feature] = STfusion_layer(scnn_feature,tcnn_feature)
    fusion_feature = zeros(size(tcnn_feature));
    for i=1:size(tcnn_feature,3)
        for j=1:size(tcnn_feature,4)
            fusion_feature(:,:,i,j) = max(scnn_feature(:,:,i,j),tcnn_feature(:,:,i,j));
        end
    end
    STfusion_feature = fusion_feature;
end
```

The following code is the fusion function on convolutional layer. The function implements 2D max pooling on the count part feature maps of the elaborately convolutional layers of spatial CNN and temporal CNN.

PROGRAMME 14.11: Convolutional feature fusion function

```
        images = zeros(NUM_WIDTH, NUM_HEIGHT, L*2, batch_size, 'single');
        tmp = video(:,:,:,range);
        for ii = 1 : size(tmp,4)
            img = single(tmp(:,:,:,ii));
            images(:,:,:,ii) = permute(img -FLOW_MEAN,[2,1,3]);
        end
        net.blobs('data').set_data(images);
        net.forward_prefilled();
        feature_conv4_3 = permute(net.blobs('conv4_3').get_data(),[2,1,3,4]);
        if isempty(FCNNFeature_conv4_3)
            FCNNFeature_conv4_3    =    zeros(size(feature_conv4_3,1),    size(feature_conv4_3,2),
    size(feature_conv4_3,3), num_images, 'single');
        end
        FCNNFeature_conv4_3(:,:,:,range) = feature_conv4_3(:,:,:,mod(range-1,batch_size)+1);
    end
end
```

The following code is the main function of final recognition. In the first place, PCA function is called, which will produce a 2KD dimensional vector. Next, a linear SVM is utilized as classifier.

PROGRAMME 14.12: Main function of fisher vector encoding and SVM classifier

```
function [cnn_feature1] = FeatureMapNormalization(cnn_feature)
    r = size(cnn_feature,1);
    c = size(cnn_feature,2);
    f = size(cnn_feature,3);
    t = size(cnn_feature,4);
    cnn_feature1 = permute(cnn_feature,[1,2,4,3]);
    cnn_feature1 = reshape(cnn_feature1,r*c*t,[]);
    cnn_feature1 = bsxfun(@rdivide,cnn_feature1,max(cnn_feature1,[],1)+eps);
    cnn_feature1 = reshape(cnn_feature1,r,c,t,f);
    cnn_feature1 = permute(cnn_feature1,[1,2,4,3]);
end


function [STfusion_feature] = STfusion_layer(scnn_feature,tcnn_feature)
    fusion_feature = zeros(size(tcnn_feature));
    for i=1:size(tcnn_feature,3)
        for j=1:size(tcnn_feature,4)
            fusion_feature(:,:,i,j) = max(scnn_feature(:,:,i,j),tcnn_feature(:,:,i,j));
        end
    end
    STfusion_feature = fusion_feature;
end
clear all;clc;
z_config;
save_layer = ['save_VIRAT20/save_VIRAT20_layerfusion',num2str(1)];
result_layer = '/home/VIRAT20/result_layer_cnn/VIRAT20_layer';
[train_vid, train_label] = textread('viratTrainTestlist_recongnition/trainlist_virat20.txt', '%s%d');
train_feat_file = cellfun(@(x){strrep(x, '.avi', '.mat')}, train_vid);
[test_vid, test_label] = textread('viratTrainTestlist_recongnition/testlist_virat20.txt', '%s%d');
test_feat_file = cellfun(@(x){strrep(x, '.avi', '.mat')}, test_vid);
tr_idx = 1:length(train_label); te_idx = 1:length(test_label);
train_feat_file = train_feat_file(tr_idx);
train_label = train_label(tr_idx);
test_feat_file = test_feat_file(te_idx);
test_label = test_label(te_idx);
TESTFILES = 1000;
NUMFEATS = 100;
params.K=256;     % num of GMMs
FISHER_RES1 = ['fisher/fisher1_virat20/'];
FISHER_RES2 = ['fisher/fisher2_virat20/'];
```

```
% pca
feat = pca_param.U * (bsxfun(@minus, train_spa_feat, pca_param.mu));
% whitening
feat = bsxfun(@rdivide, feat, sqrt(pca_param.vars));
coding_conv4_3 = vl_fisher(feat, gmm.means, gmm.cov, gmm.priors);
%power normalization f(pk) = sign(pk) |pk|
coding_conv4_3=sign(coding_conv4_3).*(abs(coding_conv4_3).^0.5);
% below is intra-normalization
for ii=1: 2 * 256
new_feature = 0;
epsilon = 0.00005;
[pca_param, gmm] = z_pca _STfusion( train_feat_file, NUMFEATS, params,save_layer,result_layer);
disp('training...');
if exist([save_layer '/trfvt1.mat'], 'file')
     load([save_layer '/trfvt1.mat']);
else
     new_feature = 1;
     trfvt_conv4_3 = [];
     for i = 1 : length(train_feat_file)
          if exist([FISHER_RES1 train_feat_file{i}], 'file')
                tmp = load([FISHER_RES1 train_feat_file{i}]);
                coding_conv4_3 = tmp.fv;
          else
                typeName = regexp(train_feat_file{i}, '/', 'split');
                typeName = typeName{1};
                dir = [FISHER_RES1 typeName];
                if ~exist(dir, 'dir')
                     mkdir(dir)
                end
                if exist(fullfile(result_layer, train_feat_file{i}), 'file')
                     tmp = load(fullfile(result_layer, train_feat_file{i}), 'result');
                else
                     continue;
                end
                fprintf('processing %d, video: %s...\n', i, train_feat_file{i});
train_spa_feat = tmp.result;
if isempty(train_spa_feat)
     coding_conv4_3 = single(zeros(32768, 1));
else
```

```
                temp=coding_conv4_3((ii-1) * 64+1:64 * ii,1);
                temp=temp / (sqrt(temp' * temp) + epsilon);
                coding_conv4_3((ii-1) * 64+1:64 * ii,1)=temp;
            end
        end
end
        trfvt_conv4_3 = [trfvt_conv4_3 coding_conv4_3];
    end
    save([save_layer '/trfvt_conv4_3.mat'], 'trfvt_conv4_3', '-v7.3');
end
disp('testing...');
if exist([save_layer '/tefvt1.mat'], 'file')
    load([save_layer '/tefvt1.mat']);
else
    new_feature = 1;
    tefvt_conv4_3 = [];
    for i = 1 : length(test_feat_file)
        if exist([FISHER_RES1 test_feat_file{i}], 'file')
            tmp = load([FISHER_RES1 test_feat_file{i}])
        else
            typeName = regexp(test_feat_file{i}, '/', 'split');
            typeName = typeName{1};
            dir = [FISHER_RES1 typeName];
            if ~exist(dir, 'dir')
                mkdir(dir)
            end
            if exist(fullfile(result_layer, test_feat_file{i}), 'file')
                tmp = load(fullfile(result_layer, test_feat_file{i}), 'result');
            else
                continue;
            end
            fprintf('processing %d, video: %s...\n', i, test_feat_file{i});
```

```
        test_spa_feat = tmp.result;
        if isempty(test_spa_feat)
                coding_conv4_3 = single(zeros(32768, 1));
        else
                % pca
                feat = pca_param.U * (bsxfun(@minus, test_spa_feat, pca_param.mu));
                % whitening
                feat = bsxfun(@rdivide, feat, sqrt(pca_param.vars));
                coding_conv4_3 = vl_fisher(feat, gmm.means, gmm.cov, gmm.priors);
                %power normalization f(pk) = sign(pk) |pk|
                coding_conv4_3=sign(coding_conv4_3).*(abs(coding_conv4_3).^0.5);
                % below is intra-normalization
                for ii=1: 2 * 256
                        temp=coding_conv4_3((ii-1) * 64+1:64 * ii,1);
                        temp=temp / (sqrt(temp' * temp) + epsilon);
                        coding_conv4_3((ii-1) * 64+1:64 * ii,1)=temp;
                end
        end
    end
    tefvt_conv4_3 = [tefvt_conv4_3 coding_conv4_3];
  end
  save([save_layer '/tefvt_conv4_3.mat'], 'tefvt_conv4_3', '-v7.3');
end
disp('svm...');
model = svmtrain(double(train_label),double(trfvt_conv4_3)',svm_option);
[pred_labels, accuracy, prob_estimates] = svmpredict(double(test_label), double(tefvt_conv4_3)', model, '-b 1');
fprintf('Final accuracy: %f.\n', accuracy);
fprintf('\n Successfully finished... \n');
```

The following code is PCA function. PCA method will reduce the dimension to 64. Then GMM is trained with K = 256.

PROGRAMME 14.13: PCA function

```
[pca_param,gmm]=z_pca _STfusion(train_feat_file,NUMFEATS,params,save_layer,result_layer)

    feats = [];

    loop = 0;

    e = 0.00001;

    if ~exist(save_layer,'dir')

        mkdir(save_layer);

    end

    parfor i = 1 : length(train_feat_file)

        fprintf('process loop %d: %s\n', i, train_feat_file{i});

        if exist(fullfile(result_layer, train_feat_file{i}), 'file')

            tmp = load( fullfile(result_layer, train_feat_file{i}), 'result');

        else

            continue;

        end


        train_spa_feat = tmp.result';

        train_spa_feat = z_subsample(train_spa_feat, NUMFEATS);

        feats = [feats train_spa_feat];

    end

    disp('pca...');

    [U, mu, vars] = pca(feats);

    x_rot = U(:, 1:64)' * (bsxfun(@minus, feats, mu));

    x_rot = bsxfun(@rdivide, x_rot, sqrt(vars(1:64)+e));

    pca_param.U = U(:, 1:64)';

    pca_param.mu = mu;

    pca_param.vars = vars(1:64) + e;

    disp('gmm...');

    [means, cov, priors] = vl_gmm(x_rot, params.K);

    gmm.means = means;

    gmm.cov = cov;

    gmm.priors = priors;

end
```

In this case, deep feature extraction function will be firstly run to get the input data of FV. Then other functions will be executed to obtain the final recognition result. The accuracy of deep learning method is more than 80%, which is superior to traditional methods, such as 55% of BOW [17], 73% of structural model [18].

References

1. Rumelhart DE, McClell JL, PDP Research Group (1986) Parallel distributed processing. Bradford Books
2. Cortes Corinna, Vapnik Vladimir (1995) Support-vector networks. Mach Learn 20(3):273–297
3. Hinton GE, Salakhutdinov RR (2006) Reducing the dimensionality of data with neural networks. Science 313:504–507
4. Lecun Y, Bottou L, Bengio Y et al (1998) Gradient-based learning applied to document recognition. Proc IEEE 86(11):2278–2324
5. Krizhevsky A, Sutskever I, Hinton GE (2012) ImageNet classification with deep convolutional neural networks. In: International conference on neural information processing systems, pp 1097–1105
6. Szegedy C, Liu W, Jia Y et al (2015) Going deeper with convolutions. In: IEEE conference on computer vision and pattern recognition, pp 1–9
7. He K, Zhang X, RenS et al (2016) Deep residual learning for image recognition. In: IEEE conference on computer vision and pattern recognition, pp 770–778
8. Zeiler MD, Fergus R (2014) Visualizing and understanding convolutional networks. In: European conference on computer vision. Springer, Cham, pp 818–833
9. Ioffe S, Szegedy C (2015) Batch normalization: accelerating deep network training by reducing internal covariate shift. In: International conference on international conference on machine learning. JMLR.org, pp 448–456
10. Simonyan K, Zisserman A (2014) Very deep convolutional networks for large-scale image recognition. arXiv:1409.1556
11. Wang X, Ji Q (2015) Video event recognition with deep hierarchical context model. In: IEEE conference on computer vision and pattern recognition, pp 4418–4427
12. Li Y, Ge R, Ji Y, Gong S, Liu C. Trajectory-pooled spatial-temporal architecture of deep convolutional neural networks for video event detection. IEEE Trans Circuits Syst Video Technol. https://doi.org/10.1109/tcsvt.2017.2759299
13. Oh S, Hoogs A, Perera A et al (2011) A large-scale benchmark dataset for event recognition in surveillance video. In: IEEE conference on computer vision and pattern recognition, pp 3153–3160
14. Wang L, Xiong Y, Wang Z et al (2015) Towards good practices for very deep two-stream convnets. arXiv:1507.02159
15. Feichtenhofer C, Pinz A, Zisserman A (2016) Convolutional two-stream network fusion for video action recognition. In: IEEE conference on computer vision and pattern recognition
16. Sánchez J, PerronninF MensinkT et al (2013) Image classification with the fisher vector: theory and practice. Int J Comput Vis 105(3):222–245
17. Jiang YG, Ngo CW, Yang J (2007) Towards optimal bag-of-features for object categorization and semantic video retrieval. In: ACM international conference on image and video retrieval, pp 494–501
18. Zhu Y, Nayak NM, Roy-Chowdhury AK (2013) Context-aware modeling and recognition of activities in video. In: IEEE conference on computer vision and pattern recognition, pp 2491–2498

Appendix
Common Evaluation Criterion

Abstract In appendix we introduce two categories of visual quality evaluation method: subjective evaluation and objective evaluation, in which we highlight structured quality evaluation methods and classification evaluation methods in objective evaluation.

Introduction

Image and video quality evaluation criterion can effectively evaluate the performance of image and video algorithm, which is significant to the application of image and video processing technology. For instance, camera designer will decide which camera can better convert the natural image into a digital image, a medical diagnostic instrument requires the quality of the image to determine the disease and the cause of the pathogeny, geological detector requires the quality of the image to determine the purity of the ore. In short, the evaluation criterion of image and video quality has developed into a research field across multiple disciplines. Its main applications as follows: (a) For image restoration, it is mainly used for evaluating the reason and degree of image distortion, which is utilized for recovering the image; (b) For image enhancement, it is mainly used to evaluate the improvement of visual effects of digital images for further processing; (c) For video behavior recognition, event recognition, pedestrian re-identification and other fields, it is mainly used to evaluate the classification results; (d) For other professional fields, the evaluation of image and video quality also has a good application prospect.

The methods for visual quality evaluation can be divided into two categories: subjective evaluation and objective evaluation. The subjective evaluation method is to evaluate the quality of the image based on the subjective perception of the perceiver. The objective evaluation method is to measure the image quality by mathematical model, and output values as the evaluation values or distortion of the quality. In the following sections, these two kinds of methods are briefly described.

Subjective Evaluation

There have been several evaluation criteria for subjective scores. For example, the subjective evaluation methods of multimedia application are stipulated in literature [1] and the subjective evaluation methods of TV image are provided in literature [2]. Both criteria listed above enact detailed and strict rules for the process and the environment of subjective evaluation, which involves factors such as test sequence, age range, distance, brightness in the environment, brightness of natural light, etc.

Table A.1 gives the five-point rating scale commonly used, including quality scale and hindrance scale. The hindrance scale is mostly adopted from the point of view of professionals, but the quality scale for the average person.

The relative evaluation is that an observer compares several images and gives a corresponding score. Its criteria shows in Table A.2. The results of the evaluation can be obtained by the average score given by a certain number of observers.

Based on the above analysis, we can see that the advantages of subjective image quality evaluation are intuitive and consistent with the observation results of human visual system while the shortcomings of subjective evaluation are obvious:

(1) Evaluation scores will vary with the observer, even for the same image, results also varies with time which generates the insufficient accuracy.
(2) It is rather harsh for the observer, evaluation environment and some objective conditions. Such as the age of the observer, distance of observation, brightness of environment and so on.
(3) Relatively high cost, relatively low efficiency, and narrow areas of application.

Table A.1 Absolute rating scale

Score	Quality scale	Hindrance scale
5	Very good	The quality of the image is not bad at all
4	Good	The quality of the image changes without interfering with the viewing
3	General	The quality of the image is bad, which is a slight hindrance to viewing
2	Bad	Interfere with the review
1	Very bad	Very serious obstruction of view

Table A.2 Comparison of relative evaluation scale and absolute evaluation scale

Score	Absolute evaluation scale	Relative evaluation scale
5	Very good	Best of all
4	Good	Better than average in that group
3	General	Average of the group
2	Bad	Below average of the group
1	Very bad	Worst of the group

In view of the shortcomings of subjective quality evaluation method, it's necessary to use mathematic models to solve the problems of subjective quality evaluation method, which leads to the emergence of many objective quality evaluation methods.

Classic Objective Quality Evaluation Methods

Objective evaluation methods are implemented by establishing models, which receive original image and distorted image as input, output a value that can reflect the quality of the distorted image. Classic objective quality evaluation methods include statistical characteristics evaluation, information content evaluation, sharpness evaluation, spectral information evaluation, and so on.

Statistical Characteristics

There is no ideal standard reference image, so the process effect of image is objectively evaluated based on the statistical characteristics of the target image and the performance index which reflects the relationship between the target image and original image. $\{x(i,j), 0 \leq i \leq M-1, 0 \leq j \leq N-1\}$ indicates a pixel value of the original image, $\{\hat{x}(i,j), 0 \leq i \leq M-1, 0 \leq j \leq N-1\}$ represents the pixel value of the target image after the corresponding compression and $\{e(i,j) = x(i,j) - \hat{x}(i,j), 0 \leq i \leq M-1, 0 \leq j \leq N-1\}$ represents error image.

(1) Average Value (AV)

The size of the mean represents the average size of the image pixel values, which is an evaluation index that belongs to the statistical characteristics. The brightness that human eye can be perceived in grayscale images in the form of grayscale, so the average value of the gray scale has a greater effect on the visual effect of the image. If the average value of the image is appropriate, the result of process is better. The average value of the image is defined as:

$$\mu = \frac{1}{M \times N} \sum_{i=0}^{M-1} \sum_{j=0}^{N-1} \hat{x}(i,j) \tag{A.1}$$

(2) Standard Deviation

The centrality or discretization of the image gray value relative to the gray scale is generally reflected by the standard deviation which reflecting the distribution of the pixel values and showing the contrast of the image. If the standard deviation of the target image is small, the contrast is small, that is, the amount of information contained therein is smaller. The larger the standard deviation, the more gray-scale distribution, the better the visual effect. The standard deviation is obtained indirectly by the average value, and the standard deviation of the image is defined as:

$$\sigma = \frac{1}{M \times N} \sqrt{\sum_{i=0}^{M-1} \sum_{j=0}^{N-1} [\hat{x}(i,j) - \mu]} \tag{A.2}$$

(3) Difference Index (DI)

The average of the ratio of the absolute value of the difference between the target image and the original image and the original image value is called difference index. In general, the smaller the difference index, the smaller the degree of target image deviation from the original, the more the original grayscale information remains. It is defined as:

$$DI = \frac{1}{M \times N} \sum_{i=0}^{M-1} \sum_{j=0}^{N-1} \frac{|x(i,j) - \hat{x}(i,j)|}{x(i,j)} \tag{A.3}$$

Ideally $DI = 0$.

(4) Degree of Distortion (DD)

DD reflects the degree of distortion of the target image relative to the original image, the smaller the value, the better the effect of the target image. It is defined as:

$$DD = \frac{1}{M \times N} \sum_{i=0}^{M-1} \sum_{j=0}^{N-1} |x(i,j) - \hat{x}(i,j)| \tag{A.4}$$

(5) Correlation Coefficient (CC)

CC reflects the correlation degree of the two spectral features of the image. Generally speaking, the closer the correlation coefficient of the image is to 1, the better the proximity of the image is, the more information is obtained from the original image, the less information, the better the processed effect. It is defined as:

$$CC = \frac{\sum_{x=0}^{N-1} \sum_{y=0}^{N-1} (f(x,y) - \mu_f)(g(x,y) - \mu_g)}{\sqrt{\sum_{x=0}^{N-1} \sum_{y=0}^{N-1} (f(x,y) - \mu_f)^2 (g(x,y) - \mu_g)^2}} \tag{A.5}$$

where μ_f, μ_g are the average of the original image $x(i,j)$ and the target image $\hat{x}(i,j)$ respectively.

(6) Mean Squared Error (MSE)

Mean Squared Error (MSE), Peak Signal to Noise Rate (PSNR), and Mean Absolute Error (MAE), etc.

MSE firstly calculates the mean square value of pixel difference value of original image and distorted image, then determines the distortion degree of the distortion image by the size of the mean square value. The mean square error is expressed as:

$$MSE = \frac{1}{MN} \sum_{i=0}^{M-1} \sum_{j=0}^{N-1} e^2(i,j) = \frac{1}{MN} \sum_{i=0}^{M-1} \sum_{j=0}^{N-1} [x(i,j) - \hat{x}(i,j)]^2 \qquad (A.6)$$

(7) Peak Signal to Noise Rate (PSNR)

Set $x_{max} = 2^K - 1$, where K represents binary number used in a pixel point, PSNR can be defined as:

$$PSNR = 10 \lg \frac{x_{max}^2}{MSE} = 10 \lg \left[\frac{x_{max}^2 MN}{\sum_{i=0}^{M-1} \sum_{j=0}^{N-1} [x(i,j) - \hat{x}(i,j)]^2} \right] (dB) \qquad (A.7)$$

In many video series and commercial image applications $K = 8$, so $x_{max} = 255$. Combining with Eq. A.7, we can get:

$$PSNR = 10 \lg \frac{255^2}{MSE} = 10 \lg \left[\frac{255^2 MN}{\sum_{i=0}^{M-1} \sum_{j=0}^{N-1} [x(i,j) - \hat{x}(i,j)]^2} \right] (dB) \qquad (A.8)$$

(8) Mean Absolute Error (MAE)

MAE can be used for evaluating coding performance, which is defined as:

$$MAE = \frac{1}{MN} \sum_{i=0}^{M-1} \sum_{j=0}^{N-1} |x(i,j) - \hat{x}(i,j)| \qquad (A.9)$$

Information Content

(1) Information entropy

Information entropy is an important indicator of the degree of abundance of image information. It is reflected degree of deviation in the image range from the peak area of the gray histogram. The larger the entropy of the target image, the more the information volume of the target image increases, the richer the image is, the better effect of the image process. Information entropy is defined as:

$$H = -\sum_{l=0}^{L} h(l) \log[h(l)] \tag{A.10}$$

where L represents the total gray level of the target image, $h(l)$ represents the ratio of the number of pixels n_l of the gray scale value l to the total number N_T of images, where $h(l) = \frac{n_l}{N_T}$, which reflects the probability distribution of the pixel with the gray value of l in the image can be regarded as the normalized histogram of the image.

(2) Joint entropy

Joint entropy is also a parameter that reflects the amount of information contained in the image. On this basis, it reflects the correlation between the original image and the processed result, and quantitatively measures the correlation between them. Similarly, the greater the joint entropy of the processed result, the larger the amount of information carried, and the better effect. It is defined as:

$$C(f : g) = -\sum_{l_2=0}^{L} \sum_{l_1=0}^{L} P_{fg}(l_1, l_2) \log P_{fg}(l_1, l_2) \tag{A.11}$$

Sharpness Evaluation

Average gradient is also called sharpness, which reflects the small detail contrast and texture change in the image, and also reflects the sharpness of the image, which can be used as an index to judge the sharpness of the processes result. It is defined as:

$$A = \frac{1}{M \times N} \sqrt{\sum_{i=0}^{M-1} \sum_{j=0}^{N-1} [G_X^2(i,j) + G_Y^2(i,j)]^{1/2}} \tag{A.12}$$

where $G_X^2(i,j)$ and $G_Y^2(i,j)$ are the difference in x and y direction, respectively. In general, the larger the average gradient of the image, the greater the clarity of the image, the better the processed effect.

The defects of classic objective quality evaluation method can be conclude as following: results of classic objective quality evaluation method often do not agree with subjective visual effect, for the reason that MSE, PSNR, etc. all reflect the global difference between original image and target image. However, it cannot reflect the truth of large gray value difference among a few pixels and small difference among most pixels. Obviously, the classic objective quality evaluation methods treat all pixels in an image using the same equation, which cannot reflect the visual characteristics of human eyes completely.

Structured Quality Evaluation Methods

Universal Quality Index

Wang and Bovic proposed UQI model [3], which firstly stated that the distortion of the image includes relevancy, brightness, and contrast. This theory is widely adopted later.

Details of UQI are as follows:

$x = \{x_i | i = 1, 2, \ldots, N\}$ indicates reference image, $y = \{y_i | i = 1, 2, \ldots, N\}$ indicates distorted image, UQI can be depicted as:

$$Q = \frac{4\sigma_{xy}\bar{x}\bar{y}}{\left(\sigma_x^2 + \sigma_y^2\right)\left[(\bar{x})^2 + (\bar{y})^2\right]} \tag{A.13}$$

where

$$\bar{x} = \frac{1}{N}\sum_{i=1}^{N} x_i \tag{A.14}$$

$$\bar{y} = \frac{1}{N}\sum_{i=1}^{N} y_i \tag{A.15}$$

$$\sigma_x^2 = \frac{1}{N-1}\sum_{i=1}^{N} (x_i - \bar{x})^2 \tag{A.16}$$

$$\sigma_y^2 = \frac{1}{N-1}\sum_{i=1}^{N} (y_i - \bar{y})^2 \tag{A.17}$$

$$\sigma_{xy} = \frac{1}{N-1}\sum_{i=1}^{N} (x_i - \bar{x})(y_i - \bar{y}) \tag{A.18}$$

From Eq. A.13, we can see that the value of UQI is between -1 and 1.

When original reference image is identical to distortion image, UQI equals 1. If $y_i = 2\bar{x} - x_i$, UQI equals -1. With proper mathematical transformation, the UQI model can be changed as:

$$Q = \frac{\sigma_{xy}}{\sigma_x \sigma_y} \times \frac{2\bar{x}\bar{y}}{(\bar{x})^2 + (\bar{y})^2} \times \frac{2\sigma_x \sigma_y}{\sigma_x^2 \sigma_y^2} \tag{A.19}$$

where, three parts in Eq. A.19 respectively represent the correlation coefficient of x and y, the similarity value of gray degree and the similarity value of contrast.

Structural Similarity

Wang et al. [4] proposed an image quality evaluation method based on structural similarity (SSIM) with the basic idea that the main function of the human eye is to extract the structure information of image background in the scope of visual field, and human visual system can adaptively complete the task, so that the image structure distortion measurement is one of the best approximation of image perception quality.

SSIM divides the evaluation index into the comparisons of brightness, contrast and structure similarity between test image and original image. Then it multiplies the three indicators of the comparison results to represent the total image quality evaluation indicator. The schematic diagram is shown as Fig. A.1.

The evaluation method can be depicted as follows:

X and Y indicate the original image and the test image respectively with the size $M \times N$. The average brightness of original image and test image are u_x and u_y, with

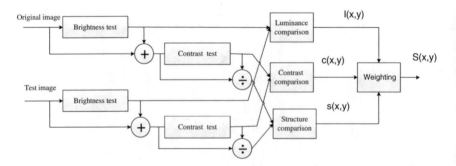

Fig. A.1 Schematic diagram of SSIM model

the standard deviation σ_x and σ_y, covariance σ_{xy}. $l(x,y)$ is the luminance comparison function while $c(x,y)$ is the contrast comparison function and $s(x,y)$ is the structure similarity comparison function. SSIM indicates evaluation value of structure distortion. The calculation equations are as follows:

$$u_x = \frac{1}{MN} \sum_{i=1}^{N} \sum_{j=1}^{M} x_{ij} \tag{A.20}$$

$$u_y = \frac{1}{MN} \sum_{i=1}^{N} \sum_{j=1}^{M} y_{ij} \tag{A.21}$$

$$\sigma_x = \left[\frac{1}{(M-1) \times (N-1)} \sum_{i=1}^{N} \sum_{j=1}^{M} (x_{ij} - u_x)^2 \right]^{\frac{1}{2}} \tag{A.22}$$

$$\sigma_y = \left[\frac{1}{(M-1) \times (N-1)} \sum_{i=1}^{N} \sum_{j=1}^{M} (y_{ij} - u_y)^2 \right]^{\frac{1}{2}} \tag{A.23}$$

$$\sigma_{xy} = \frac{1}{(M-1) \times (N-1)} \sum_{i=1}^{N} \sum_{j=1}^{M} (x_{ij} - u_x)(y_{ij} - u_y) \tag{A.24}$$

$$l(x,y) = \frac{2u_x u_y + C_1}{u_x^2 + u_y^2 + C_1} \tag{A.25}$$

$$c(x,y) = \frac{2\sigma_x \sigma_y + C_2}{\sigma_x^2 + \sigma_y^2 + C_2} \tag{A.26}$$

$$s(x,y) = \frac{\sigma_{xy} + C_3}{\sigma_x \sigma_y + C_3} \tag{A.27}$$

$$SSIM = [l(x,y)]^{\alpha}[c(x,y)]^{\beta}[s(x,y)]^{\gamma} \tag{A.28}$$

where, α, β, $\gamma > 0$ are weight coefficients used for adjusting brightness, contrast and structural similarity. In general, $\alpha = \beta = \gamma = 1$. In Eqs. A.26, A.27, A.28, C_1, C_2 and C_3 are constants where $C_1 = (K_1 L)^2$, $C_2 = (K_2 L)^2$, $C_3 = C_2/2$.

By comparing with the UQI model, in Eqs. A.26, A.27, A.28, where $C_1 = C_2 = C_3 = 0$, SSIM = UQI. The range of SSIM is 0–1, while the range of UQI is -1 to 1. Therefore, the reason that SSIM makes the improvement on UQI model is that it can make the evaluation results more stable and convergent.

Information Fidelity Criterion

Sheikh et al. [5] proposed a new image quality evaluation model of IFC in 2005, in which the distorted image is transformed by the original image through a distortion channel.

IFC used a new mathematical model to represent the original image and distorted image. For the original reference image, the wavelet decomposition is made to decompose it into a number of subbands in space index of M in the first place. Each subband indicates a scalar Gaussian Mixture Model. The scalar coefficient is a model of a random variable:

$$X = Z \times U = \{Z_i \times U_i, i \in M\} \tag{A.29}$$

The mathematical model is:

$$X(i, k) = Z(i, k)U(i, k) \tag{A.30}$$

where $Z(i, k)$ is an random field (RF) of positive scalars, $U(i, k)$ is Gaussian scalar RF with mean zero and variance. The distortion model can be expressed as:

$$Y(i, k) = \beta(i, k)X(i, k) + V(i, k) \tag{A.31}$$

where $\beta(i, k)$ is a deterministic scalar attenuation field and $V(i, k)$ is a stationary additive zero-mean Gaussian noise RF with variance $\sigma_y(k)^2$. According to information entropy equation in information theory, IFC model can be represented as follows under the condition of $Z(i, k)$:

$$IFC[X(i, k), Y(i, k)] = I[X(i, k), Y(i, k)|Z(i, k)] \tag{A.32}$$

$$= \frac{1}{2}log_2\left(\frac{\beta(i, k)^2 Z(i, k)^2 + \sigma_y(k)^2}{\sigma_y(k)^2}\right) \tag{A.33}$$

From another perspective, IFC uses completely different mathematical model to describe image quality, which is an evaluation model. The IFC model is slightly better than SSIM in terms of performance, but computation complexity of IFC model is more complicated. What's more, IFC model also has the same shortcomings as the UQI model, stability and convergence of which are not guaranteed.

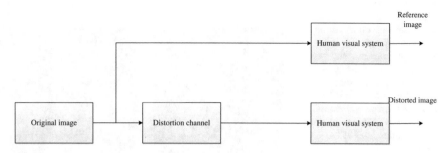

Fig. A.2 Schematic diagram of VIF model

Visual Information Fidelity

In 2006, Sheikh proposed a new model VIF [6] adding human visual system on the basis of the IFC, which is the best image quality evaluation model by far, as shown in Fig. A.2.

The mathematical model of VIF is expressed as follows:

$$Y(i,k) = \beta(i,k)X(i,k) + V(i,k) + W(i,k) \tag{A.34}$$

$$\chi(i,k) = X(i,k) + W(i,k) \tag{A.35}$$

where $W(i,k)$ is Gaussian scalar RF with variance of k, which depicts the neural noise in the human vision system. $\chi(i,k)$ is original reference image filtered through the human eye system and $Y(i,k)$ is distorted image filtered through the human eye system. VIF model is expressed as follows:

$$VIF[X(i,k), Y(i,k)] = \frac{I[X(i,k), Y(i,k)|Z(i,k)]}{I[\chi(i,k), Y(i,k)|Z(i,k)]} \tag{A.36}$$

$$= \frac{\frac{1}{2}\log_2\left(\frac{\beta(i,k)^2 Z(i,k)^2 + \sigma_y(k)^2 + k}{\sigma_y(k)^2 + k}\right)}{\frac{1}{2}\log_2\left(\frac{Z(i,k)^2 + k}{k}\right)} \tag{A.37}$$

where, k is set to 0.01 in literature [6].

Case Analysis

In this section, we will take the classical image Lena as an example, and analyze and compare the methods above. Figure A.3a is the original image of Lena. Figure A.3b is a distorted image with salt-and-pepper noise 0.05. Figure A.3c is a distorted image with salt-and-pepper noise 0.1. Figure A.3d is a blurred image with

Fig. A.3 Lena original image and its distorted images. **a** Original image of Lena. **b** Distorted image with salt-and-pepper noise 0.05. **c** Distorted image with salt-and-pepper noise 0.1. **d** Blurred image with caustic radius 5. **e** Blurred image with caustic radius 10. **f** JPEG compressed image with quantization factor 5. **g** JPEG compressed image with quantization factor 10

caustic radius 5. Figure A.3e is a blurred image with caustic radius 10. Figure A.3f is a JPEG compressed image with quantization factor 5. Figure A.3g is a JPEG compressed image with quantization factor 10.

Table A.3 The comparison of evaluation methods of Lena original image and its distorted images

Evaluation method	Image (a)	Image (b)	Image (c)	Image (d)	Image (e)	Image (f)	Image (g)
PSNR	Inf	18.4802	15.4426	26.2586	23.0846	29.8234	26.7117
MSE	0	6.3657	12.6516	30.3304	46.0259	25.2299	44.7597
UQI	1.0	0.2671	0.1542	0.4532	0.2514	0.4889	0.3555
SSIM	1.0	0.5418	0.3996	0.8329	0.6706	0.8818	0.7878
IFC	77.2479	1.0504	0.7226	1.4405	0.5561	1.5462	0.9149
VIF	1.0	0.1971	0.1362	0.1898	0.0654	0.2779	0.1676

Table A.3 is the comparison result of evaluation methods of Lena original image and its distorted ones. From Table A.3 we can see that traditional objective quality evaluation method is not effective. The three groups of images with the counterpart of visual effects in Fig. A.3b–g, differentiation of PSNR is little, the value of MSE and the visual effects of images appear opposite results. Several objective quality evaluation indexes of structured methods are ideal, of which the VIF model is the best. In conclusion, we can see that the SSIM and VIF models are the improvement of UQI and IFC respectively.

PROGRAMME A.1: Structured quality evaluation methods

```
%%%% main program
clc;
clear all;
img_in1 = imread('D:\imgdb\img1.bmp');
img_in_d1 = double(img_in1);
net = feedforwardnet(10);
for i=1:8
    img_in2 = imread(['D:\imgdb\img',num2str(i),'.bmp']);
    img_in_d2 = double(img_in2);
    UQI(i) = uqi(img_in_d1, img_in_d2);
SSIM(i) = ssim(img_in_d1,img_in_d2);
    VIF(i) = vifvec(img_in_d1,img_in_d2);
    IFC(i) = ifcvec(img_in_d1, img_in_d2);
    PSNR(i) = psnr(img_in2,img_in1);
    MSE(i) = mse(net,img_in2,img_in1,'normalization','none');
end
function [quality, quality_map] = uqi(img1, img2, block_size)
    if (nargin == 1 | nargin > 3)
    quality = -Inf;
    quality_map = -1*ones(size(img1));
    return;
end
```

```
if (size(img1) ~= size(img2))
    quality = -Inf;
    quality_map = -1*ones(size(img1));
    return;
end
if (nargin == 2)
    block_size = 8;
end
N = block_size.^2;
sum2_filter = ones(block_size);
img1_sq    = img1.*img1;
img2_sq    = img2.*img2;
img12 = img1.*img2;
img1_sum    = filter2(sum2_filter, img1, 'valid');
img2_sum    = filter2(sum2_filter, img2, 'valid');
img1_sq_sum = filter2(sum2_filter, img1_sq, 'valid');
img2_sq_sum = filter2(sum2_filter, img2_sq, 'valid');
img12_sum = filter2(sum2_filter, img12, 'valid');
img12_sum_mul = img1_sum.*img2_sum;
img12_sq_sum_mul = img1_sum.*img1_sum + img2_sum.*img2_sum;
numerator = 4*(N*img12_sum - img12_sum_mul).*img12_sum_mul;
denominator1 = N*(img1_sq_sum + img2_sq_sum) - img12_sq_sum_mul;
denominator = denominator1.*img12_sq_sum_mul;
quality_map = ones(size(denominator));
index = (denominator1 == 0) & (img12_sq_sum_mul ~= 0);
quality_map(index) = 2*img12_sum_mul(index)./img12_sq_sum_mul(index);
index = (denominator ~= 0);
quality_map(index) = numerator(index)./denominator(index);
quality = mean2(quality_map);
function [mssim, ssim_map] = ssim(img1, img2, K, window, L)
if (nargin < 2 || nargin > 5)
    mssim = -Inf;
    ssim_map = -Inf;
    return;
end
if (size(img1) ~= size(img2))
    mssim = -Inf;
    ssim_map = -Inf;
    return;
end
```

```matlab
[M N] = size(img1);
if (nargin == 2)
    if ((M < 11) || (N < 11))
        mssim = -Inf;
        ssim_map = -Inf;
        return
    end
    window = fspecial('gaussian', 11, 1.5);   %
    K(1) = 0.01;                        % default settings
    K(2) = 0.03;
    L = 255;
end
if (nargin == 3)
    if ((M < 11) || (N < 11))
        mssim = -Inf;
        ssim_map = -Inf;
        return
    end
    window = fspecial('gaussian', 11, 1.5);
    L = 255;
    if (length(K) == 2)
        if (K(1) < 0 || K(2) < 0)
            mssim = -Inf;
            ssim_map = -Inf;
            return;
        end
    else
        mssim = -Inf;
        ssim_map = -Inf;
        return;
    end
end
if (nargin == 4)
    [H W] = size(window);
    if ((H*W) < 4 || (H > M) || (W > N))
        mssim = -Inf;
        ssim_map = -Inf;
        return
    end
    L = 255;
```

```
    if (length(K) == 2)
        if (K(1) < 0 || K(2) < 0)
                mssim = -Inf;
            ssim_map = -Inf;
             return;
         end
    else
            mssim = -Inf;
        ssim_map = -Inf;
            return;
    end
end
if (nargin == 5)
    [H W] = size(window);
    if ((H*W) < 4 || (H > M) || (W > N))
        mssim = -Inf;
        ssim_map = -Inf;
         return
    end
    if (length(K) == 2)
        if (K(1) < 0 || K(2) < 0)
                mssim = -Inf;
            ssim_map = -Inf;
             return;
          end
    else
            mssim = -Inf;
        ssim_map = -Inf;
            return;
    end
end
img1 = double(img1);
img2 = double(img2);
% automatic downsampling
f = max(1,round(min(M,N)/256));
%downsampling by f
%use a simple low-pass filter
if(f>1)
    lpf = ones(f,f);
    lpf = lpf/sum(lpf(:));
```

```
        img1 = imfilter(img1,lpf,'symmetric','same');
        img2 = imfilter(img2,lpf,'symmetric','same');
        img1 = img1(1:f:end,1:f:end);
        img2 = img2(1:f:end,1:f:end);
end
C1 = (K(1)*L)^2;
C2 = (K(2)*L)^2;
window = window/sum(sum(window));
mu1     = filter2(window, img1, 'valid');
mu2     = filter2(window, img2, 'valid');
mu1_sq = mu1.*mu1;
mu2_sq = mu2.*mu2;
mu1_mu2 = mu1.*mu2;
sigma1_sq = filter2(window, img1.*img1, 'valid') - mu1_sq;
sigma2_sq = filter2(window, img2.*img2, 'valid') - mu2_sq;
sigma12 = filter2(window, img1.*img2, 'valid') - mu1_mu2;
if (C1 > 0 && C2 > 0)
    ssim_map = ((2*mu1_mu2 + C1).*(2*sigma12 + C2))./((mu1_sq + mu2_sq + C1).*(sigma1_sq +
    sigma2_sq + C2));
else
    numerator1 = 2*mu1_mu2 + C1;
    numerator2 = 2*sigma12 + C2;
     denominator1 = mu1_sq + mu2_sq + C1;
    denominator2 = sigma1_sq + sigma2_sq + C2;
    ssim_map = ones(size(mu1));
    index = (denominator1.*denominator2 > 0);
    ssim_map(index)                                                                 =
    (numerator1(index).*numerator2(index))./(denominator1(index).*denominator2(index));
    index = (denominator1 ~= 0) & (denominator2 == 0);
    ssim_map(index) = numerator1(index)./denominator1(index);
end
mssim = mean2(ssim_map);
return
function ifc=ifcvec(imorg,imdist)
M=3;
subbands=[4 7 10 13 16 19 22 25];
[pyr,pind] = buildSpyr(imorg, 4, 'sp5Filters', 'reflect1'); % compute transform
org=ind2wtree(pyr,pind); % convert to cell array
[pyr,pind] = buildSpyr(imdist, 4, 'sp5Filters', 'reflect1');
dist=ind2wtree(pyr,pind);
```

```
% calculate the parameters of the distortion channel
[g_all,vv_all]=distsub_est_M(org,dist,subbands,M);
% calculate the parameters of the reference image
[ssarr, larr, cuarr]=refparams_vecgsm(org,subbands,M);
% reorder subbands. This is needed since the outputs of the above functions
% are not in the same order
vvtemp=cell(1,max(subbands));
ggtemp=vvtemp;
for(kk=1:length(subbands))
    vvtemp{subbands(kk)}=vv_all{kk};
    ggtemp{subbands(kk)}=g_all{kk};
end
% compute reference and distorted image information from each subband
for i=1:length(subbands)
    sub=subbands(i);
    g=ggtemp{sub};
    vv=vvtemp{sub};
    ss=ssarr{sub};
    lambda = larr(sub,:);,
    cu=cuarr{sub};
    % how many eigenvalues to sum over. default is 1.
    neigvals=1;
    % compute the size of the window used in the distortion channel estimation, and use it to calculate
the offset from subband borders
    % we do this to avoid all coefficients that may suffer from boundary
    % effects
    lev=ceil((sub-1)/6);
    winsize=2^lev+1; offset=(winsize-1)/2;
    offset=ceil(offset/M);
    % select only valid portion of the output.
    g=g(offset+1:end-offset,offset+1:end-offset);
    vv=vv(offset+1:end-offset,offset+1:end-offset);
    ss=ss(offset+1:end-offset,offset+1:end-offset);
    %IFC
    temp1=0; temp2=0;
    for j=1:length(lambda)
        temp1=temp1+sum(sum((log2(1+g.*g.*ss.*lambda(j)./(vv+1e-10))))); % IFC for the i'th
subband. tolerence for zero variance
    end
    num(i)=temp1;
```

```
end
% compuate IFC and normalize to size of the image
ifc=sum(num)/prod(size(imorg));
function vif=vifvec(imorg,imdist)
M=3;
subbands=[4 7 10 13 16 19 22 25];
sigma_nsq=0.4;
% Do wavelet decomposition.
[pyr,pind] = buildSpyr(imorg, 4, 'sp5Filters', 'reflect1'); % compute transform
org=ind2wtree(pyr,pind); % convert to cell array
[pyr,pind] = buildSpyr(imdist, 4, 'sp5Filters', 'reflect1');
dist=ind2wtree(pyr,pind);
% calculate the parameters of the distortion channel
[g_all,vv_all]=vifsub_est_M(org,dist,subbands,M);
% calculate the parameters of the reference image
[ssarr, larr, cuarr]=refparams_vecgsm(org,subbands,M);
% reorder subbands. This is needed since the outputs of the above functions
% are not in the same order
vvtemp=cell(1,max(subbands));
ggtemp=vvtemp;
for(kk=1:length(subbands))
    vvtemp{subbands(kk)}=vv_all{kk};
    ggtemp{subbands(kk)}=g_all{kk};
end
% compute reference and distorted image information from each subband
for i=1:length(subbands)
    sub=subbands(i);
    g=ggtemp{sub};
    vv=vvtemp{sub};
    ss=ssarr{sub};
    lambda = larr(sub,:);
    cu=cuarr{sub};
    % how many eigenvalues to sum over. default is all.
    neigvals=length(lambda);
    % compute the size of the window used in the distortion channel estimation, and use it to calculate
    the offset from subband borders
    lev=ceil((sub-1)/6);
    winsize=2^lev+1; offset=(winsize-1)/2;
    offset=ceil(offset/M);
    % select only valid portion of the output.
```

```
   g=g(offset+1:end-offset,offset+1:end-offset);
   vv=vv(offset+1:end-offset,offset+1:end-offset);
   ss=ss(offset+1:end-offset,offset+1:end-offset);
   %VIF
   temp1=0; temp2=0;
   for j=1:length(lambda)
       temp1=temp1+sum(sum((log2(1+g.*g.*ss.*lambda(j)./(vv+sigma_nsq)))));    %    distorted
   image information for the i'th subband
       temp2=temp2+sum(sum((log2(1+ss.*lambda(j)./(sigma_nsq)))));    %    reference    image
   information
   end
   num(i)=temp1;
   den(i)=temp2;
end
% compuate VIF
vif=sum(num)./sum(den);
```

Classification Evaluation Methods

Positive Samples and Negative Samples

For all supervised learning methods, they need positive samples and negative samples. Taking a human face detector as an example, face images are positive samples, images without face are negative samples.

For a group of positive samples and negative samples, after testing, they can be in one of four statuses, shown as Table A.4.

Possible scenarios generated by positive samples include:

1. TP (true positive), that is to say positive sample is determined as target by the detector.
2. FN (false negative), that is to say positive sample is determined as non-target by the detector.

Table A.4 Prediction result of sample

	Actual positive	Actual negative
Predicted positive	True positives (TP)	False positives (FP)
Predicted negative	False negatives (FN)	True negatives (TN)

Possible scenarios generated by negative samples include:
3. TN (true negative), that is to say negative sample is determined as non-target by the detector.
4. FP (false positive), that is to say negative sample is determined as target by the detector.

Precision, Recall, Accuracy, and F1

In Machine learning (ML), Natural language processing (NLP), information retrieval (IR) and other fields, evaluation is a necessary work, and its evaluation indexes tend to have the following aspects: Accuracy, Precision, Recall and F1-Measure.

Now let's assume a specific scenario as an example: if a class has 80 black sheep and 20 white sheep, 100 in total. The goal is to find out all the white ones. If we pick out 50 sheep, 20 of whom are white, other 30 black ones are chosen wrongly. The next task is to assess this work (evaluation).

Precision is the proportion of all correctly retrieved items (TP) accounts for all actually retrieved items (TP + FP). Its equation is:

$$Precision = \frac{TP}{TP + FP} \tag{A.38}$$

In the above example, we want to know the ratio of right one (white) to all. That is, precision = 20/(20 + 30) = 40% (20 whites/(20 whites + 30 blacks error judged)).

Recall is the proportion of all correctly retrieved items (TP) accounts for all should be retrieved items (TP + FN). Its equation is:

$$Recall = \frac{TP}{TP + FN} \tag{A.39}$$

In the above example, it's the ratio of retrieved whites to all. That is, recall =20/(20 + 0) = 100% (20 whites/(20 whites + 0 whites who misjudged to be black)).

Accuracy is the proportion of all correctly classified samples accounts for all samples. Its equation is:

$$Accuracy = \frac{TP + TN}{TP + FP + FN + TN} \tag{A.40}$$

In the above example, Accuracy = (20 + 50)/100 = 70%.

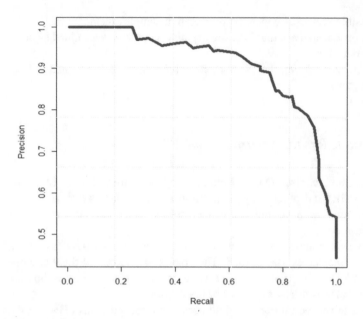

Fig. A.4 PR curve diagram

F-Measure is weighted harmonic average of precision and recall [7]:

$$F1 = \frac{(a^2 + 1)Precion * Recall}{a^2(Precion + Recall)} \tag{A.41}$$

When $a = 1$, $F1$ is transformed as:

$$F1 = \frac{2Precion * Recall}{Precion + Recall} \tag{A.42}$$

In the above example, F1-measure can be calculated as: F1 = 2 * 0.4 * 1/(0.4 + 1) = 57.14%.

PR Curves

PR curve is often used in information retrieval. PR curve traverses all thresholds, and draws different points of precision and recall. A PR curve diagram is shown as Fig. A.4.

ROC Curves

A good classifier should be as close to the top left of the graph as possible, and a random prediction model should be located on the main diagonal of the connection point $(TPR = 0, FPR = 0)$ and $(TPR = 1, FPR = 1)$. TPR and FPR are calculated as following:

$$TPR = \frac{TP}{TP + FN} \tag{A.43}$$

$$FPR = \frac{FP}{FP + TN} \tag{A.44}$$

Receiver Operating Characteristic (ROC) curve is a graphical plot that illustrates the diagnostic ability of a binary classifier system as its discrimination threshold is varied [8]. Area under the ROC Curve (AUC) provides another way to evaluate the average performance of the model. If the model is perfect, then its AUC = 1. If the model is a simple random prediction one, then its AUC = 0.5. If a model is better than another, its area below the curve is relatively larger. A ROC curve diagram is shown as Fig. A.5.

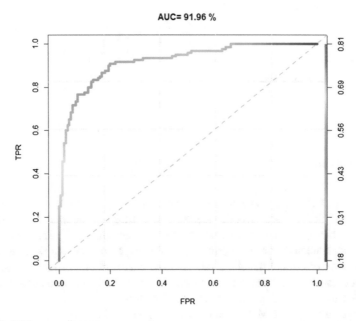

Fig. A.5 ROC curve diagram

The main function of ROC curve is as Programme A.2.

PROGRAMME A.2: ROC curve instance

```
function auc = computeROC(predict, ground_truth)
    pos_num = sum(ground_truth==1);
    neg_num = sum(ground_truth==0);
     m = size(ground_truth,1);
    [pre,Index]   = sort(predict);
    ground_truth = ground_truth(Index);
     x = zeros(m+1,1);
    y = zeros(m+1,1);
    auc   = 0;
    x(1) = 1;
    y(1) = 1;
     for i = 2:m
         TP = sum(ground_truth(i:m)==1);
         FP = sum(ground_truth(i:m)==0);
          x(i) = FP/neg_num;
          y(i) = TP/pos_num;
           auc = auc+(y(i)+y(i-1))*(x(i-1)-x(i))/2;
    end;
     x(m+1) = 0;
    y(m+1) = 0;
    auc      = auc+y(m)*x(m)/2;
    plot(x,y);
end
```

CMC Curves

Cumulative Matching Characteristic (CMC) curve is a widely used evaluation index in the field of pedestrian re-identification [9]. For a common matching and sorting problem, in order to analyze the performance of the algorithm visually, it can be accumulated for the correct matching rate. A drawing diagram of the accumulation of correct matching rate is CMC curve. In order to clearly show the matching results in each rank, top-n matching rate in CMC curve is often shown in a list. A CMC curve diagram is shown as Fig. A.6.

The main function of CMC curve is as Programme A.3.

Fig. A.6 CMC curve diagram

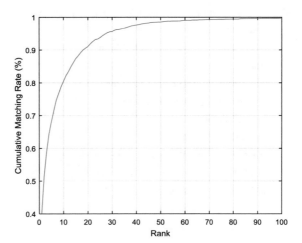

PROGRAMME A.3: CMC curve instance

```
function [cmc] = computeCMC( dist, probeID, galleryID,numRanks)
    % --------------------------------------------------------------------
    % Inputs:
    %      dist       : m*n distance matrix
    %      probeID    : 1*m, ID of probe image
    %      galleryID : 1*n, ID of gallery image
    %      numRanks   : top numRanks value of CMC
    %
    % Outputs:
    %      cmc        : cumulative matching rate, i.e. matching rates of each rank!
    % --------------------------------------------------------------------
    if ~isrow(galleryID)
        galleryID = galleryID';
    end
    if ~iscolumn(probeID)
        probeID = probeID';
    end
    eqFlag = bsxfun(@eq, probeID, galleryID);
```

```
% check whether all probe samples belong to the gallery
if any( all(eqFlag == false, 2) );
     error('Gallery set is not a closed-set!');
end
%% get the matching rank of each probe
[~, sortedIndex] = sort(dist,2,'ascend');
eqFlag = double(eqFlag);
[~, maxIndex]    = max(eqFlag,[],2);
rankMatrix = bsxfun(@eq, sortedIndex, maxIndex);
rankRate = sum(rankMatrix,1)/sum(rankMatrix(:));
cmc = cumsum(rankRate );
cmc = cmc(1:numRanks );                    % cumulative matching curve
plot(cmc);          % plot figure
end
```

Confusion Matrix

Confusion matrix, in the field of artificial intelligence, is a particular table for visually representing, especially used in supervised learning [10].

Each column of confusion matrix represents an instance of the predicted sample, and each line represents an instance of the actual sample. By observing the confusion matrix, we can clearly know whether the classification system correctly distinguishes two different categories. In other words, we also can determine the effect of a classifier classification by confusion matrix.

As a simple example to understand confusion matrix and its implications. Given 16 sample data, which are divided into four categories: class 1, class 2, class 3, class 4, and 4 samples per class. The predicted results by the classifier are shown in Table A.5.

Table A.5 Predicted samples and actual samples

		Predicted samples			
		Class 1	Class 2	Class 3	Class 4
Actual samples	Class 1	2	1	1	0
	Class 2	0	3	1	0
	Class 3	0	0	4	0
	Class 4	1	0	0	3

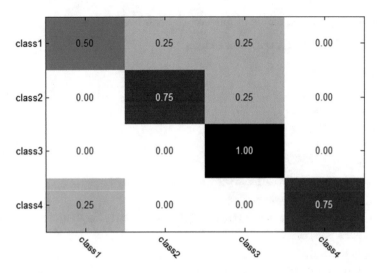

Fig. A.7 Confusion matrix

The meanings of each row and column are as follows:

The first row: in the 4 samples of class 1, 2 samples are divided into class 1, 1 sample is divided into class 2, 1 sample is divided into class 3, and 0 sample is divided into class 4.

The second row: in the 4 samples of class 2, 0 sample is divided into class 1, 3 samples are divided into class 2, 1 sample is divided into class 3, and 0 sample is divided into class 4.

The rest can be done in the same manner. Examples on the main diagonal of confusion matrix are the cases of correctly classified, such as 2, 3, 4, 3 in Table 6.5. By observing the confusion matrix in Table 6.5, we can calculate the classification accuracy and error rate.

For multi-class classification problem, each category may be assigned to the other categories, but their own category should be the most, thus the calculated percentage forms a matrix. If classification accuracy is high, the values on the diagonal should be high. The confusion matrix is shown in Fig. A.7.

The following code is the main function of confusion matrix.

PROGRAMME A.4: Confusion matrix

```
clc;   clear;
close all;
addpath('./ConfusionMatrices')
pred_labels = [1 2 1 3 2 2 3 2 3 3 3 3 4 4 1 4];
num=[4 4 4];
name=cell(1,4);
name{1}='class1';name{2}='class2';name{3}='class3';name{4}='class4';
compute_cm(pred_labels,num,name)
function [confusion_matrix]=compute_cm (predict_label,num_in_class,name_class)
    num_class=length(num_in_class);
    num_in_class=[0 num_in_class];
    confusion_matrix=size(num_class,num_class);
    for ci=1:num_class
        for cj=1:num_class
            summer=0;
            c_start=sum(num_in_class(1:ci))+1;
            c_end=sum(num_in_class(1:ci+1));
            summer=size(find(predict_label(c_start:c_end)==cj),2);
            confusion_matrix(ci,cj)=summer/num_in_class(ci+1);
        end
    end
    draw_cm2(confusion_matrix,name_class,num_class);
end
function draw_cm2(mat,tick,num_class)
    %%%
    %   Matlab code for visualization of confusion matrix;
    %   Parameters£ºmat: confusion matrix;
    %   tick: name of each class, e.g. 'class_1' 'class_2'...
    %   num_class: number of class
    %%%
    imagesc(1:num_class,1:num_class,mat);                    %# in color
    colormap(flipud(gray));    %# for gray; black for large value.
    textStrings = num2str(mat(:),'%0.2f');
    textStrings = strtrim(cellstr(textStrings));
    [x,y] = meshgrid(1:num_class);
    hStrings = text(x(:),y(:),textStrings(:), 'HorizontalAlignment','center');
    midValue = mean(get(gca,'CLim'));
```

```
    textColors = repmat(mat(:) > midValue,1,3);
    set(hStrings,{'Color'},num2cell(textColors,2));    %# Change the text colors
    set(gca,'xticklabel',tick,'XAxisLocation','bottom');
    set(gca, 'XTick', 1:num_class, 'YTick', 1:num_class);
    set(gca,'yticklabel',tick);
    set(gcf,'color','w');
    set(gca,'Position',[.17 .2 .8 .7]);
    rotateXLabels(gca,315);    % rotate the x tick
    F=getframe(gcf);
end
```

Image Quality and Fusion Evaluations

Although the image fusion method is numerous, the technology is also endless, the purpose is nothing more than to improve the picture quality or increase the content of the image information, which is the effect of the evaluation of the fundamental starting point. For different levels of fusion, the evaluation of the effect of indicators is not the same. In terms of the underlying fusion, generally the visual effects can be compared and analyzed, and the higher the level, the greater the degree of demand satisfaction. In theory, the fusion of image should be to preserve the effective information in two or more images and to synthesize them into an image. Therefore, the evaluation of the fusion effect should include two aspects: the improvement level and the continuation level.

For image observers, the meaning of the image mainly includes two aspects: one is the fidelity of the image, the other is the image of the comprehensibility. The existing methods of image fusion performance evaluation can be divided into: objective and subjective evaluation of fusion quality. The former by virtue of observation, which depends largely on the observer's subjective consciousness, with as well as the difference and variation, will change with the application area, where the situation, personal preferences and other changes, the latter is a quantitative calculation, through the value to judge, in general, it has a certain relevance to subjective evaluation.

Subjective Evaluation of Image Fusion

In the evaluation of image fusion effect, subjective evaluation mainly from the following aspects:

(1) Registration accuracy evaluation. If the degree of registration deviation is small, ghosting will occur, if the deviation is large, there will be serious dislocation.

(2) Color distribution evaluation. If the color distribution is reasonable, the naked eye will feel comfortable; if the distribution is unreasonable, the whole image color distribution is uneven, visual impact will increase.

(3) Sharpness evaluation. If the sharpness is close to or improved with the original image, the fused image is clear; if the sharpness is reduced, the fused image will appear to a certain extent blurred.

(4) Brightness and contrast evaluation. If this two are inappropriate, the fused image will have patches or fog and other noise-like parts.

(5) Texture information evaluation. If the texture information is sufficient, the fused image will look plumper, if there a loss in the fusion process, it will become dull and lack of hierarchy.

In term of this aspect of evaluation, there are common international 5-point evaluation criteria, see section 'Subjective Evaluation'.

Objective Evaluation of Image Fusion

For the subjective evaluation, the human eye can only see the obvious changes, the small differences are not sensitive, and subjective judgments will be affected by many factors and always vary. Therefore, a quantitative evaluation method with a uniform standard is indispensable. Now according to the evaluation principle, the objective evaluation method can be divided into statistical characteristics evaluation, information content evaluation, sharpness evaluation, signal to noise ratio (SNR) evaluation and spectral information evaluation. The following is a brief introduction to the main method. In the following evaluation indicators, the original image is $f(x, y)$, the fused image is $g(x, y)$, the ideal image is $i(x, y)$ and the size of image is $N \times N$.

1. Evaluation based on statistical characteristics

There is no ideal standard reference image, so the fusion effect of image is objectively evaluated based on the statistical characteristics of the fusion image and the performance index which reflects the relationship between the fusion image and original image.

(1) Average Value (AV) of image

The size of the mean represents the average size of the image pixel values, which is an evaluation index that belongs to the statistical characteristics. The brightness that human eye can be perceived in grayscale images in the form of grayscale, so the average value of the gray scale has a greater effect on the visual effect of the image.

If the average value of the image is appropriate, the result of fusion is better. The average value of the image is defined as:

$$\mu = \frac{1}{N \times N} \sum_{x=0}^{N-1} \sum_{y=0}^{N-1} g(x, y) \tag{A.45}$$

(2) Standard Deviation

The centrality or discretization of the image gray value relative to the gray scale is generally reflected by the standard deviation which reflecting the distribution of the pixel values and showing the contrast of the image. If the standard deviation of the fusion image is small, the contrast is small, that is, the amount of information contained therein is smaller. The larger the standard deviation, the more gray-scale distribution, the better the visual effect. The standard deviation is obtained indirectly by the average value, and the standard deviation of the image is defined as:

$$\sigma = \frac{1}{N \times N} \sqrt{\sum_{x=0}^{N-1} \sum_{y=0}^{N-1} [g(x, y) - \mu]} \tag{A.46}$$

(3) Root mean square error

The root mean square error can be used to detect the degree of deviation between the image to be detected and the ideal image, which can be evaluated using the known ideal image. The smaller the deviation between the fusion result and ideal image, the better the fusion effect. It is defined as:

$$RMSE = \sqrt{\frac{1}{N \times N} \sum_{x=0}^{N-1} \sum_{y=0}^{N-1} [g(x, y) - i(x, y)]^2} \tag{A.47}$$

2. Objective evaluation based on information content

(1) Information entropy

Information entropy is an important indicator of the degree of abundance of image information. It is reflected degree of deviation in the image range from the peak area of the gray histogram. The larger the entropy of the fused image, the more the information volume of the fused image increases, the richer the image is, the better effect of the image fusion. it is defined as:

$$H = -\sum_{l=0}^{L} h(l) \log[h(l)] \tag{A.48}$$

where L represents the total gray level of the fused image F, $h(l)$ represents the ratio of the number of pixels n_l of the gray scale value l to the total number N of images, that is $h(l) = \frac{n_l}{N}$, which reflects the probability distribution of the pixel with the gray value of l in the image can be regarded as the normalized histogram of the image.

(2) Joint entropy

Joint entropy is also a parameter that reflects the amount of information contained in the image. On this basis, it reflects the correlation between the original image and the fusion result, and quantitatively measures the correlation between them. Similarly, the greater the joint entropy of the fusion result, the larger the amount of information carried, and the better effect. It is defined as:

$$C(f : g) = -\sum_{l_2=0}^{L} \sum_{l_1=0}^{L} P_{fg}(l_1, l_2) \log P_{fg}(l_1, l_2) \tag{A.49}$$

3. Objective evaluation based on the sharpness

(1) Average gradient

The average gradient is also called sharpness, which reflects the small detail contrast and texture change in the image, and also reflects the sharpness of the image, which can be used as an index to judge the sharpness of the fusion result. It is defined as:

$$A = \frac{1}{N \times N} \sqrt{\sum_{x=0}^{N-1} \sum_{y=0}^{N-1} [G_X^2(x, y) + G_Y^2(x, y)]^{1/2}} \tag{A.50}$$

where $G_X^2(x, y)$ and $G_Y^2(x, y)$ are the difference in x and y direction, respectively. In general, the larger the average gradient of the image, the greater the clarity of the image, the better the fusion effect.

4. Objective evaluation based on spectral information

(1) Correlation Coefficient

The correlation coefficient reflects the correlation degree of the two spectral features of the image. Generally speaking, the closer the correlation coefficient of the image is to 1, the better the proximity of the image is, the more information is obtained from the original image, the less information, the better the fusion effect. It is defined as:

$$CC = \frac{\sum_{x=0}^{N-1} \sum_{y=0}^{N-1} (f(x,y) - \mu_f)(g(x,y) - \mu_g)}{\sqrt{\sum_{x=0}^{N-1} \sum_{y=0}^{N-1} (f(x,y) - \mu_f)^2 (g(x,y) - \mu_g)^2}} \tag{A.51}$$

where μ_f, μ_g are the average of the original image $f(x,y)$ and the fused image $g(x,y)$ respectively.

(2) Structure Similarity

The structural similarity is calculated as (A.52).

$$SSIM(f,g) = l(f,g)c(f,g)s(f,g) = \frac{(2\mu_f\mu_g + c_1)(2\sigma_f\sigma_g + c_2)(\sigma_{fg} + c_3)}{(\mu_f^2 + \mu_g^2 + c_1)(\sigma_f^2 + \sigma_g^2 + c_2)(\sigma_f\sigma_g + c_3)} \tag{A.52}$$

where $l(f,g) = \frac{2\mu_f\mu_g + c_1}{\mu_f^2 + \mu_g^2 + c_1}$, $c(f,g) = \frac{2\sigma_f\sigma_g + c_2}{\sigma_f^2 + \sigma_g^2 + c_2}$, $s(f,g) = \frac{\sigma_{fg} + c_3}{\sigma_f\sigma_g + c_3}$ is brightness comparison, contrast comparison and structural comparison, respectively; $\mu_f, \mu_g, \sigma_f^2, \sigma_g^2, \sigma_{fg}$ represent the average, variance and covariance of the original image and the fusion image, respectively.

5. Objectively evaluation based on signal to noise ratio (SNR)

(1) Signal to Noise Ratio (SNR)

In the process of image fusion, the noise from the sensor that acquires the image is also a key factor to consider. Therefore, the signal to noise ratio has been applied, the greater the value and the better fusion effect. It is defined as:

$$SNR = 10 \lg \frac{\sum_{x=0}^{N-1} \sum_{y=0}^{N-1} (g(x,y))^2}{\sum_{x=0}^{N-1} \sum_{y=0}^{N-1} (f(x,y) - g(x,y))^2} \tag{A.53}$$

(2) Difference Index (DI)

The average of the ratio of the absolute value of the difference between the fusion image and the original image and the original image value is called difference index. In general, the smaller the difference index, the smaller the degree of fusion image deviation from the original, the more the original grayscale information remains. It is defined as:

$$D = \frac{1}{N \times N} \sum_{x=0}^{N-1} \sum_{y=0}^{N-1} \frac{|f(x,y) - g(x,y)|}{f(x,y)} \tag{A.54}$$

Ideally $DI = 0$.

(3) Peak Signal to Noise Ratio (PSNR)

PSNR is achieved by assuming that the difference between the fused and original image is caused by noise, and the original image is treated as useful information to evaluate quality of the fused image. The larger its value, the closer relation between fusion image and the original image. It is defined as:

$$PSNR = 10 \lg \frac{N \times N \times [\max g(x, y) - \min(g(x, y))]}{\sum_{x=0}^{N-1} \sum_{y=0}^{N-1} [f(x, y) - g(x, y)]^2} \tag{A.55}$$

(4) Degree of Distortion (DD)

DD reflects the degree of distortion of the fused image relative to the original image, the smaller the value, the better the effect of fusing the image. It is defined as:

$$DD = \frac{1}{N \times N} \sum_{x=0}^{N-1} \sum_{y=0}^{N-1} |f(x, y) - g(x, y)| \tag{A.56}$$

In addition to the above several indicators of image quality evaluation, there are some other evaluation indicators, such as general indictors of image quality evaluation, indictors of weighted fusion evaluation. Although the above list of indicators in most cases can accurately evaluate the quality of the image, but the exception of events has also occurred, that is why subjective evaluation is the main evaluation and objective evaluation is auxiliary in the practical application. Therefore, it is one of the hot issues in the study to process a general objective evaluation index which can accurately reflect the quality of the image.

References

1. P.910:Subjective video quality assessment methods for multimedia applications, ITU-T Recommendation P.910 (2008)
2. BT.500-13: Methodology for the subjective assessment of the quality of television pictures, Recommendation ITU-R BT.500-13 (2012)
3. Wang Z, Bovik AC (2002) A universal image quality index. IEEE Signal Process Lett 9 (3):81–84
4. Wang Z, Bovik AC, Sheikh HR, Simoncelli EP (2004) Image quality assessment: from error visibility to structural similarity. IEEE Trans Image Process 13(4):600–612
5. Sheikh HR, Bovik AC, Veciana GD (2005) An information fidelity criterion for image quality assessment using natural scene statistics. IEEE Trans Image Process 14(12):2117–2128

6. Sheikh HR, Bovik AC (2006) Image information and visual quality. IEEE Trans Image Process A Publ IEEE Signal Process Soc 15(2):430–444
7. Powers DMW (2011) Evaluation: from precision, recall and F-Measure to ROC, informedness, markedness & correlation. J Mach Learn Technol 2(1):37–63
8. https://en.wikipedia.org/wiki/Receiver_operating_characteristic
9. Gray D, Brennan S, Tao H (2007) Evaluating appearance models for recognition, reacquisition, and tracking. In: International workshop on performance evaluation for tracking and surveillance (PETS), pp 41–47
10. Remus JJ, Collins LM (2005) Expediting the identification of impaired cochlear implant acoustic model channels through confusion matrix analysis. In: IEEE EMBS conference on neural engineering, pp 418–421

Bibliography

1. Prathusha P, Jyothi S (2018) A novel edge detection algorithm for fast and efficient image segmentation
2. Wei X, Phung SL, Bouzerdoum A (2014) Object segmentation and classification using 3-D range camera. J Visual Commun Image Rep 25(1):74–85
3. Yu H, Zhang X, Wang S et al (2013) Context-based hierarchical unequal merging for SAR image segmentation. IEEE Trans Geosci Remote Sens 51(2):995–1009
4. Minetto R, Spina TV, Falcão AX et al (2012) IFTrace: video segmentation of deformable objects using the image foresting transform. Comput Vis Image Understand 116(2):274–291
5. Yu H, Xian M, Qi X (2014) Unsupervised co-segmentation based on a new global GMM constraint in MRF. In: IEEE international conference on image processing (ICIP), pp 4412–4416
6. Chen J, Benesty J, Huang Y et al (2006) New insights into the noise reduction Wiener filter. IEEE Trans Audio Speech Lang Process 14(4):1218–1234
7. Koller D, Daniilidis K, Nagel HH (1993) Model-based object tracking in monocular image sequences of road traffic scenes. Int J Comput 10(3):257–281
8. Yilmaz A, Javed O, Shah M (2006) Object tracking: a survey. ACM Comput Surv 38(4)
9. Zhong Y, Jain AK, Dubuisson-Jolly MP (1998) Object tracking using deformable templates. In: International conference on computer vision. IEEE Computer Society, p 440
10. Milan A, Roth S, Schindler K (2014) Continuous energy minimization for multitarget tracking. IEEE Trans Pattern Anal Mach Intell 36(1):58–72
11. Andriyenko A, Schindler K (2011) Multi-target tracking by continuous energy minimization. In: IEEE conference on computer vision and pattern recognition. IEEE Computer Society, pp 1265–1272
12. Kwolek B (2013) Multi-object tracking using particle swarm optimization on target interactions. In: Advances in heuristic signal processing and applications, pp 96–121. Springer, Berlin Heidelberg
13. Bosch A, Zisserman A (2006) Scene classification via pLSA. In: European conference on computer vision. Springer, pp 517–530
14. Boutell MR, Luo J, Shen X et al (2004) Learning multi-label scene classification. Pattern Recogn 37(9):1757–1771
15. Derpanis KG, Lecce M, Daniilidis K et al (2012) Dynamic scene understanding: the role of orientation features in space and time in scene classification. In: IEEE computer vision and pattern recognition. IEEE, pp 1306–1313
16. Thériault C, Thome N, Cord M (2013) Dynamic scene classification: learning motion descriptors with slow features analysis. In: IEEE computer vision and pattern recognition. IEEE, pp 2603–2610

17. Yeo BL, Yeung MM (1997) Classification, simplification, and dynamic visualization of scene transition graphs for video browsing. In: Photonics West'98 Electronic Imaging. DBLP, 1997:60–70
18. Vasudevan AB, Muralidharan S, Chintapalli SP et al (2013) Dynamic scene classification using spatial and temporal cues. In: IEEE International Conference on Computer Vision Workshops. IEEE, pp 803–810
19. Liu C, Lin H, Chen N et al (2015) Dynamic scene classification method based on topic model, CN 104268546A
20. Irie G, Satou T, Kojima A et al (2010) Affective audio-visual words and latent topic driving model for realizing movie affective scene classification. IEEE Trans Multimed 12(6):523–535
21. Wang L, Gong S, Liu C et al (2015) Multi-scale feature learning for dynamic scene classification. In: International conference on cyberspace technology. IET, pp 1–6
22. Zhu Q, Zhong Y, Zhang L et al (2017) Scene classification based on the fully sparse semantic topic model. IEEE Trans Geosci Remote Sens (99):1–14

Printed in the United States
By Bookmasters